New Trends in Integrability and Partial Solvability

NATO Science Series

A Series presenting the results of scientific meetings supported under the NATO Science Programme.

The Series is published by IOS Press, Amsterdam, and Kluwer Academic Publishers in conjunction with the NATO Scientific Affairs Division

Sub-Series

I. Life and Behavioural Sciences	IOS Press
II. Mathematics, Physics and Chemistry	Kluwer Academic Publishers
III. Computer and Systems Science	IOS Press
IV. Earth and Environmental Sciences	Kluwer Academic Publishers
V. Science and Technology Policy	IOS Press

The NATO Science Series continues the series of books published formerly as the NATO ASI Series.

The NATO Science Programme offers support for collaboration in civil science between scientists of countries of the Euro-Atlantic Partnership Council. The types of scientific meeting generally supported are "Advanced Study Institutes" and "Advanced Research Workshops", although other types of meeting are supported from time to time. The NATO Science Series collects together the results of these meetings. The meetings are co-organized bij scientists from NATO countries and scientists from NATO's Partner countries – countries of the CIS and Central and Eastern Europe.

Advanced Study Institutes are high-level tutorial courses offering in-depth study of latest advances in a field.
Advanced Research Workshops are expert meetings aimed at critical assessment of a field, and identification of directions for future action.

As a consequence of the restructuring of the NATO Science Programme in 1999, the NATO Science Series has been re-organised and there are currently Five Sub-series as noted above. Please consult the following web sites for information on previous volumes published in the Series, as well as details of earlier Sub-series.

http://www.nato.int/science
http://www.wkap.nl
http://www.iospress.nl
http://www.wtv-books.de/nato-pco.htm

Series II: Mathematics, Physics and Chemistry – Vol. 132

New Trends in Integrability and Partial Solvability

edited by

A.B. Shabat

Landau Institute for Theoretical Physics,
Russian Academy of Science, Moscow, Russia

A. González-López

Departamento de Física Teórica II,
Universidad Complutense de Madrid, Spain

M. Mañas

Departamento de Física Teórica II,
Universidad Complutense de Madrid, Spain

L. Martínez Alonso

Departamento de Física Teórica II,
Universidad Complutense de Madrid, Spain

and

M.A. Rodríguez

Departamento de Física Teórica II,
Universidad Complutense de Madrid, Spain

Kluwer Academic Publishers

Dordrecht / Boston / London

Published in cooperation with NATO Scientific Affairs Division

Proceedings of the NATO Advanced Research Workshop on
New Trends in Integrability and Partial Solvability
Cadiz, Spain
2–16 June 2002

A C.I.P. Catalogue record for this book is available from the Library of Congress.

ISBN 1-4020-1835-5

Published by Kluwer Academic Publishers,
P.O. Box 17, 3300 AA Dordrecht, The Netherlands.

Sold and distributed in North, Central and South America
by Kluwer Academic Publishers,
101 Philip Drive, Norwell, MA 02061, U.S.A.

In all other countries, sold and distributed
by Kluwer Academic Publishers,
P.O. Box 322, 3300 AH Dordrecht, The Netherlands.

Printed on acid-free paper

Printed in the Netherlands.

Contents

Preface

The Advanced Research Workshop "New Trends in Integrability and Partial Solvability" (ARW.978791) took place in the beautiful setting of the Faculty of Medicine of Cádiz University's main room on June 13–15, 2002. Although the number of participants was 30, the lectures were attended by more than one hundred researchers from around the world who were also attending the NEEDS 2002 meeting. The aim of the organizers was to take advantage of these events to bring together researchers from the field of integrable systems and/or from the particular subject of partial integrability, in view of the current interest in combining methods and ideas arising from both areas. A wide variety of topics were covered in the talks and the subsequent discussions, including the analysis of reductions and solutions of integrable nonlinear partial differential equations and dynamical systems, new methods for the analysis of initial-boundary value problems for linear partial differential equations, quasi-exactly solvable Bose systems, the geometric theory of ordinary differential equations, exactly and partially solvable spin models, the theory of nonlocal symmetries of differential equations, and superintegrable systems. The workshop revealed the growing importance of the theory of integrable system as well as the emerging theory of partially solvable systems.

The present volume contains a series of invited contributions describing the background and recent developments of the main subjects discussed in the workshop. Special emphasis has been laid on providing self-contained and detailed presentations of the theory.

M.J. Ablowitz and J. Villarroel give a detailed description of the inverse scattering for the KP equation, a keystone in the theory of integrable systems. The authors investigate how the inverse scattering method should be applied to obtain solutions decaying off a background line, multi-pole lump soliton solutions and slowly decaying solutions.

F. Calogero presents some recent work on dynamical systems such that all solutions with initial conditions in a certain region of the phase space are periodic and have the same period. Many examples of this kind of systems, which the author calls "partially superintegrable", are discussed in detail.

A new method for dealing with initial-boundary value problems for linear PDEs, the *analyticity approach*, is systematically presented in the contribution by A. Degasperis, S.V. Manakov and P.M. Santini. The method is applied to several classical problems (Dirichlet, Neumann, mixed, periodic) for second- and third-order evolutionary PDEs in $1+1$ and in $n+1$ dimensions.

S.N. Dolya and O.B. Zaslavskii consider the extension of the concept of quasi-exact solvability to Bose systems. They find conditions for an algebraization of part of the spectrum and, in some cases, explicit formulae for several energy levels of an anharmonic Bose oscillator. Two important aspects of this contribution are, on the one hand, that the results do not depend on perturbation techniques and, hence, are valid in the strong coupling regime and, on the other hand, that they can be extended to many-particle Bose systems with interaction.

V. Dryuma focuses on the study of the geometric notions underlying certain types of second-order scalar ordinary differential equations. A family of second-order differential equations polynomial in the first derivative is studied by relating it to the equations of the geodesics of a four-dimensional Riemannian metric of a certain kind. The so-called dual equation of an arbitrary second-order differential is also investigated, and its relation to three-dimensional Einstein–Weyl spaces is elucidated.

A general method for constructing and classifying exactly or quasi-exactly solvable Calogero-Sutherland models is discussed in the article by F. Finkel, D. Gómez-Ullate, A. González-López, M.A. Rodríguez and R. Zhdanov, whose main ingredient is the use of new types of Dunkl operators. The authors obtain, in particular, several families of quantum spin models with elliptic interaction.

The contribution by V.M. Goncharenko and A.P. Veselov describes some of the authors' latest results on the quantum Yang–Baxter equation. The road for these new findings departs from the inverse scattering theory of the matrix Korteweg–de Vries equation. In particular, the authors show how previous formulae for the two soliton solutions of the matrix Korteweg–de Vries equation can be extended to give new Yang–Baxter maps on the Grasmannian $G(k, n)$ and on the Cartesian product of two Grasmannians $G(k, n) \times G(n-k, n)$, and find the corresponding Lax pairs for such maps.

The theory of nonlocal symmetry algebras is the subject of the contribution presented by P.J. Olver. A new and consistent framework for the latter theory is developed including, in particular, a description of the calculus of "ghost vector fields". In this way the apparent outstanding paradox of the violation of the Jacobi identity by certain nonlocal vector fields is resolved.

S.N.M. Ruijsenaars' contribution uncovers the deep structure of the BC_N relativistic Calogero–Moser model, whose basic operators are analytic difference operators of D_4 or D_8 type, depending on the hyperbolic or elliptic character of the model. The author develops a "relativistic" version of the hypergeometric functions possessing most of the usual properties of their classical counterparts.

A.B. Shabat and L. Martínez Alonso present a prolongation of a hierarchy of hydrodynamic chains which exhibits a universal character, as it reduces to many of the standard integrable hierarchies. Several new integrable nonlinear models are derived and the properties of the differential reductions of this enlarged hierarchy are studied.

P. Winternitz presents some new insights into the theory of superintegrable systems, specially in relation to the existence and practical computation of third-order integrals of motion.

The workshop was mainly financed by a grant of the Cooperative Science and Technology Sub-Programme of the NATO SCIENCE PROGRAMME. We also received financial support from the Complutense University of Madrid and the University of Cádiz. We would like to thank the authorities of the University of Cádiz, where the workshop took place, for generously making several of the university's facilities available to the participants.

Finally we would like to thank Ms. Asunción Iglesias for her excellent work with the administrative arrangements during the workshop.

Madrid, June 2003

The Editors

List of Contributors

M.J. Ablowitz
Department of Applied Mathematics University of Colorado, Boulder 80309-0526, USA

F. Calogero
Dipartimento di Fisica, Università di Roma "La Sapienza", 00185 Roma, Italy, and Istituto Nazionale di Fisica Nucleare, Sezione di Roma
francesco.calogero@roma1.infn.it

A. Degasperis
Dipartimento di Fisica, Università di Roma "La Sapienza", Roma, Italy
Istituto Nazionale di Fisica Nucleare, Sezione di Roma
antonio.degasperis@roma1.infn.it

S.N. Dolya
B. Verkin Institute for Low Temperature Physics and Engineering, 47 Lenin Prospekt, Kharkov 61164, Ukraine
dolya@ilt.kharkov.ua

Valerii Dryuma
Institute of Mathematics and Informatics, AS RM, 5 Academiei Street, 2028 Kishinev, Moldova
valery@gala.moldova.su

F. Finkel
Dpto. de Física Teórica II, Universidad Complutense, 28040 Madrid, Spain
federico@ciruelo.fis.ucm.es

D. Gómez-Ullate
Centre de recherches mathématiques
Université de Montréal, C.P. 6128, succ. Centre-Ville,
Montréal, QC, H3C 3J7, Canada
ullate@crm.umontreal.ca

V.M. Goncharenko
Chair of Mathematics and Financial Applications, Financial Academy, Leningradsky prospect, 49, Moscow, Russia
mike-gonchar@mtu-net.ru

A. González-López
Dpto. de Física Teórica II, Universidad Complutense, 28040 Madrid, Spain
artemio@fis.ucm.es

S. V. MANAKOVmanakov@itp.ac.ru
Landau Institute for Theoretical Physics, Moscow, Russia

L. MARTÍNEZ ALONSO
Departamento de Física Teórica II, Universidad Complutense
E28040 Madrid, Spain
luism@fis.ucm.es

PETER J.OLVER
Department of Mathematics, University of Minnesota,
Minneapolis, MN, USA 55455
olver@math.umn.edu

M.A. RODRÍGUEZ
Dpto. de Física Teórica II, Universidad Complutense, 28040 Madrid, Spain
rodrigue@fis.ucm.es

S.N.M. RUIJSENAARS
Centre for Mathematics and Computer Science,
P.O. Box 94079, 1090 GB Amsterdam, The Netherlands

P. M. SANTINI
Dipartimento di Fisica, Università di Roma "La Sapienza", Roma, Italy
Istituto Nazionale di Fisica Nucleare, Sezione di Roma
paolo.santini@roma1.infn.it

A.B. SHABAT
Landau Institute for Theoretical Physics, RAS, Moscow 117 334, Russia

A.P. VESELOV
Department of Mathematical Sciences, Loughborough
University, Loughborough, Leicestershire, LE 11 3TU, UK
Landau Institute for Theoretical Physics, Kosygina 2, Moscow, 117940, Russia
A.P.Veselov@lboro.ac.uk

J.VILLARROEL
Universidad de Salamanca, Facultad de Ciencias, 37008 Plaza de la Merced, Salamanca, Spain

P. WINTERNITZ
Centre de recherches mathématiques et
Département de mathématiques et statistique,
Université de Montréal, C.P. 6128, succ. Centre-Ville,
Montréal, QC, H3C 3J7, Canada
wintern@crm.umontreal.ca

O.B. Zaslavskii
Department of Mechanics and Mathematics, Kharkov
V.N. Karazin's National University, Svoboda Sq. 4,
Kharkov 61077, Ukraine
aptm@kharkov.ua

R. Zhdanov
Dpto. de Física Teórica II, Universidad Complutense, 28040 Madrid, Spain

INITIAL VALUE PROBLEMS AND SOLUTIONS OF THE KADOMTSEV–PETVIASHVILI EQUATION

M.J. ABLOWITZ
Department of Applied Mathematics University of Colorado, Boulder 80309-0526, USA

J. VILLARROEL
Universidad de Salamanca, Facultad de Ciencias, 37008 Plaza de la Merced, Salamanca, Spain

Abstract. Initial value problems and solutions associated with the Kadomtsev–Petviashvili equation are analyzed. The discussion includes the inverse scattering transform for suitably decaying data, solutions decaying off a background line, multi-pole lump soliton solutions and solutions which are slowly decaying. Existence and uniqueness of the associated eigenfunctions are discussed in terms of natural functional norms.

1. Introduction

The Kadomtsev–Petviashili (KP) equation

$$(u_t + u_{xxx} + 6uu_x)_x + 3\epsilon^2 u_{yy} = 0\,, \tag{1.1}$$

with both $\epsilon^2 = 1$ or -1, is a physically significant equation which arises in the study of small amplitude, long, two dimensional surface water waves with surface tension, which vary slowly in the transverse direction to the wave propagation [1]. It is also an important equation in plasma physics [2]. Corresponding to $\epsilon^2 = -1$ we speak of KPI, while KPII corresponds to taking $\epsilon^2 = 1$. The KP equation is an extension to $2+1$ dimensions of the prototype equation integrable by the inverse scattering transform (IST) in $1+1$ dimensions: the Korteweg–deVries (KdV) equation, which arises under similar assumptions. Its importance is highlighted by the fact that it is the best known prototype of a $2+1$ dimensional integrable equation. This means that the solution of the corresponding Cauchy problem with decaying data in the plane can be obtained by means of the inverse scattering transform. In [3–5] the relevant extension of IST to the multidimensional case was developed and used to find the solution of the corresponding Cauchy problem

A.B. Shabat et al. (eds.), New Trends in Integrability and Partial Solvability, 1–47.

with decaying data for eqs. (1); we note that IST has been henceforth employed to solve the initial value problem (i.v.p.) corresponding to decaying data for a number of important nonlinear evolution equations appearing in physics, see [6–8].

KP is the compatibility of the linear spectral equations $L\psi(x,y) = M\psi(x,y) = 0$, where the operator L is either the nonstationary Schrödinger (NS) equation when $\epsilon^2 = -1$ or the heat operator with $\epsilon^2 = 1$ defined as

$$L\psi \equiv [\epsilon\partial_y + \partial_{xx} + u(x,y)]\psi(x,y)\,, \tag{1.2}$$

and

$$M\psi \equiv (\partial_t + 4\partial_{xxx} + 6u\partial_x + 3u_x - 3\epsilon(\partial_y\partial_x^{-1}u) + h(k))\psi = 0 \tag{1.3}$$

where $h(k)$ is an arbitrary function [9].

The solution of the Cauchy problem with decaying data for equations (1.1) is connected to the solution of the spectral problem (1.2) as we explain below. The IST for KPII involves using a $\overline{\partial}$ problem [4]. The relevant ideas are developed in Section 2. Another feature of KPII is the appearance of line solitons: real and localized solutions that decay exponentially at infinity everywhere except along a line. From a dynamical perspective these are also asymptotically free objects moving with uniform velocities in which the only effect of interaction is a certain translation, which is usually termed a "phase shift". These configurations seem to be generic and stable under KPII evolution. We note that although line solitons were known as early as 1976 [10], they are not recovered by IST formulations in [4] as they are off the class studied. To explain the appearance of line solitons of KPII, new ideas are necessary that we discuss in Section 3.

In Section 4 we consider initial value problems (i.v.p.) for KPI. The IST involves using a nonlocal Riemann problem [3]. Despite the fact that KPI also has line soliton solutions, they are unstable (cf. [1, 2, 7]). However, it supports multi-lump solutions; these are localized configurations that decay rationally everywhere which possess simple dynamics. A class of such solutions is constructed in [11]. This class corresponds to asymptotically free objects that move with uniform velocities and decay rationally (we call this an N-lump solution if the result consists of N of these objects). The dynamics is characterized by an elastic interaction between the lumps, which is the typical behavior for localized solitonic solutions of integrable equations. From a spectral perspective, the N-lump solution is associated with wave functions that have simple poles in the spectral variable [3]. Section 5 develops the relevant ideas.

Localized real valued solutions with rational decay but exhibiting nontrivial asymptotic dynamics have been found for KPI and several other integrable equations [12]–[15]. The connection with the discrete spectrum of the relevant spectral problem has been discussed in detail for the case of the NS and KPI equations [16]–[18]. Spectrally, these potentials correspond to wave functions that have higher order poles and have an associated winding number Q, an integer greater than one; while standard noninteracting lumps correspond to $Q = 1$ (see also [19, 20] regarding direct methods used to examine the properties of these NS and KPI solutions, and [21, 22] in connection with the Davey–Stewartson eqs.). We develop the relevant ideas in Section 5.

Section 6 discusses the rigorous aspects of the direct and inverse problems associated with both KPI and KPII. Conditions for existence and uniqueness are given in terms of some natural functional norms. Finally in Section 7 we consider the role of the constraints with regard to the initial value problem. We find that, with data not satisfying these constraints, a KP evolution is possible but the corresponding solutions decay weakly, irrespective of the decay of the initial data.

2. The Cauchy problem for KPII with decaying potentials

As mentioned above, the KPII equation

$$(u_t + u_{xxx} + 6uu_x)_x + 3u_{yy} = 0 \tag{2.1}$$

is associated to the the heat operator with a potential defined as

$$L\psi \equiv [-\partial_y + \partial_{xx} + u(x,y)]\psi(x,y) = 0 \tag{2.2}$$

and

$$M\psi \equiv (\partial_t + 4\partial_{xxx} + 6u\partial_x + 3u_x + 3(\partial_y\partial_x^{-1}u) + h(k))\psi = 0\,. \tag{2.3}$$

We assume that the initial data $u(x, y, t = 0)$ is real, nonsingular and has suitable decay as $x^2 + y^2 \to \infty$ and that $\int dx\, u(x, y, t = 0) = 0$. Consider a new function $\mu(x, y, k)$ via $\psi(x, y) = \mu(x, y, k)\exp(ikx - k^2y)$ that includes the complex spectral parameter $k \equiv k_R + ik_I$ so μ satisfies

$$(-\partial_y + \partial_{xx} + 2ik\partial_x)\mu(x,y,k) = -u(x,y)\mu(x,y,k)\,. \tag{2.4}$$

We consider an eigenfunction of (2.4) that is bounded and tends to 1 as $x^2 + y^2 \to \infty$. Under this proviso, (2.4) can be converted into the integral equation

$$\mu(x,y,k) = 1 + \int dx'dy' G_(x - x', y - y', k)u\mu(x', y', k)\,. \tag{2.5}$$

$G(x, y, k)$ is an appropriate Green's function for (2.4), i.e., it solves

$$(-\partial_y + \partial_{xx} + 2ik\partial_x)G(x,y,k) = -\delta(x)\delta(y) \tag{2.6}$$

selected by requiring that the eigenfunction $\mu(k)$ is bounded. By Fourier transformation we find that

$$G(x,y,k) = \frac{1}{(2\pi)^2}\int_{-\infty}^{\infty} dp \int_{-\infty}^{\infty} dq \exp\left[i(px+iqy)\right]/(p^2+2pk+iq)\,. \quad (2.7)$$

Note that the integrand has integrable singularities at the points $p=q=0$ and $p=-2k_R, q=4k_Rk_I$; hence the Green's function is well defined. Note also that they depend separately on both real and imaginary parts of the spectral parameter. This property also holds for the Green's function, which is therefore analytic nowhere. Actually, upon integration one obtains

$$\begin{aligned} G(x,y,k) &= \frac{1}{2\pi}\int dp\, e^{ipx-y(p^2+2kp)}\Big[\theta(y)(\chi_{A_+}\theta(k_R)+\chi_{A_-}\theta(-k_R)) \\ &\quad -\theta(-y)(\chi_{A_+^c}\theta(k_R)+\chi_{A_-^c}\theta(-k_R)\Big], \end{aligned} \quad (2.8)$$

where we define the sets

$$A_+ \equiv (-\infty,-2k_R)\cup(0,\infty);\;\; A_- \equiv (-\infty,0)\cup(-2k_R,\infty), \quad (2.9)$$

$A_\pm^c$ stands for:

$$A_+^c \equiv I\!R - A_+ = [-2k_R, 0], \quad A_-^c \equiv I\!R - A_- = [0,-2k_R],$$

and the indicator function χ_A of the set $A\subset I\!R$ is defined as

$$\chi_{A_\pm}(p) = \begin{cases} 1, & p\in A_\pm, \\ 0, & p\notin A_\pm\,. \end{cases}$$

Note that, as a function of k, $G(x,y,k)$ is not holomorphic anywhere. Besides, it has different representations according to whether $k_R>0$, or $k_R<0$. However, it is easily seen to be continuous across $k_R=0$ (see below (2.11)).

We next list a few useful properties of the Greens function relevant in this connection.

i) The first one is the departure from holomorphicity measured by $\frac{\partial G}{\partial \bar k}(k)$. By direct calculation on formula (2.8) and using $\partial/\partial\bar k = \frac{1}{2}(\partial/\partial k_R + i\partial/\partial k_I)$ we find that

$$\frac{\partial G}{\partial \bar k}(x,y,k) = \lambda e^{iqy+ipx}\,. \quad (2.10)$$

where $\lambda \equiv -\frac{\operatorname{sign} k_R}{2\pi}$; $\quad p=-2k_R, \quad q\equiv 4k_Rk_I$.

ii) G is continuous across $k_R=0$ and

$$G(x,y,k=ik_I) = \frac{1}{\sqrt{4\pi y}} e^{k_Ix-k_I^2y-\frac{x^2}{4y}}\theta(y)\,. \quad (2.11)$$

iii) G satisfies the symmetry relationship

$$G(x, y, k) = e^{iqy+ipx} G(x, y, -\bar{k}) . \tag{2.12}$$

iv)

$$\lim_{|k|\to\infty} G(x, y, k) = 0 . \tag{2.13}$$

The integral equation (2.5) is of Fredholm type. Existence and uniqueness of its solutions can be studied by iteration [4, 23]. We prove in chapter 6 that existence and uniqueness is guaranteed when the initial data $u_0 \equiv u(x, y, 0)$ has L_1 and L_2 norms that satisfy $\|u\|_1 \|u\|_2^2 < 0.26$, or more concretely

$$54\|u\|_1\|u\|_2^2 C^2 < 1 \quad \text{where} \quad C \equiv \frac{1}{(2\pi)^2} \int_0^{2\pi} \frac{d\varphi}{\sqrt{|\sin\varphi|}} .$$

See (6.11) below.

If the latter condition is not satisfied, solutions ϕ to the homogeneous version of (2.5) might exist at some points k_α (the eigenvalues). This would be similar to what happens for KPI (see Section 5); however the existence of real, nonsingular, localized modes has never been proven. In this regard we note that corresponding to $k_R = 0$ it is easy to establish that no such modes exist; indeed, in this case, (2.11) shows that the integral equation (2.5) is of Volterra type, precluding the existence of eigenvalues.

The next step of the IST transformation is the so called inverse problem. It involves reconstructing the eigenfunction $\mu(k) \equiv \mu(x, y, k)$ from appropriate data. As has already been commented, $\mu(k)$ generically is not holomorphic anywhere. We evaluate its departure from holomorphicity by using the $\bar{\partial}$ derivative of the Greens function. By direct calculation, and using (2.10), we find the following integral equation for $\frac{\partial\mu}{\partial \bar{k}}$:

$$\frac{\partial\mu}{\partial\bar{k}}(x, y, k) = F(k)e^{iqy+ipx} + \int dx'dy'\, G(x - x', y - y', k)u\frac{\partial\mu}{\partial\bar{k}}(x', y', k), \tag{2.14}$$

where

$$F(k) \equiv \lambda \int dxdye^{-iqy-ipx} u\mu(x, y, k) \tag{2.15}$$

defines the bidimensional scattering data of the problem. Using (2.12) it then follows that μ and its $\bar{\partial}$ derivative are related as follows:

$$\frac{\partial\mu}{\partial\bar{k}}(x, y, k) = e^{iqy+ipx} F(k)\mu(x, y, -\bar{k}) . \tag{2.16}$$

Once the departure from holomorphicity of the function μ is evaluated, the inverse problem is formulated using the normalization

$$\mu(k) = 1 + \frac{\mu_1}{k} + O(1/k^2), \quad |k| \to \infty, \quad \text{where } 2ik\partial_x\mu_1 = -u , \tag{2.17}$$

which follows from (2.5), (2.13), and the generalized Cauchy formula

$$\mu(k) = \frac{1}{2\pi i}\int_{D_\infty} \frac{\mu(z)}{z-k}dz + \frac{1}{2\pi i}\int_D \frac{\frac{\partial\mu}{\partial\bar{z}}}{z-k}dz\wedge d\bar{z}\,, \tag{2.18}$$

where D is a large disk in the complex plane and D_∞ its boundary. In terms of the scattering data described above, we have

$$\mu(x,y,k) = 1 + \frac{1}{2\pi i}\int_C F(z)\mu(x,y,-\bar{z})\frac{e^{2iz_R(2z_I y - x)}}{z-k}dz\wedge d\bar{z}\,. \tag{2.19}$$

Eq.(2.17) implies that the potential is obtained from

$$u = \frac{1}{\pi}\frac{\partial}{\partial x}\int_C F(z)\mu(x,y,-\bar{z})e^{2iz_R(2z_I y - x)}dz\wedge d\bar{z}\,. \tag{2.20}$$

We next determine the temporal evolution of the scattering data. To this end, recall that $\psi(x,y,k)$ must also solve (2.3). Requiring the normalization $\psi(x,y) \to \exp(ikx - k^2 y)$ as $x^2+y^2 \to \infty$ to hold for all time implies, using the fact that the potentials vanish at ∞, that $h(k) = 4ik^3$. Note also that $\partial_{\bar{k}}\psi$ also solves $M\partial_{\bar{k}}\psi = 0$, since M does not depend on $\bar{k}$: $\partial_{\bar{k}}M\psi = M\partial_{\bar{k}}\psi = 0$. Inserting (2.16) in the latter equation one obtains that the data $F(k)$ must evolve in time via

$$\frac{\partial}{\partial t}F(k,t) = -4iF(k,t)(\bar{k}^3 + k^3)\,, \tag{2.21}$$

i.e.,

$$F(k,t) = F(k,0)e^{-4i(\bar{k}^3+k^3)t}\,. \tag{2.22}$$

Hence the solution $u(x,y,t)$ of the KPII equation is obtained from eqs. (2.19), (2.20) after the time dependence of F is inserted. In Section 6 rigorous results associated with the IST analysis (direct and inverse scattering) for the KPII eq. are described.

3. Initial value problem for KPII with potentials supported along a line

Natural interesting solutions of KPII, which seem to be generic and stable under KPII evolution, are the line solitons which are real and localized solutions that decay exponentially at infinity everywhere except along a line. We note that they are not included in the IST formulation of the latter section as they are off the class studied. The Cauchy problem in the plane for eq. (2.1) corresponding to physical data which is real, nonsingular and decays at infinity everywhere except along a line involves new ideas that have been developed recently [24]. Here we summarize our work regarding this situation. More precisely, we consider general initial data $u(x,y,t=0)$ that is real, nonsingular and decays at infinity everywhere except along a line $\mathcal{L}$ on the plane: $\mathcal{L} = \{(x,y)|\, x - vy = x_0\}$ (here x_0, v are the parameters defining the line) i.e.: $\lim_{x,y\to\infty} u(x,y) = 0$ for $(x,y) \notin \mathcal{L}$ and if $(x,y) \in \mathcal{L}$, $\lim_{x,y\to\infty} u(x,y) = u_\infty(x_0)$. Here $u_\infty(x_0)$ is an arbitrary given function that we assume is smooth and decaying: $\lim_{x_0\to\infty} u_\infty(x_0) = 0$. It may contain, in particular, a superposition of solitons. The potential is decomposed into an asymptotic nondecaying component along a line and a decaying contribution: $u(x,y) = u_\infty(x-vy) + U(x,y)$. (We note that other interesting studies regarding the inverse scattering associated with the heat operator with one soliton potential plus a decaying background have been published (cf. [25]- [30]).

Important features of these results are:

i) The inverse problem is a combination of a $\overline{\partial}$ contribution due to $U(x,y)$ plus a Riemann–Hilbert (RH) contribution arising from the reflection coefficient obtained from $u_\infty(x)$ plus pole contributions that are due to the discrete eigenvalues associated with $u_\infty(x)$. Generically, the spectrum is a combination of these three terms.

ii) If u_∞ vanishes the formalism reduces to the $\overline{\partial}$ solution of the last section.

iii) The Green's function is bounded in the spectral plane and is continuous away from the discrete eigenvalues; in particular, it has no line discontinuities.

iv) Completeness relations for the time independent Schrödinger operator are important in establishing the properties of the Green's function.

3.1. DIRECT AND INVERSE PROBLEMS

Given eq. (2.2), consider a new function $\mu(x,y,k)$ via

$$\psi(x,y,k) = \mu(x,y,k)\exp\Big[-ik(x-vy) - \frac{v}{2}x + \Big(\frac{v^2}{4} - k^2\Big)y\Big] \tag{3.1}$$

that includes the spectral parameter k, and new coordinates $\tilde{x} \equiv x - vy$, $\tilde{y} \equiv y$. In the new frame, and after dropping tildes, we find that the spectral problem (2.2) associated to KPII is conveniently transformed into

$$\begin{aligned} \hat{L}\mu(x,y,k) &= -U(x,y)\mu(x,y,k); \\ \hat{L} &\equiv -\partial_y + \partial_{xx} - 2ik\partial_x + u_\infty(x)\,. \end{aligned} \tag{3.2}$$

Following [24], we next discuss how to construct a Green's function $G(x, x', y, k)$ for the operator $\hat{L}$:

$$(-\partial_y + \partial_{xx} - 2ik\partial_x + u_\infty(x))G(x, x', y, k) = -\delta(x - x')\delta(y) . \qquad (3.3)$$

Define solutions $\phi_\pm(x, k), \psi_\pm(x, k)$ to the equations

$$\begin{aligned}(\partial_{xx} + k^2 + u_\infty(x))\phi_\pm(x, k) &= 0;\\ (\partial_{xx} + k^2 + u_\infty(x))\psi_\pm(x, k) &= 0\end{aligned} \qquad (3.4)$$

by requiring the conditions

$$\phi_\pm(x, k) \overset{x \to -\infty}{=} e^{\mp ikx} ; \qquad \psi_\pm(x, k) \overset{x \to \infty}{=} e^{\pm ikx} . \qquad (3.5)$$

(Note that $\psi_\pm(x, k)$ is not directly related to the wave function $\psi(x, y, k)$.) If $u_\infty(x)$ satisfies $\int(1 + |x|^2)|u_\infty(x)|dx < \infty$, the former functions exist and are analytic functions of $k \equiv k_R + ik_I$ on $C_\pm$ (the upper/lower half k planes), having limits on the boundary $\{k_I = 0\}$ (cf. [31, 32]). These limits satisfy the following relation:

$$\phi_+(x, k) = a(k)\psi_-(x, k) + b(k)\psi_+(x, k), \qquad k \in \mathbb{R} , \qquad (3.6)$$

for certain functions $a(k), b(k)$. The function $a(k)$ can be proven to be analytic in k on the upper half plane, having a denumerable set of (simple) zeros $\{k_j \equiv i\chi_j, \chi_j \in \mathbb{R}^+\}_{j=1 \ldots N}$. If $\psi_j(x) \equiv \psi_+(x, k_j)$ then $\phi_+(x, k_j) = \beta_j\psi_j(x)$ for some complex constant β_j. We call $a_+(k) \equiv a(k)$.

Define next

$$G_c(x, x', y, k) = G_{c+}(x, x', y, k)\theta(k_I) + G_{c-}(x, x', y, k)\theta(-k_I) , \qquad (3.7)$$

where

$$\begin{aligned} G_{c\pm} &= \frac{1}{2\pi} \int_{C'_\pm} dp\, e^{ik(x-x')-y(p^2+2kp)} g_\pm(x, x', p + k)\\ &\quad \times [\theta(y)(\chi_{A_+}\theta(k_R) + \chi_{A_-}\theta(-k_R))\\ &\qquad - \theta(-y)(\chi_{A_+^c}\theta(k_R) + \chi_{A_-^c}\theta(-k_R))] , \end{aligned} \qquad (3.8)$$

$$g_+(x, x', k) \equiv \frac{\phi_+(x, k)\psi_+(x', k)}{a_+(k)} , \qquad k_I > 0 \qquad (3.9)$$

$$g_-(x, x', k) \equiv -\frac{\phi_-(x', k)\psi_-(x, k)}{a_-(k)} , \qquad k_I < 0 , \qquad (3.10)$$

and $A_\pm$ are defined in (2.9). $C'_\pm$ is the contour along the real axis with a small semicircular indentation below (−)/above (+) the point $p = -k_R$ (corresponding to the zeros of $a(p + k)$ when $p + k = i\chi_j$). Let also

$$G_d(x, x', y, k) = i \sum_{|k_j| \geq |k_I|} C_j e^{ik(x-x')+(k^2+\chi_j^2)y} \psi_j(x)\psi_j(x')\theta(-y) \qquad (3.11)$$

(the discrete part of the Green's function). Finally we consider

$$G(x,x',y,k) = G_c(x,x',y,k) + G_d(x,x',y,k)\,. \tag{3.12}$$

It can be verified that the Green's function satisfies (3.3). Note that in the case that $u_\infty = 0$

$$\phi_\pm(x,k) = e^{\mp ikx}; \quad \psi_\pm(x,k) = e^{\pm ikx},\ a(k) = 1,\ \{k_j\} = \emptyset\,,$$

and then $G_d = 0$, $G_+ = G_-$, and the Green's function reduces (upon letting $k \to -k$) to that of the latter section. We also note that G is bounded and continuous everywhere except at the *points* $k = i\chi_j$, where it has limits from both sides $k_I \to \chi_j^+, k_I \to \chi_j^-$ but it has a jump.

We consider a solution $\mu(x,y,k)$ to (3.2) via

$$\mu(x,y,k) = h(x,k) + \int dx'dy' G(x,x',y-y',k) U \mu(x',y',k)\,, \tag{3.13}$$

where

$$\mu = \begin{cases} \mu_+(x,k), & k_I > 0 \\ \mu_-(x,k), & k_I < 0; \end{cases} \tag{3.14}$$

$$h(x,k) = \begin{cases} e^{ikx}\frac{\phi_+}{a}(x,k), & k_I > 0 \\ e^{ikx}\psi_-(x,k), & k_I < 0\,. \end{cases} \tag{3.15}$$

Operating with $\frac{\partial}{\partial \bar{k}}$ on equations (3.8), (3.13) we find the following integral equation for $\frac{\partial \mu_\pm}{\partial \bar{k}}$ corresponding to $k \in C_\pm$:

$$\begin{aligned} \frac{\partial \mu_\pm}{\partial \bar{k}}(k) &= \pi \sum_{j=1}^{N} C_j e^{-\chi_j x} \psi_j(x) \delta(k_R) \delta(k_I - \chi_j) \\ &\quad + e^{i(qy-px)} F_\pm(k) h_\pm(x,-\bar{k}) \\ &\quad + \int dx'dy' G_\pm(x,x',y-y',k) U \frac{\partial \mu_\pm(k)}{\partial \bar{k}}\,, \end{aligned} \tag{3.16}$$

where $\lambda \equiv -\frac{\operatorname{sign} k_R}{2\pi}$, $q \equiv 4k_R k_I$, $p = -2k_R$. The functions

$$\begin{aligned} F_+(k) &\equiv \lambda \int dx dy e^{-ikx-iqy} U \mu_+(x,y,k) \psi_+(x,-\bar{k}) \\ F_-(k) &\equiv -\frac{\lambda}{a_-(-\bar{k})} \int dx dy\, e^{-ikx-iqy} U \mu(x,y,k) \phi_-(x,-\bar{k})\,. \end{aligned} \tag{3.17}$$

define the bidimensional scattering data of the problem. It then follows that

$$\begin{aligned} \frac{\partial \mu_\pm}{\partial \bar{k}} &= \pi \sum_{j=1}^{N} C_j e^{-2\chi_j x} \mu_-(-i\chi_j) \delta(k_R) \delta(k_I - \chi_j) \\ &\quad + e^{i(qy-px)} F_\pm(k) \mu_\pm(-\bar{k})\,. \end{aligned} \tag{3.18}$$

The function $\mu(x,y,k)$ has a jump on the real axis that satisfies

$$[\mu_+ - \mu_-](x,y,k) = \rho(k)e^{2ikx}\ \mu_-(x,y,-k)\,. \tag{3.19}$$

Therefore, the global function $\mu(k)$ satisfies

$$\frac{\partial\mu}{\partial\bar{k}} = \pi\sum_1^N C_j e^{-2\chi_j x}\mu_-(-i\chi_j)\delta(k_R)\delta(k_I-\chi_j) \tag{3.20}$$

$$+\, e^{i(qy-px)}\sum_{\pm} F_\pm(k)\mu_\pm(-\bar{k})\theta(\pm k_I) - \frac{\rho(k)}{2i}e^{2ikx}\mu_-(-k)\delta(k_I)\,. \tag{3.21}$$

Once the departure from holomorphicity of the function μ is evaluated, the inverse problem follows from the generalized Cauchy formula (2.18)

$$\begin{aligned}\mu(k) = 1\ &+ \sum_1^N C_j e^{-2\chi_j x}\frac{\mu_-(-i\chi_j)}{k-i\chi_j}\\ &+ \frac{1}{2\pi i}\sum_{\pm}\int_{C_\pm} F_\pm(z)\mu_\pm(-\bar{z})\frac{e^{i(qy+2z_R x)}}{z-k}dz\wedge d\bar{z}\\ &+ \frac{1}{2\pi i}\int_{-\infty}^{\infty} dz_R\frac{\rho(z_R)e^{2izx}\mu_-(-z_R)}{z_R-k}\,.\end{aligned} \tag{3.22}$$

From (3.22) we obtain a coupled system of integral equations after appropriately evaluating the terms for $k\in C_\pm$. Then, the potential is obtained from

$$u\ =\ \frac{\partial}{\partial x}\Bigg[2i\sum_1^N C_j e^{-2\chi_j x}\mu_-(-i\chi_j)$$

$$-\frac{1}{\pi}\sum_{\pm}\int_{C_\pm} F_\pm(z)\mu_\pm(-\bar{z})e^{i(qy+2z_R x)}dz\wedge d\bar{z} \tag{3.23}$$

$$-\frac{1}{\pi}\int_{-\infty}^{\infty} dz_R\rho(z_R)e^{2iz_R x}\mu_-(-z_R)\Bigg], \tag{3.24}$$

and, after restoring the initial coordinates, u is given by (3.24) letting $x\to x-vy$.

The fact that ν_+ has simple poles at $k=i\chi_j$, $j=1,2,\ldots,n$, suggests that the inverse problem (3.22) has a double pole, viz. a nonintegrable singularity. However, the singularity is less severe. Indeed letting $\mu_+ = a(k)\mu_1$ implies that $F_+ = (\lambda F_1(k))/a(k)$, where

$$F_1(k) \equiv \int dxdye^{-iqy}U\mu_1(x,y,k)\psi_+(x,-\bar{k})\,.$$

Evaluating this at $k = ik_I \neq i\chi_j$, using eq. (3.2) and employing the boundary conditions one can prove that $\lim_{k_I \to \kappa_j} F_1(k_R = 0, k_I) = 0$. Actually, the singularity is at most a simple pole; one can prove the following:

$$\lim_{k_I \to \kappa_j} \frac{\partial F_1}{\partial k}(k_R = 0, k_I) = \lim_{k_I \to \kappa_j} \frac{\partial F_1}{\partial \bar{k}}(k_R = 0, k_I) = 0\,.$$

3.2. TEMPORAL EVOLUTION AND LINE SOLITONS

The temporal evolution is obtained using again (2.3). Corresponding to our boundary conditions we have, for all time, the normalization

$$\psi(x, y) \to \exp\Big[-ik(x - vy) - \frac{v}{2}x + \Big(\frac{v^2}{4} - k^2\Big)y\Big], \quad x^2 + y^2 \to \infty\,. \quad (3.25)$$

Hence we take $h(k) = 4(\frac{v}{2} + ik)^3$. This implies that

$$\frac{\partial}{\partial t} F_\pm(k, t) = F_\pm(k, t)(h(-\bar{k}) - h(k))\,, \quad (3.26)$$

and hence

$$F_\pm(k, t) = F_\pm(k, 0)e^{4[(\frac{v}{2} - i\bar{k})^3 - (\frac{v}{2} + ik)^3]t}\,. \quad (3.27)$$

The rest of the scattering data are found to satisfy

$$\frac{\partial}{\partial t}\rho(k, t) = 4\Big[\Big(\frac{v}{2} - ik\Big)^3 - \Big(\frac{v}{2} + ik\Big)^3\Big]\rho \quad (3.28)$$

$$\frac{\partial}{\partial t}C_j(t) = 4\Big[\Big(\frac{v}{2} + \chi_j\Big)^3 - \Big(\frac{v}{2} - \chi_j\Big)^3\Big]C_j(t)\,, \quad (3.29)$$

so that

$$\begin{aligned} \rho(k, t) &= \rho(k, 0)e^{4[(\frac{v}{2} - ik)^3 - (\frac{v}{2} + ik)^3]t}\,; \\ C_j(t) &= C_j(0)e^{4[(\frac{v}{2} + \chi_j)^3 - (\frac{v}{2} - \chi_j)^3]t}\,. \end{aligned} \quad (3.30)$$

Hence the solution $u(x, y, t)$ of the KPII equation is obtained from eqs. (3.19, 3.20) after the time dependence of $F_\pm, \rho, C_j$ is inserted.

When the continuous scattering data are all zero: $F_+(k) = F_-(k) = \rho(k) = 0$, the equations to recover the eigenfunction μ in modified coordinates (reminding the reader that $\tilde{x} = x - vy$) are

$$\zeta_l = C_l e^{-\chi_l \tilde{x}} - \sum_1^N C_l \frac{e^{-(\chi_j + \chi_l)\tilde{x}}}{i(\chi_l + \chi_j)}\zeta_j\,, \quad (3.31)$$

where we call $C_j\mu(-i\chi_j)e^{-\chi_j\tilde{x}} \equiv \zeta_j$. Solving this system in the same way as for KdV, restoring the original coordinates and introducing the temporal evolution we obtain that the solution is given in terms of this data by

$$u(x,y,t) = 2\frac{d^2}{dx^2}\log\det F(x-vy,t)\,, \tag{3.32}$$

where the $N\times N$ matrix $(F_{lj})_{N\times N}$ is defined by

$$F_{lj}(x,t) = \delta_{lj} - iC_l\,\frac{e^{-(\chi_j+\chi_l)x+8\chi_j(\chi_j^2+3\frac{v^2}{4})t}}{\chi_j+\chi_l}\,. \tag{3.33}$$

The solutions are line solitons, all of them moving along the same direction, and with properties similar to that of the standard soliton. In particular with $N=1$, $\chi_1\equiv\chi$

$$u(x,y,t) = 2\chi^2\mathrm{sech}^2\chi[(x-vy)-(4\chi^2+3v^2)t-x_0]\,, \tag{3.34}$$

where $x_0 = \frac{1}{2\chi}\log\left(\frac{C_1(0)}{2i\chi}\right)$.

4. KPI equation with decaying initial data

We now consider the i.v.p. for the KPI equation given by (1.1) with $\epsilon = i$ corresponding to decaying initial data $u(x,y,0)$. The relevant operator is now the nonstationary Schrödinger operator (1.2). We follow [3]. As above, we define a new eigenfunction $\mu(x,y,k) = \psi(x,y)e^{-ikx+ik^2y}$, that solves

$$0 = (i\partial_y + \partial_{xx} + 2ik\partial_x + u)\mu(x,y,k) \equiv L\mu\,. \tag{4.1}$$

We consider solutions μ to (4.1) that correspond to regular and decaying potentials $u(x,y)$ tending to 1 as $x^2+y^2\to\infty$. Eq. (4.1) can be converted into the integral equation

$$\mathcal{G}\mu(x,y,k) = 1\,, \tag{4.2}$$

where the operator $\mathcal{G}(k)$ is as follows

$$\begin{aligned}\mathcal{G}f(x,y) &= f(x,y) - \int_{-\infty}^{\infty}dx'\int_{-\infty}^{\infty}dy' G(x-x',y-y',k)(uf)(x',y')\,,\\ G(x,y,k) &= \frac{1}{(2\pi)^2}\int_{-\infty}^{\infty}dp\int_{-\infty}^{\infty}dq\exp\left[i(px+iqy)\right]/((p^2+2pk+q)\,.\end{aligned}$$

$G(x,y,k)$ is an appropriate Green's function selected by requiring that the eigenfunction $\mu(k)$ admits an analytic extension to the complex k plane. We denote the analytic extensions to the upper $(+)$/lower $(-)$ half planes by

$G_\pm(k)$. Note that the integrand has an integrable singularity at the point $p = q = 0$ if $k_I \neq 0$, while there is an entire parabola of singularities at the points $(p, -p^2 - 2k_R p)$ if $k_I = 0$. It follows that if $k_I \neq 0$ the Green's function is well defined. Upon integration one obtains

$$\begin{aligned} G_\pm(x,y,k) &= -\frac{1}{2\pi i}\int_{-\infty}^{\infty} dp \exp\left[i(px - (p^2 + 2kp)y)\right] \\ &\qquad \times \left(\theta(y)\theta(\mp p) - \theta(-y)\theta(\pm p)\right). \end{aligned} \tag{4.3}$$

A few useful properties of this function are the following:

$$\begin{aligned} G_+ - G_- &= \frac{1}{2\pi i}\int_{-\infty}^{\infty} \operatorname{sign}(p) \exp\left[i(px - (p^2 + 2kp)y)\right], \qquad (4.4) \\ &\operatorname{Im} k = 0\,; \\ \frac{\partial G(x,y,k)}{\partial k} &= -i(x - 2ky)G(x,y,k) - \frac{1}{2\pi i}\operatorname{sign}(\operatorname{Im} k), \qquad (4.5) \\ &\operatorname{Im} k \neq 0\,; \\ \bar{G}(x,y,\bar{k}) &= G(-x,-y,k)\,, \qquad (4.6) \end{aligned}$$

where the bar denotes complex conjugation.

Likewise, the analytic extensions to the upper $(+)$/lower $(-)$ half planes of $\mu(k)$, that we denote $\mu_\pm$, satisfy

$$\mu_\pm(x,y,k) = 1 - \int_{-\infty}^{\infty} dx' \int_{-\infty}^{\infty} dy' G(x - x', y - y', k)(u\mu_\pm)(x', y', k). \tag{4.7}$$

Existence of solutions to this equation is discussed in Section 6.

In order to determine the RH problem governing the inverse problem we next evaluate the jump $\Delta(k) \equiv \Delta(x,y,k) \equiv \mu_+(k) - \mu_-(k)$ that $\mu(k)$ has across the real axis. Note that Δ solves

$$\begin{aligned} \Delta = &\int_{-\infty}^{\infty} dx' \int_{-\infty}^{\infty} dy' (G_+ - G_-)(x - x', y - y', k)(u\mu_+)(x', y', k) \\ &+ \int_{-\infty}^{\infty} dx' \int_{-\infty}^{\infty} dy' G_-(x - x', y - y', k)(u\Delta)(x', y'). \end{aligned} \tag{4.8}$$

Hence using (4.4) we have

$$\begin{aligned} \Delta = &\int_{-\infty}^{\infty} T_+(k,l) e^{\beta(x,y,k,l)} dl \\ &+ \int_{-\infty}^{\infty} dx' \int_{-\infty}^{\infty} dy' G_-(x - x', y - y', k)(u\Delta)(x', y'), \end{aligned} \tag{4.9}$$

where

$$\begin{aligned} \beta(x,y,k,l) &\equiv e^{i[(l-k)x - (l^2 - k^2)y]} \\ T_+(k,l) &= \frac{\operatorname{sign}(l-k)}{2\pi i}\int_{-\infty}^{\infty} dx \int_{-\infty}^{\infty} dy (u\mu_+)(x,y,k) e^{\beta(x,y,k,l)}\,. \end{aligned}$$

It follows that

$$\Delta(x,y,k) \equiv \mu_+(k) - \mu_-(k) = \int_{-\infty}^{\infty} T_+(k,l)N(x,y,k,l)\,dl\,, \tag{4.10}$$

where $N(x,y,k,l), k,l \in I\!R$ solves

$$\begin{aligned} N(x,y,k,l) &= e^{\beta(x,y,k,l)} \\ &+ \int_{-\infty}^{\infty} dx' \int_{-\infty}^{\infty} dy' G_-(x-x',y-y',k)(uN)(x',y',k,l)\,. \end{aligned} \tag{4.11}$$

Note that the eigenfunctions $N(x,y,k,l), \mu_-(x,y,k)$ are related by

$$\mu_-(x,y,k) = N(x,y,k,k)\,. \tag{4.12}$$

We next establish the relationship between $N(x,y,k,l)$ and $\mu_-(x,y,k)$. Define $\hat{f}(x,y,k) = f(x,y,k)e^{i(kx-k^2y)}$; then note that $\hat{N}(x,y,k,l) = N(x,y,k,l)e^{i(kx-k^2)}$ solves

$$\begin{aligned} \hat{N}(x,y,k,l) &= e^{i(lx-l^2y)} \\ &+ \int_{-\infty}^{\infty} dx' \int_{-\infty}^{\infty} dy' \hat{G}_-(x-x',y-y',k)(u\hat{N})(x',y',k,l)\,. \end{aligned} \tag{4.13}$$

Hence $e^{-i(kx-k^2y)}\partial_k \hat{N}(x,y,k,l) \equiv \pi(x,y,k,l)$ solves

$$\begin{aligned} \pi(x,y,k,l) &= F(k,l) \\ &+ \int_{-\infty}^{\infty} dx' \int_{-\infty}^{\infty} dy' G_-(x-x',y-y',k)(u\pi)(x',y',k,l)\,, \end{aligned} \tag{4.14}$$

which implies that $\pi(x,y,k,l) = F(k,l)\mu_-(x,y,k)$. Here we define

$$F(k,l) = \frac{1}{2\pi i} \int_{-\infty}^{\infty} dx \int_{-\infty}^{\infty} dy(uN)(x,y,k,l)\,. \tag{4.15}$$

By integration (taking (4.12) into account) one obtains

$$\begin{aligned} N(x,y,k,l) &= \mu_-(x,y,l)e^{\beta(x,y,k,l)} \\ &\quad + \int_l^k F(p,l)\mu_-(x,y,p)e^{\beta(x,y,k,p)}\,p\,, \end{aligned} \tag{4.16}$$

which is the desired relationship. Eq. (4.16) expresses $N(x,y,k,l)$ in terms of $\mu_-(x,y,k)$ and of the data $F(p,l)$. Hence $N(x,y,k,l)$ must be determined directly from (4.11), and then $F(k,l)$ follows from (4.15).

It turns out that $T_+(k,l)$ and $N(x,y,k,l)$ do not show directly in the final representation of the jump. Only $F(k,l)$ and $\mu_-(k)$ do so. Indeed, using (4.10) and (4.16) we obtain (see [3])

$$\mu_+(x,y,k) - \mu_-(x,y,k) = \int_{-\infty}^{\infty} dl f(k,l)(u\mu_-)(x,y,l)e^{\beta(x,y,k,l)}\,, \tag{4.17}$$

and $f(k,l) \equiv \mathrm{sign}(k-l)F(k,l)$.

Alternatively, one can obtain the data $F(k,l)$ from the following integral equation (use (5.2.29) of ref. [8]):

$$F(k,l) = S(k,l) + \int_l^k S(k,p)F(p,l)\,dp\,, \tag{4.18}$$

where we define

$$S(k,l) = \frac{1}{2\pi i}\int_{-\infty}^{\infty} dx \int_{-\infty}^{\infty} dy\,(u\mu_-)(x,y,l)e^{\beta(x,y,k,l)}\,. \tag{4.19}$$

The determination of the jump is thereby reduced to solving the linear integral eqs. (4.7), (4.17) and either (4.11) or (4.18).

The inverse problem equation and the potential are obtained using the formulae (2.17), (2.18), (4.17). After some transformations one obtains

$$\mu(x,y,k) = 1 + \frac{1}{2\pi i}\int_{-\infty}^{\infty}\int_{-\infty}^{\infty} f(p,l)e^{\beta(x,y,p,l)}\mu_-(x,y,l)\frac{dp\,dl}{p-k+i0} \tag{4.20}$$

$$u(x,y) = -\frac{1}{\pi}\frac{\partial}{\partial x}\int_{-\infty}^{\infty}\int_{-\infty}^{\infty} f(p,l)e^{\beta(x,y,p,l)}\mu_-(x,y,l)\,dp\,dl\,. \tag{4.21}$$

Solutions to KPI are obtained after inserting the relevant time dependence, which in this case reads

$$F(k,l,t) = F(k,l,0)\exp 4i(l^3-k^3)t\,. \tag{4.22}$$

Notes

1. We assume that $\mu_\pm$ exist and have analytic extensions to the upper (+)/lower (−) half planes, respectively, and that they do not have singularities. (In Section 6 we discuss conditions that guarantee these properties. The addition of pole singularities is discussed in Section 5).
2. Formula (4.20) for the inverse problem was first derived in [5], although no connection to initial value problems was drawn; i.e., it remained to establish how the inverse problem and the scattering data follow from the direct side and to connect to the i.v.p.. This is accomplished from (4.11) and (4.15) (or from (4.18)).

5. Discrete modes for the KPI equation with decaying initial data

5.1. CHARACTERIZATION OF THE DISCRETE SPECTRUM OF THE KPI EQUATION

In the previous section we considered real, nonsingular potentials that decay at infinity, which correspond to solutions μ to (1.2) that satisfy (4.2). Existence and uniqueness of the Fredholm-like eq. (4.2) is obtained (see [33, 34]) when the initial data $u(x,y,0)$ satisfy the small norm condition

$$\int\int |\hat{u}(p,y,0)|dydp < 1\,, \qquad \hat{u}(p,y,0) \equiv \int dx e^{-ipx} u(x,y,0)\,. \tag{5.1}$$

Other conditions directly in terms of norms on the physical space are derived in Section 6 below.

When condition (5.1) does not hold, $\mu(k)$ may have singularities that we assume to be poles. The case when both the pole's multiplicity and a certain integer (referred to as the charge or index, see below (5.10)) are unity, as well as the necessary modification of (4.20), was discussed in [3]. In this section we discuss the situation when higher-order pole multiplicities and/or higher charges, are allowed. It seems that the notion of index is absent for one-dimensional problems and only appears in multidimensions; see also [45] regarding multiple poles for the Dirac operator in one dimension. Around any pole k_1 we have the Laurent expansion

$$\mu(k) = \sum_{r=0}^{\infty} \frac{\rho_r}{r!}(k-k_1)^r + \sum_{r=1}^{M} \frac{\Psi_r}{(k-k_1)^r} = \nu(k) + \frac{\Phi}{(k-k_1)} + \sum_{r=2}^{M} \frac{\Psi_r}{(k-k_1)^r}\,, \tag{5.2}$$

where $\Psi_1 \equiv \Phi$, $\nu(k_1) = \rho_0$ and

$$\nu(k) = \sum_{r=0}^{\infty} \frac{\rho_r}{r!}(k-k_1)^r \tag{5.3}$$

is the regular part of $\mu(k)$ at k_1, while $\mu(k) - \nu(k)$ is the singular part of $\mu(k)$ at k_1. The next result establishes the connection between the pole structure of $\mu(k)$ and the discrete spectrum.

5.1.1. *Result I*

Assume an eigenfunction $\mu(k)$ with a meromorphic representation exists:

$$\mu(k) = 1 + \sum_{\alpha=1}^{n} \left\{ \frac{\Phi_\alpha}{k-k_\alpha} + \sum_{r=2}^{M_\alpha} \frac{\Psi_{r,\alpha}}{(k-k_\alpha)^r} \right\}, \tag{5.4}$$

where $\Phi_\alpha, \Psi_{r,\alpha}$ depend only on x, y. Note that

$$\nu(k_\alpha) = 1 + \sum_{\beta=1,\beta\neq\alpha}^{n} \left\{ \frac{\Phi_\beta}{k_\alpha - k_\beta} + \sum_{r=2}^{M_\beta} \frac{\Psi_{r,\beta}}{(k_\alpha - k_\beta)^r} \right\}. \tag{5.5}$$

At any point k_1 the (negative) Laurent coefficients of $\mu(k)$ around k_1:

$$\{\nu(k_1), \Phi_1, \Psi_{r,1}\}, \qquad r = 2, \ldots, M, \quad M \geq 2\,,$$

satisfy the integral equations

$$\begin{aligned}\{\mathcal{G}\Psi_{M,1}\}_{k=k_1} &= \left\{\mathcal{G}\Psi_{M-1,1} + \frac{\partial \mathcal{G}}{\partial k}\Psi_{M,1}\right\}_{k=k_1} = \cdots \\ &= \left\{\mathcal{G}\Phi_1 + \frac{\partial \mathcal{G}}{\partial k}\Psi_{2,1} + \cdots + \frac{1}{(M-1)!}\frac{\partial^{M-1}\mathcal{G}}{\partial k^{M-1}}\Psi_{M,1}\right\}_{k=k_1} = 0\,,\end{aligned}$$

$$\left\{\mathcal{G}\nu + \frac{\partial \mathcal{G}}{\partial k}\Phi_1 + \frac{1}{2}\frac{\partial^2 \mathcal{G}}{\partial k^2}\Psi_{2,1} + \cdots + \frac{1}{M!}\frac{\partial^M \mathcal{G}}{\partial k^M}\Psi_{M,1}\right\}_{k=k_1} = 1. \tag{5.6}$$

Proof.
Take the limit $k \to k_1$ in eq. (4.2) and set the coefficients of

$$1/(k-k_1)^j, \qquad j = 0, 1, \ldots, M\,,$$

equal to zero.

If (5.1) is not met, then eigenfunctions or solutions $\omega(x, y)$ to the homogeneous version

$$\begin{aligned}\mathcal{G}_{k_1}\omega &\equiv \omega(x, y) \\ &\quad - \int dx' \int dy' G(x - x', y - y', k_1) u(x', y') \omega(x', y') = 0\end{aligned} \tag{5.7}$$

may exist at some points k_1 (the eigenvalues), where we denote $\mathcal{G}_{k_1} \equiv \mathcal{G}(k = k_1)$. These are called homogeneous solutions of the operator $\mathcal{G}_{k_1}$. The span (vector space) of all such functions is denoted $\operatorname{Ker} \mathcal{G}_{k_1}$. They define the discrete spectrum of the nonstationary Schrödinger operator (4.2), and are responsible for the appearance of the lumps, i.e., real, nonsingular potentials that decay rationally at infinity. The relevant theory for standard lumps was obtained in [3]. The general situation is however much more elaborated, and was developed in the articles [16]–[18].

We next review the relevant ideas (the detailed proofs of the main result can be found in ref. [17]). Call $\rho(x, y)$ the homogeneous adjoint solution, which satisfies

$$\mathcal{G}^{\dagger}_{k_1}\rho \equiv \rho(x, y) - \int dx' \int dy' \bar{G}(x' - x, y' - y, k_1) u(x, y) \rho(x', y') = 0\,. \tag{5.8}$$

We reiterate that certain properties of these operators, e.g. compactness and decay of their solutions, is assumed in this discussion.

5.1.2. *Result II*

At any point k_1 the number of linearly independent homogeneous solutions to $\mathcal{G}_{k_1}\omega_1 = 0$, $\mathcal{G}^{\dagger}_{k_1}\rho = 0$ and $\mathcal{G}_{\bar{k}_1}\omega_{\bar{1}} = 0$ is the same. Result I implies that the poles of $\mu(k)$ are eigenvalues, i.e., they are points of the discrete spectrum of (4.2), while $\Psi_{M,\alpha}$ are eigenfunctions of (4.2) at the eigenvalue k_α. By Result II we have that the discrete spectrum is an even dimensional set $\{k_j, \bar{k}_j\}_{j=1,\dots,N}$. As a consequence, pure meromorphic functions μ have the structure (5.4), where the poles occur in pairs $(k_j, k_{j+N}) \equiv (k_j, \bar{k}_j)$, and hence without loss of generality their position can be arranged such that $\mathrm{Im}\,(k_j) > 0; j = 1, 2, \dots, N$.

Suppose μ corresponds to the discrete spectrum of the Schrödinger operator and has 2N poles $\{k_j, \bar{k}_j\}_{j=1,\dots,N}$ with multiplicities $\{M_j, M_{\bar{j}}\}_{j=1,\dots,N}$, and $\mu \to 1$ as $k \to \infty$. Such an eigenfunction is given by

$$\mu(k) = 1 + \sum_{j=1}^{N} \left\{ \frac{\Phi_j}{k - k_j} + \frac{\Phi_{\bar{j}}}{k - \bar{k}_j} + \sum_{r=2}^{M_j} \frac{\Psi_{r,j}}{(k - k_j)^r} + \sum_{r=2}^{M_{\bar{j}}} \frac{\Psi_{r,\bar{j}}}{(k - \bar{k}_j)^r} \right\}, \tag{5.9}$$

where we have used $k_{j+N} = \bar{k}_j$ in (5.4).

Fundamental objects in our development are certain quantities that we call indices (winding numbers) or charges. At any pole $k_\alpha, \alpha = 1, \dots, 2N$ we define the index of the pole as

$$Q_\alpha \equiv \frac{1}{2\pi i} \,\mathrm{sign}(\mathrm{Im}\, k_\alpha) \int\!\!\int u\Phi_\alpha \, dx dy\,, \tag{5.10}$$

where Φ_α is the residue of μ at the pole. We also introduce the secondary indices

$$Q_{r,\alpha} \equiv \frac{1}{2\pi i} \,\mathrm{sign}(\mathrm{Im}\, k_\alpha) \int\!\!\int u\Psi_{r,\alpha} \, dx dy, \qquad 2 \le r \le M_\alpha\,. \tag{5.11}$$

Note: to stress the importance of the charges $Q_{1,\alpha}$, we drop the subscript and write $Q_\alpha \equiv Q_{1,\alpha}$.

5.1.3. *Result III*

Let $\mu(k)$ be given by eq. (5.9) with discrete spectrum $\{k_j,\ \bar{k}_j\}_{j=1}^N$. Then the indices Q_α have the following properties:

i)

$$\bar{Q}_\alpha = Q_{\bar{\alpha}} \quad \text{and} \quad \bar{Q}_{r,j} = Q_{r,\bar{j}}\,. \tag{5.12}$$

ii) Eigenfunctions corresponding to eigenvalues k_1, k_2 with $k_1 \neq \bar{k}_2$ are orthogonal with respect to the inner product

$$\langle f, g \rangle \equiv \frac{1}{\pi} \int \bar{f}_x \, g \, dx dy \tag{5.13}$$

If $k_1 = \bar{k}_2$ their scalar product equals the "charge" Q.

iii) The charge Q can take only positive integers values.

iv) Q has a topological interpretation as

$$Q = \text{winding number } H \tag{5.14}$$

where H is related to the residue Φ of the eigenfunction μ satisfying eq. (4.2) by

$$\Phi = i\frac{\partial}{\partial f}\log H + O(1/r^2), \qquad r^2 \equiv x^2 + y^2 \to \infty\,, \tag{5.15}$$

where $f \equiv x - 2k_1 y$, and the derivative is taken along a closed contour at infinity Γ_∞ surrounding the origin once. In fact we find $H = f^Q + O(1/r)$ as $r \to \infty$.

v) Solutions of the nonstationary Schrödinger operator (eq. (1.2),(4.2)) can have higher-order poles.

vi) Characterization of the class of localized potentials requires the following information: the position and multiplicity M_j of the poles at k_j, the associated indices Q_j and the dimension $r_j \geq 1$ of the null space of the corresponding eigenfunctions. In this regard we note that we have found that the Fredholm conditions yield at k_0 the following relationship between these numbers to hold at any eigenvalue

$$M_0 + M_{\bar{0}} + r_0 - 2 = Q_0\,, \tag{5.16}$$

and hence in particular $M_0 + M_{\bar{0}} - 1 \leq Q_0$, where $M_0, M_{\bar{0}}$ are the order of the poles of the eigenfunction at $k_0, \bar{k}_0$ resp. We have in fact found from direct methods for all these solutions that the following "two sided" inequality (see ref. [20]) holds:

$$M_0 + M_{\bar{0}} - 1 \leq Q_0 \leq M_0 M_{\bar{0}}\,. \tag{5.17}$$

vii) Meromorphic wave functions correspond to real, nonsingular and decaying potentials of the form $u(x,y) = 2\frac{\partial^2}{\partial x^2}\log F(x,y,\gamma,\delta,\ldots)$ where F is a certain polynomial and $\gamma, \delta, \ldots$ constants. They also correspond to localized, real, nonsingular, decaying solutions of the KPI equation when time dependence is inserted. The dynamical evolution of these configurations is nontrivial, as opposed to the situation for standard lumps or solitons.

Remark.

1. Standard lumps potentials of KPI correspond to having simple poles for which the index $Q = 1$ [3]. As we have just seen, in the general situation poles of any multiplicity are allowed, as well as higher order values of the index.
2. The inner product (5.13) had also appeared in [35].

5.2. EQUATIONS DETERMINING THE EIGENFUNCTION

In this section we discuss a method for determining equations which fix the eigenfunctions $\mu(k)$ satisfying eq. (4.2) with certain assumed analytic structure. Once these equations have been determined, we show in Section 5.3 how they can be used to solve for particular cases of eigenfunctions and related reflectionless multipole potentials which are real, nonsingular and decaying at infinity. A linear relationship is obtained.

5.2.1. *Simple poles*

We begin by analyzing simple pole solutions of eq. (4.2), i.e., special cases of eq. (5.4). Assume that the solution to eq. (4.2) has the following structure around $k = k_1$:

$$\mu(k) = \nu(k) + \frac{\Phi}{k - k_1}, \tag{5.18}$$

where $\nu(k)$ is regular at k_1 and $\nu(k) \underset{k\to\infty}{\longrightarrow} 1$. Define recursively

$$J_n(f, y), \qquad n = 0, \dots, \infty,$$

by

$$J_0 = 1, \qquad \partial_f J_n = iJ_{n-1}, \qquad (i\partial_y + \partial_{ff})J_n = 0\,, \tag{5.19}$$

where $f \equiv f(k_1) = x - 2k_1 y$ (see (5.15)). Note that, by (5.19), the first few J_n's are given by

$$\begin{aligned} J_0 &= 1, \quad J_1 = i(f + \gamma_s), \quad J_2 = \frac{1}{2}(J_1^2 - \delta_s - 2iy)\,, \\ J_3 &= \frac{1}{3!}(J_1^3 - 3J_1(\delta_s + 2yi) + \epsilon_s)\,, \end{aligned} \tag{5.20}$$

with $\gamma_s, \delta_s, \epsilon_s$ arbitrary constants (here the subscript s stands for a simple pole). Note that when we consider the KPI equation they become functions of time determined by the associate eq. (1.3)). We also write

$$H_n(x, y) \equiv J_n(x, y, \gamma_s = 0, \delta_s = 0, \epsilon_s = 0, \cdots).$$

Thus

$$H_0 = 1, \quad H_1 = if, \quad H_2 = -\frac{1}{2}(2iy + f^2), \quad H_3 = \frac{1}{3!}(6yf - if^3)\,.$$

One has the following:

5.2.2. *Result IV*
Let ρ be the solution of the adjoint equation at k_1, and n be the greatest integer for which the integrals

$$\int H_{j-1}\bar{\rho}(x,y)dxdy,\ j=0,\ldots,n,$$

exist.Suppose also that $\operatorname{Dim}\operatorname{Ker}\mathcal{G}_{k_1} = 1$. In what follows we define the index $Q = Q(k_1) \equiv Q_1$. Then:

i) $Q = n$.

ii) The Laurent coefficients are related by

$$J_0\rho_{n-1} + \cdots + J_{n-2}\rho_1 + J_{n-1}\rho_0 + J_n\Phi = 0. \tag{5.21}$$

iii) As $r^2 \equiv x^2 + y^2 \to \infty$ we have

$$\Phi \approx i\frac{\partial}{\partial f}\log H_n = i\frac{\partial}{\partial f}\log f^n = i\frac{n}{f},$$

and

$$Q = \text{ winding number } H_n. \tag{5.22}$$

Hence Q is a topological invariant for the Schrödinger operator that counts the number of zeros that $H_n(f,y)$ has as a function of f.

iv) The functions H_n and J_n of (5.19) (5.32) are given by

$$H_n(f,y) = \sqrt{\frac{(2iy)^n}{n!}}h_n\left(\sqrt{\frac{i}{2y}}x\right), \tag{5.23}$$

where the h_n are the standard Hermite polynomials. J_n are linear combinations of the former.

The results of [3] correspond to the case $n = 1$. This is the case of the well known lumps associated with KPI.

5.2.3. *Result V. Double poles*
Assume that the solution to (4.2) has the following structure around $k = k_1$:

$$\mu(k) = \nu(k) + \frac{\Phi}{k-k_1} + \frac{\Psi}{(k-k_1)^2}, \tag{5.24}$$

where $\nu(k)$ is regular at k_1 and tends to 1 as $k \to \infty$. Note that, as compared to the general formula (5.4), we simply write $\Psi \equiv \Psi_{2,1}$. Suppose also that $\Psi \neq 0$ and that $\operatorname{Dim}\operatorname{Ker}\mathcal{G}_{k_1} = 1$. Then

i) If $u = O(1/r^2)$ then $Q_2 = 0$, where

$$Q_2 = \frac{\text{sign}(\text{Im}\, k_1)}{2\pi i} \int\int u\Psi dx dy.$$

Let $Q \geq 2$.

ii) In the cases below one has the following relationships between μ and its Laurent coefficients:

$$\Phi + J_1 \Psi = 0\,, \tag{5.25}$$

and also if (with the coefficients ρ_r from (5.3)):

1. $Q = 2 \Rightarrow$

$$\rho_0 + J_{2,d}\Psi = 0\,, \tag{5.26}$$

where

$$J_1 \equiv J_{1,d} \equiv i(f + \gamma_d),\ J_{2,d} \equiv -\frac{1}{2}(J_1^2 + 2iy + \delta_d)\,,$$

and γ, δ_d are arbitrary constants (here d stands for double pole). We find it convenient to use a similar notation to that of the simple pole case.

2. If $Q = 3$ then

$$\rho_1 + (J_1 + 2\sigma)\rho_0 + \frac{1}{3} J_{3,d}\Psi = 0\,, \tag{5.27}$$

where

$$J_{3,d} \equiv -(J_1 + \sigma)^3 + 3\sigma^2(J_1 + \sigma) - 6i\sigma y + i\beta\,, \tag{5.28}$$

and β, δ_d, σ are arbitrary constants.

iii) If Φ, Ψ are related with μ by the above equations then as $r^2 \equiv x^2 + y^2 \to \infty$ we have

$$\Phi = i\frac{\partial}{\partial f_1} \log h_{n,d} + O(1/r^2),$$

and

$$Q = \frac{1}{2\pi i} \int_{\Gamma_\infty} \frac{\partial_f J_{n,d}}{J_{n,d}} df = \text{winding number of } J_{n,d}\,. \tag{5.29}$$

5.3. SCHRÖDINGER POTENTIALS

In this section we construct several families of decaying potentials for the nonstationary Schrödinger operator. Given an assumed pole structure (5.9) of μ around both k_1 and $\bar{k}_1$, the value of all the relevant charges is determined. The results of the last section show that under the proviso $\mathrm{Dim\,Ker}\,\mathcal{G}_{k_1} = 1$ a system of linear equations, which serve to fix the coefficients of the poles and thus $\mu(k)$, is obtained. We begin by considering the simplest possibility.

5.3.1. *1+1 poles*

Assume that μ is meromorphic with simple poles at points $k_1 \equiv a + ib, \bar{k}_1$:

$$\mu(k) = 1 + \frac{\Phi_1}{k - k_1} + \frac{\Phi_{\bar{1}}}{k - \bar{k}_1}\,. \tag{5.30}$$

Then the charges defined by eq. (5.10) are given by

i) $Q_{\bar{1}} = Q_1 = 1$. (We use the notation $Q_{\bar{1}} = Q(\bar{k}_1)$, $Q_1 = Q(k_1)$).

ii) The Laurent coefficients satisfy the system of linear algebraic equations (5.21) with $n = 1 : \rho_{0\alpha} = -i(f_\alpha + \gamma_\alpha)\Phi_\alpha, \quad \alpha = 1, \bar{1}\,.$

iii) Let $\gamma_2 = \bar{\gamma}_1, \quad z = x' - 2ay', \quad x' = x - x_0, \quad y' = y - y_0,$

$$x_0 = \frac{\gamma_I\, a - \gamma_R\, b}{b}, \quad y_0 = \frac{\gamma_I}{2b}, \tag{5.31}$$

where a, b, γ_R, γ_I are arbitrary real constants. Then the function (5.30) solves (4.18) with the Schrödinger potential $u(x, y) = 2\partial_{xx} \log F$, with

$$F = z^2 + 4b^2 y'^2 + \frac{1}{4b^2}\,. \tag{5.32}$$

The well known one lump solution is obtained from

$$u(x, y, t) = 2\frac{\partial^2}{\partial x^2} \log F.$$

We say that the solution has charge (index) one.

5.3.2. *Simple + double poles*

Assume that μ has the following structure:

$$\mu = 1 + \frac{\Phi_1}{k - k_1} + \frac{\Psi_2}{(k - k_1)^2} + \frac{\Phi_{\bar{1}}}{k - \bar{k}_1}\,. \tag{5.33}$$

Suppose also that $\mathrm{Dim\,Ker}\,\mathcal{G}_{k_1} = \mathrm{Dim\,Ker}\,\mathcal{G}_{\bar{k}_1} = 1$. Then

i) $Q_1 = 2$.

ii) The Laurent coefficients satisfy the following system of linear algebraic equations

a) At $k = k_1$ (see eqs. (5.25) and (5.26))

$$\Phi_1 + J_1 \Psi_2 = 0, \quad \rho_0 + J_{2,d} \Psi_2 = 0. \tag{5.34}$$

b) At $k = \bar{k}_1$ (see eq. (5.21))

$$\rho_1 + J_1 \rho_0 + J_2 \Phi_{\bar{1}} = 0. \tag{5.35}$$

Note that here we simply write $\Psi_{2,1} \equiv \Psi_2$.

iii) The function (5.33) solves (4.2) with the Schrödinger potential $u(x, y) = 2\partial_{xx} \log F$, where

$$\begin{aligned} F &= \left(z^2 - 4b^2 y'^2 + \delta_R\right)^2 + \left(2y'(1 + 2bz) + \frac{\gamma_I}{b} - \delta_I\right)^2 \\ &\quad + \frac{1}{b^2}\left((z - \frac{1}{2b})^2 + 4b^2 y'^2 + \frac{1}{4b^2}\right), \end{aligned} \tag{5.36}$$

δ_R, δ_I being arbitrary real constants.

5.3.3. *Two double poles*
Assume that μ has the following structure:

$$\mu = 1 + \frac{\Phi_1}{k - k_1} + \frac{\Psi_2}{(k - k_1)^2} + \frac{\Phi_{\bar{1}}}{k - \bar{k}_1} + \frac{\Psi_{\bar{2}}}{(k - \bar{k}_1)^2}, \tag{5.37}$$

where $\operatorname{Dim} \operatorname{Ker} \mathcal{G}_{k_1} = 1$. Then

i) $Q_1 = 3$.

ii) At $k = k_\alpha$, $\alpha = 1, \bar{1}$, the Laurent coefficients satisfy the following system of linear algebraic equations (obtained from (5.27),(5.28)):

$$\Phi_\alpha = -J_{1,\alpha} \Psi_{2,\alpha}; \tag{5.38}$$

$$\rho_{1,\alpha} + (J_{1,\alpha} + 2\sigma_\alpha)\nu_\alpha + \tfrac{1}{3} J_{3.\alpha,d} \Psi_{2,\alpha} = 0, \tag{5.39}$$

where the notation is defined as follows:

$$2.1 \equiv 2, \quad 2.\bar{1} \equiv \bar{2}, \quad 3.1 \equiv 3, \quad 3.\bar{1} \equiv \bar{3}.$$

iii) If

$$k_{\bar{1}} = \bar{k}_1, \quad \beta_{\bar{1}} = \bar{\beta}_1, \quad \gamma_{\bar{1}} = \bar{\gamma}_1 + 2i\bar{\sigma}_1, \quad \sigma_{\bar{1}} = \bar{\sigma}_1$$

the function (5.37) solves (4.2) with the Schrödinger potential determined by

$$\begin{aligned} F = & \left(z^3 - 12b^2y'^2z + 3(\sigma_R^2 - \sigma_I^2)z + 12b\sigma_I\sigma_R y' - 6\sigma_R y' + \hat{\beta}_R\right)^2 \\ & + \left(6bz^2y' - 8b^3y'^3 - 6\sigma_I\sigma_R z + 6b(\sigma_R^2 - \sigma_I^2)y' + 6\sigma_I y' - \hat{\beta}_I\right)^2 \\ & + \frac{9}{4b^2}\left\{\left(z - \sigma_I - \frac{1}{2b}\right)^2 + (2by' - \sigma_R)^2 + \frac{1}{4b^2}\right\} \\ & \quad \times \left\{\left(z + \sigma_I + \frac{1}{2b}\right)^2 + (2by' + \sigma_R)^2 + \frac{1}{4b^2}\right\}, \end{aligned} \tag{5.40}$$

where the coordinates z, x', y' are the same as those in eq. (5.31) except

$$x_0 = \frac{a}{b}(\gamma_I - \sigma_R) - (\gamma_R + \sigma_I), \quad y_0 = \frac{\gamma_I - \sigma_R}{2b},$$

and $\hat{\beta}$ is a minor redefinition of the original free parameter β. This solution depends on 8 parameters.

5.4. KPI SOLUTIONS

The above class of potentials is related to localized rationally decaying solutions of the KPI equation when the appropriate temporal dependence is inserted. Requiring $\mu(x, y, t, k)$ to solve also (1.3) implies a particular time evolution for the constants $\gamma, \beta, \ldots$ entering in the potential. Here we present the solutions corresponding to potentials derived earlier and then we study their dynamics.

Solutions to KPI are obtained when the constants of (5.20), (5.28) "evolve" as $\dot{k}_\alpha(t) = 0$ and

$$\begin{aligned} \dot{\gamma}_\alpha(t) = 12k_\alpha^2, \quad \dot{\delta}_\alpha(t) = -24ik_\alpha, \quad \dot{\epsilon}_\alpha(t) = 24i \\ \dot{\sigma}(t) = 0, \quad \dot{\beta}_\alpha(t) = 12(1 + 6k_\alpha\sigma), \quad \alpha = 1, \ldots, 2N. \end{aligned} \tag{5.41}$$

5.4.1. *Simple poles*

The simplest case corresponds to (5.32). Using (5.41) the KPI solution $u(x, y, t)$ is obtained by setting

$$x \to x - 12(a^2 + b^2)t, \qquad y \to y - 12at \tag{5.42}$$

in (5.32).

Corresponding to $n = 2N$ simple poles the resulting solution KPI solution, called the N lump solution, can be obtained. It describes a configuration that asymptotically is made up of N-humps all of them moving uniformly with distinct velocities $v_{jx} = 12(a_j^2 + b_j^2)$, $v_{jy} = 12a_j$ (here $k_j \equiv a_j + ib_j$). Upon interaction there is neither a phase shift nor a deflection, hence the asymptotic dynamics of these objects is trivial [11].

5.4.2. *Simple + double poles*
The KPI solution follows from (5.36) and (5.41) by taking

$$x \to x - 12(a^2 + b^2)t, \qquad y \to y - 12at, \qquad \delta_R \to \delta_R - 24bt\,. \tag{5.43}$$

Study of this KP solution shows that for long times it has two maxima or humps, each of which moves with distinct asymptotic velocities and diverge from one another, proportional to $|t|^{1/2}$ (this "anomalous scattering" was first discussed in [13] with particular choice of some of the relevant parameters, and the solution itself was first considered in [12]).

As $t \to -\infty$ the two maxima (+ denotes fast, − denotes slower hump) are located at

$$\begin{cases} x_\pm \sim 12(a^2 + b^2)t \pm \sqrt{\dfrac{-24a^2 t}{b}} + x_0 - \dfrac{1}{2b} \\ y_\pm \sim 12at \pm \sqrt{-\frac{6t}{b}} + y_0\,, \end{cases} \tag{5.44}$$

and at $t \to \infty$ the two maxima are located at

$$\begin{cases} x = 12(a^2 + b^2)t \pm \sqrt{24bt} + 0(t^{-1/2}) \\ y = 12at + y_0\,. \end{cases} \tag{5.45}$$

To describe the dynamics of this motion we consider a frame with the origin at (x_0, y_0) which moves with velocities $v_x = 12(a^2 + b^2), v_y = 12a$. With respect to this Galilean frame as $t \to -\infty$ one of the humps is located in the first quadrant while the other is the mirror image in the third quadrant. As $t \to \infty$ both of them travel on the x axis, the first one moving left, the second right. The asymptotic amplitude is $u(x_\pm, y_\pm) = 16b^2$ for both maxima as $|t| \to \infty$. The angle Ω the humps get deflected is given by:

$$\Omega = \arctan \frac{1}{2|a|}\,. \tag{5.46}$$

By properly choosing the free parameter a we can obtain a solution that scatters with any desired angle. This is in contrast to the situation for standard lumps where this angle is zero (i.e. there is no phase shift).

5.4.3. *Two double poles*
Solutions to KPI are obtained by taking

$$x \to x - 12(a^2+b^2)t, \quad y \to y - 12at, \quad \hat{\beta} \to \hat{\beta} + 12(1+6ib\sigma)t \qquad (5.47)$$

and $\sigma = \sigma_R + i\sigma_I$ constant in (5.40). This solution depends on 8 real parameters.

The expression for the solution simplifies if $\sigma = 0$, in which case

$$F = (z^3 - 12b^2y'^2z + 12t + \hat{\beta}_R)^2 + (8b^3y'^3 - 6bz^2y' + \hat{\beta}_I)^2 + \frac{9}{4b^2}\left((z - \frac{1}{2b})^2 + 4b^2y'^2 + \frac{1}{4b^2}\right)\left((z + \frac{1}{2b})^2 + 4b^2y'^2 + \frac{1}{4b^2}\right). \qquad (5.48)$$

This corresponds to a three humped structure with the position of the humps given asymptotically as $|t| \to \infty$ by

$$x_j \sim 12(a^2+b^2)t + x_0 + (12t)^{1/3}\left(\cos\left(\frac{2\pi j}{3}\right) - \frac{a}{b}\sin\left(\frac{2\pi j}{3}\right)\right) + o(1) \qquad (5.49)$$

$$y_j \sim 12at + y_0 - \frac{(12t)^{1/3}}{2b}\sin\left(\frac{2\pi j}{3}\right) + o(1), \quad j = 0, 1, \ldots . \qquad (5.50)$$

Thus the humps are located on the conic $(x' - 2ay')^2 + 4b^2y'^2 = (12t)^{2/3}$ moving with uniform speed $v_x = 12(a^2+b^2), v_y = 12a$ and whose semiaxes are first decreasing and then increasing at a rate $|t|^{1/3}$. Moreover in the moving frame the humps translate along three straight lines. After collision the formulae (5.49)–(5.50) show that the particles follow their original trajectories without any scattering or phase shift. Thus there is no scattering although there is nontrivial interaction.

Another subcase of interest is that given by $\sigma = i/(6b)$, which corresponds to

$$F = \left(z^3 - 12b^2y'^2z - \frac{z}{12b^2} + \hat{\beta}_R\right)^2 + \left(8b^3y'^3 - 6bz^2y' - 5\frac{y'}{6b} + \hat{\beta}_I\right)^2 + \frac{9}{4b^2}\left\{\left(z + \frac{2}{3b}\right)^2 + 4b^2y'^2 + \frac{1}{4b^2}\right\}\left\{\left(z - \frac{2}{3b}\right)^2 + 4b^2y'^2 + \frac{1}{4b^2}\right\}. \qquad (5.51)$$

We note that the temporal dependence drops out: (5.51) corresponds to a three-humped KP solution that is stationary (cf. also refs. [13, 14]) which moves with uniform velocity $v_x = 12(a^2+b^2) > 0$, $v_y = 12a$, and hence can be regarded as a solitary wave. Thus this solution, unlike the others considered which depend on all the variables z, y', t, also solves the Boussinesq equation. Solutions that behave asymptotically as a nonlinear superposition of these multipeaked solitary waves can be obtained in a straightforward way (see ref. [17]).

6. Rigorous results for the KP equation with decaying initial data

The analysis of the previous sections assumed the existence and uniqueness of solutions to the relevant integral eqs. (2.5) and (4.2).In this section we comment on some rigorous results to the direct problem associated with the KP equation with decaying initial data.

6.1. RIGOROUS RESULTS FOR THE KPII EQUATION WITH DECAYING INITIAL DATA

We consider eq. (2.5), and recall some standard notation in analysis. Denote by $L_\infty(I\!R^2), L_p(I\!R^2, dxdy) \equiv L_p(dxdy)$, or $L_p(I\!R^2, \rho dxdy) \equiv L_p(\rho dxdy)$ the set of all measurable functions $f(x, y)$ on the plane with the norms below finite; the relevant spaces and norms are defined as

$$
\begin{aligned}
L_\infty(I\!R^2) &= \{f \in I\!R^2 : \|f\|_\infty \equiv \sup_{x,y \in I\!R^2} |f(x,y)| < \infty\} \\
L_p(dxdy) &= \{f \in I\!R^2 : \|f\|_p \equiv \left(\int_{I\!R^2} |f(x,y)|^p dxdy\right)^{\frac{1}{p}} < \infty\}; \\
\tilde{L}_p &\equiv L_p(\rho dxdy) = \{f \in I\!R^2 : \\
& \qquad \|f\|_{\tilde{p}} \equiv \left(\int_{I\!R^2} |f(x,y)|^p \rho(x,y) dxdy\right)^{\frac{1}{p}} < \infty\}.
\end{aligned}
$$

Here $p \in I\!R^+$ and $\rho(x, y)$ is any measurable function on the plane.

6.1.1. *Result I*

i) The Green's function can be split as

$$G(x, y, k) = G_\infty(x, y, k) + G_2(x, y, k), \tag{6.1}$$

where

$$G_\infty(x, y, k) \in L_\infty(I\!R^2), \quad G_2(x, y, k) \in L_2(dxdy). \tag{6.2}$$

ii) Assume that $u(x, y, 0) \in L_2(dxdy) \cap L_1(dxdy)$ and that the condition

$$M(k) \equiv \|G_\infty\|_\infty \|u\|_1 + \|G_2\|_2 \|u\|_2 < 1 \tag{6.3}$$

is satisfied. Then there exists a unique solution μ to (2.5) on $L_\infty(I\!R^2)$ and

$$\|\mu(x, y)\|_\infty \leq 1/(1 - M). \tag{6.4}$$

iii) Assume that $u(x,y,0) \in L_\infty(I\!R^2) \cap L_1(dxdy)$. Define

$$\tilde{\mu} \equiv \mu(x,y)\sqrt{|u(x,y)|},$$

$$K(x,x',y,y',k) \equiv G(x-x',y-y',k)\sqrt{|u(x,y)|}\sqrt{|u(x',y')|}, \tag{6.5}$$

and consider the transformed equation.

$$\tilde{\mu}(x,y,k) = \sqrt{|u(x,y)|} + \int dx'dy' K(x,x',y,y',k)\tilde{\mu}(x',y',k). \tag{6.6}$$

iii.1) the kernel $K(x,x',y,y',k)$ is compact, with L_2 norm bounded as

$$\|K\|_2^2 \le \int dxdy\, dx'dy'\, |K(x,x',y,y',k)|^2 \le \tilde{M}^2 \tag{6.7}$$

$$\tilde{M}^2 \equiv 2\left(\|G_\infty\|_\infty^2 \|u\|_1^2 + \|G_2\|_2^2 \|u\|_1 \|u\|_\infty\right). \tag{6.8}$$

iii.2) If the condition $\tilde{M} < 1$ is satisfied, there exists a unique solution $\tilde{\mu}$ on $L_2(I\!R^2)$ to (6.6), and hence there also exists a unique solution $\mu(x,y)$ to (2.5) in the weighted space

$$L_2\Big(I\!R^2, \sqrt{u(x,y)}dxdy\Big) \equiv \tilde{L}_2$$

and

$$\|\tilde{\mu}(x,y)\|_2 = \|\mu(x,y)\|_{\tilde{2}} \le \|u\|_1^{\frac{1}{2}}/(1-\tilde{M}). \tag{6.9}$$

iii.3) If $\tilde{M} < \infty$ and no homogeneous modes exist in this space, there exists a unique solution $\mu(x,y)$ to (2.5) in the space $\tilde{L}_2$, regardless of the size of $\tilde{M}$.

Proof. We sketch the main ideas. Recall (2.7)

$$G(x,y,k) = \frac{1}{(2\pi)^2}\int_{-\infty}^{\infty} dp \int_{-\infty}^{\infty} dq\, \hat{G}(p,q)\exp[i(px+iqy)],$$

where

$$\hat{G}(p,q) \equiv \frac{1}{p^2+2pk+iq} \equiv \frac{1}{p^2+2pk+iq}\left(\chi_D + (1-\chi_D)\right)$$

$$\equiv \hat{G}_\infty(p,q) + \hat{G}_2(p,q).$$

Here D is a convenient region enclosing the points $p=q=0$ and $p=-2k_R$, $q = 4k_Rk_I$, $D^c \equiv R - D$, and by χ_D we denote the characteristic function of the set D (which is defined under eq. (2.9)). Note that the integrand has singularities at these points, but the singularities are integrable. Note also that $1/|p^2 + 2kp + iq|^2 = O(|p|^4 + |q|^2)^{-1}$ is integrable away from the singularities, hence $\hat{G}_\infty(p,q) \in L_1(dpdq)$, $\hat{G}_2(p,q) \in L_2(dpdq)$, and

$$G_\infty \equiv \frac{1}{(2\pi)^2}\int_D \frac{dpdqe^{i(px+iqy)}}{p^2+2pk+iq}; \qquad \|G_\infty\|_\infty \le \frac{1}{(2\pi)^2}\|\hat{G}_\infty\|_1$$

$$G_2 \equiv \frac{1}{(2\pi)^2}\int_{D^c} \frac{dpdqe^{i(px+iqy)}}{p^2+2pk+iq}; \qquad \|G_2\|_2 = \frac{1}{2\pi}\|\hat{G}_2\|_2$$

satisfy (6.2).

Next, if $\mathcal{G}\mu$ stands for the second term on the RHS of (2.5) one has

$$\|\mathcal{G}\mu\|_\infty \equiv \sup_{x,y\in I\!R^2}\left|\int G(x-x',y-y',k)u\mu(x',y',k)dx'dy'\right|$$
$$\le \|\mu\|_\infty \sup_{x,y\in I\!R^2}\int dx'dy'\left|u(x',y')(G_\infty+G_2)(x-x',y-y')\right|$$
$$\le \|\mu\|_\infty\Bigg[\|G_\infty\|_\infty\int dx'dy'|u(x',y')|$$
$$+\left(\int |G_2|^2(x,y)dxdy\int |u^2(x',y')|dx'dy'\right)^{\frac{1}{2}}\Bigg] \equiv \|\mu(x,y)\|_\infty M,$$

and the operator $\mathcal{G}$ is a contraction on L_∞if $M<1$.

To prove (6.7) notice that if $r \equiv (x,y)$, $dr \equiv dxdy$, etc., we have

$$\int drdr'|K(r,r')|^2 = \int drdr'|(G_\infty+G_2)|^2(r-r')|u(r)||u(r')|$$
$$\le 2\int drdr'|u(r)||u(r')|[|G_\infty(r-r')|^2+|G_2(r-r')|^2]$$
$$\le 2\|G_\infty\|^2_\infty dr'\int drdr'|u(r)||u(r')| + 2\|u\|_\infty\int dr|G_2(r)|^2\int dr'|u(r')|$$
$$\equiv \tilde{M}^2\,.$$

Assuming that $u(x,y)\in L_\infty(I\!R^2)\cap L_1(dxdy)$, it follows that the bound (6.7) applies and the kernel $K(x,x',y,y')$ is compact on $L_2(dxdy)$; if, besides, $\tilde{M}<1$, the RHS of (6.6) is a contraction on $L_2(dxdy)$. The rest follows from standard properties of compact or contraction operators.

Remark.

1. The basic idea of the proof, namely, the decomposition (6.1), appears in [4]. With this, the rest follows easily. The results of iii) were also pointed out in [4], and solvability on $L_\infty(I\!R^2)$ with the bound (6.4) in [23].

2. As the proof shows, the functions $M, \tilde{M}$ of (6.3) — and hence the bounds (6.4), (6.9) — depend on k. The following result gets around this difficulty:

6.1.2. *Result II*

i) $M(k)$ is bounded:

$$M(k) \leq \left(54\|u\|_1\|u\|_2^2C^2\right)^{\frac{1}{3}}, \quad C \equiv \frac{1}{(2\pi)^2}\int_0^{2\pi}\frac{d\varphi}{\sqrt{|\sin\varphi|}}. \tag{6.10}$$

If the condition

$$54\|u\|_1\|u\|_2^2C^2 < 1 \tag{6.11}$$

is satisfied, there exists a unique solution μ to (2.5) on $L_\infty(I\!R^2)$, and

$$\|\mu(x,y)\|_\infty \leq \left[1 - (54\|u\|_1\|u\|_2^2C^2)^{\frac{1}{3}}\right]^{-1}.$$

ii) $\tilde{M}(k)$ is bounded:

$$\tilde{M}(k) \leq 3^{\frac{1}{2}}2^{\frac{5}{6}}\left(\|u\|_\infty\|u\|_1^2C^2\right)^{\frac{1}{3}}. \tag{6.12}$$

If the condition

$$3^{\frac{1}{2}}2^{\frac{5}{6}}\left(\|u\|_\infty\|u\|_1^2C^2\right)^{\frac{1}{3}} < 1 \tag{6.13}$$

is satisfied, there exists a unique solution $\mu(x,y)$ to (2.5) in the weighted space $L_2(I\!R^2, \sqrt{u(x,y)}dxdy) \equiv \tilde{L}_2(I\!R^2)$, and

$$\|\mu(x,y)\|_{\tilde{2}} \leq \|u\|_1^{\frac{1}{2}}\left[1 - 3^{\frac{1}{2}}2^{\frac{1}{6}}(\|u\|_\infty\|u\|_1^2C^2)^{\frac{1}{3}}\right]^{-1}.$$

iii) If $\|u\|_\infty$, $\|u\|_1 < \infty$, and no homogeneous modes exist in the weighted space $\tilde{L}_2(I\!R^2)$, there exists a unique solution $\mu(x,y)$ to (2.5) in the space $\tilde{L}_2$.

Estimates (6.10), (6.11) are obtained by choosing the region D in a suitable way. We skip the proof. See also [36] regarding the non existence of exponentially decaying homogeneous modes, and [37] for another bound on $M(k)$.

The solvability of the inverse problem was considered in [23]. The basic result is the following:

6.1.3. *Result III*

Assume that (6.2) is satisfied. Then:

i) There exists $g(k)$ such that

$$\|g\|_\infty < \infty \quad \text{and} \quad \lim_{k_R\to\infty} g(k) = 0\,;$$

$$|F(k)| \leq \tfrac{\|u\|_1}{1-g(k)(\|u\|_1+\|u\|_2)}\,. \tag{6.14}$$

ii) If at the initial time u has two derivatives in

$$L_1(dxdy)\cap L_2(dxdy)\cap L_\infty,$$

then

$$F(k)\in L_2(k_I dk_R dk_I)\cap L_\infty\,,$$

and the solution to (2.5) also leads to a potential (2.20) which solves the KPII eq. (2.1).

iii) More generally, if at time 0 u has n derivatives in

$$L_1(dxdy)\cap L_2(dxdy)\cap L_\infty,$$

$$\partial^{(\gamma_1,\gamma_2)}u \in L_1(dxdy)\cap L_2(dxdy)\cap L_\infty \tag{6.15}$$

for some multi-index (γ_1,γ_2), then

$$F(k) = O\left(\frac{1}{1+|k_R k_I|^{\gamma_2}+|k_I|^{\gamma_1}}\right), \tag{6.16}$$

and the solution to (2.5) also leads to a potential (2.20) which solves the KPII eq. (2.1).

The reader may wish to consult [23] for the details of the proof.

6.2. RIGOROUS RESULTS FOR THE KPI EQUATION WITH DECAYING INITIAL DATA

We now consider the KPI equation (4.2).

Following [33], we introduce a a function $\hat{\nu}(p,y)$ that satisfies the allied integral equation

$$\hat{\nu}(p,y) = \hat{H}(p,y) + \hat{\mathcal{G}}(\hat{\nu})\,, \tag{6.17}$$

where $\hat{H}(p,y)$ and the integral operator $\hat{\mathcal{G}}(\hat{\nu})$ are defined as

$$\hat{H}(p,y) \equiv \textstyle\int dy'\hat{u}(p,y')\Xi(p,y-y')\exp\left[i(y-y')(p^2+2kp)\right]$$

$$\hat{\mathcal{G}}(\hat{\nu}) \equiv \textstyle\int dy'\Xi(p,y-y')\exp\left[i(p^2+2kp)(y-y')\right)]\int dl\hat{u}(l,y)\hat{\nu}(p-l,y)$$

$$\Xi(p,y) \equiv (\theta(y)\theta(\mp p) - \theta(-y)\theta(\pm p))\,. \tag{6.18}$$

Note that the above equation is formally obtained from (4.2) letting $\hat{\nu}$ be the Fourier transform of $\nu \equiv \mu - 1$ and $\hat{u}(p,y)$ that of $u(x,y)$:

$$\hat{u}(p,y) \equiv \int dx e^{-ipx} u(x,y); \quad \hat{\nu}(p,y) \equiv \int dx e^{-ipx} \nu(x,y). \tag{6.19}$$

To study (6.17) we define the norm

$$\|\hat{\nu}\|_{1,\infty} \equiv \sup_y \int dp |\, \hat{\nu}(p,y)|\,. \tag{6.20}$$

6.2.1. *Result IV*
Assume that $\hat{u}(p,y) \in L_1(dpdy)$ has small norm, concretely

$$M \equiv \int\int dydp\, |\hat{u}(p,y)| < 1\,. \tag{6.21}$$

Then a solution $\hat{\nu} \in L_1(dpdy)$ to eq. (6.17) exists, is unique and satisfies the bound

$$\|\hat{\nu}\|_{1,\infty} \leq \frac{M}{1-M}\,. \tag{6.22}$$

We refer to [33] for the proof. This result has been sharpened, with estimates for the derivatives given, in [34].

Note. The results of [33] imply that the Fourier transform of $\hat{\nu}(p,y)$,

$$\nu(x,y) \equiv \frac{1}{2\pi} \int dp e^{ipx} \hat{\nu}(p,y)\,,$$

exists and is bounded:

$$\|\nu(x,y)\|_\infty \leq \frac{M}{2\pi(1-M)}\,. \tag{6.23}$$

However it does not necessarily follow that $\nu(x,y)$ defined by (6.2.1) solves eq. (4.2). The following result fills this gap and establishes existence to solutions to both (6.17) and (4.2):

6.2.2. *Result V*
Consider the eq. (6.17), with $\hat{u}(p,y)$ given. Let

$$\begin{gathered} u(x,y) \equiv \tfrac{1}{2\pi} \int dp e^{ipx} \hat{u}(p,y) \\ I_1(x) \equiv \int dy |u(x,y)|, \quad I_\infty(x) \equiv \sup_y |u(x,y)|\,. \end{gathered} \tag{6.24}$$

Assume that:

i) $u(x, y)$ is in $L_1(dx)$ for every y and in $L_1(dy)$ for every x.

ii) $I_1 I_\infty \in L_{\frac{1}{2}}(dx)$, i.e.,

$$\int dx \sqrt{I_1(x) I_\infty(x)} < \infty . \tag{6.25}$$

iii) Eq. (6.21) holds.

Let $\hat{\nu}$ be the unique solution to (6.17). Then its Fourier transform $\nu(x, y)$ solves (4.2), i.e., there exists a solution to (4.2) in the subspace of $L_\infty(I\!R^2)$ consisting of those functions having a Fourier transform that belongs to $L_1(dxdy)$. Moreover, $\nu(x, y)$ satisfies the estimate (6.23).

The latter result establishes existence to solutions to both (6.17) and (4.2). This suggests that one can get around the need to consider the allied integral eq. (6.17), and work only in physical space. Indeed, one has the following:

6.2.3. *Result VI*

Consider eq. (4.2) with $u(x, y)$ given and satisfying ii) above with (6.25) sufficiently small: $\int dx \sqrt{I_1(x) I_\infty(x)} \ll 1$. Then there exists a unique solution to (4.2) with bounded norm $\|\nu(x, y)\|_\infty$.

Note. Conditions i), ii) above hold if $u(x, y) \in L_\infty(I\!R^2) \cap L_1(dxdy)$ and $I_\infty(x) \in L_1(dx)$. Hence existence and uniqueness to (4.2) are guaranteed if

$$\int dxdy |u(x, y)| \ll 1, \qquad \int dx \sup_y |u(x, y)| \ll 1 . \tag{6.26}$$

The proof will be skipped.

The rigorous theory of the inverse problem was considered in [34]. The basic result reads as follows:

6.2.4. *Result VII*

Assume that the conditions of result IV hold, and that $\hat{u}(p, y)$ has enough derivatives in $L_1(dp\, dy)$. Then the equations of the direct and inverse problems (6.17) and (4.20) have both a unique solution, and the potential (4.21) with time dependence (4.22) inserted is real.

The following results improves on the solvability of the inverse problem (for both KPI and KPII) [37]:

6.2.5. *Result VIII*
Assume that the potential $u(x,y)$ is real with the relevant norm bounded (but not necessarily small). Then, the inverse problem (for both KPI and KPII) is always solvable.

7. On constraints for KP and the associated initial value problem

In this section we study and comment on the difficulties inherent to the KP equations (both for KPI and KPII) due to its nonlocal nature. To understand the onset of these difficulties, we write (1.1) in evolution form, i.e.,

$$u_t + 6uu_x + u_{xxx} + 3\epsilon^2 \partial_x^{-1} u_{yy} = 0\,, \tag{7.1}$$

whereupon it follows naturally that if u and its derivatives decay at infinity then the solution should satisfy the condition

$$\chi_1(y,t) = \partial_{yy} \int_{-\infty}^{\infty} u(x,y,t)dx = 0. \tag{7.2}$$

Indeed, (7.2) has so far been assumed for the initial data $u(x,y,t=0)$.

Note that (7.2) is not preserved in time, and consequently further constraints (obtained by differentiation of (7.2) with time) are necessary, e.g.

$$\chi_2(y,t) = \partial_{yy} \int_{-\infty}^{\infty} dx \left(\int_{-\infty}^{x} dx' - \int_{x}^{\infty} dx' \right) u(x',y,t) = 0. \tag{7.3}$$

This procedure yields a hierarchy of constraints $\chi_n(y,t) = 0$, $n = 1, \ldots, \infty$, which many researchers impose at $t = 0$. For a more complete discussion on this matter and the relation with conservation laws, we refer the reader to [38]–[40].

We note that in the physical reduction of KP $\chi_1(y,0)$ is generically not zero, i.e, $\chi_1(y,0) \neq 0$ corresponds to the physical situation. It is important to decide to what extent are the constraints $\chi_n(y,t) = 0$, $n = 1, \ldots, \infty$, necessary, and whether there exists a solution corresponding to initial data $u(x,y,t=0)$ not satisfying (7.2). This delicate problem was addressed in [41]. Note that the difficulty is already present at the linear level, corresponding to dropping the term uu_x of (7.1). As the linear theory is already substantial enough and encompasses the main difficulties of the problem, we shall concentrate on it henceforth. In this context the nonlinear problem is reducible to the linear one as for such boundary conditions, the linear problem has all the critical features. Note also that similar issues arise in the context of one-dimensional problems, e.g. the Sine-Gordon equations in light-cone coordinates ($u_{xt} = \sin u$). Such i.v.p.'s with initial time discontinuity are physically important and generic. It was found in [41] that, even if

$\chi_1(y,0) \neq 0$, nevertheless $\chi_1(y,t) = 0$ for any $t > 0$, and the corresponding KP evolution exists. A related issue which will also be discussed is that of determining the natural choice for ∂_x^{-1}. It was proven in [41] (see (7.14)) that for $t > 0$ all choices are equivalent, but they produce different results at $t = 0$. In [42] it was shown how the latter discontinuity is resolved in a physical context (e.g. from water waves) and for more general KP systems.

7.1. THE LINEAR THEORY

A solution to the linear version to (7.1) may be derived formally by using Fourier transforms. We find

$$u(x,y,t) = \fint_{-\infty}^{\infty} dp \int_{-\infty}^{\infty} dq\, e^{i[px+qy+(p^3-3\epsilon^2 q^2/p)t]}\phi(p,q,0), \tag{7.4}$$

where $\fint_{-\infty}^{\infty}$ denotes the Cauchy principal value integral, and the choice $\partial_x^{-1} = \int_{-\infty}^{x} dx' - \int_{x}^{\infty} dx'$ is made to eliminate a delta function $\delta(p)$ in the exponent of (7.4).

In order to guarantee that (7.4) is an actual solution of (7.1), the following conditions are imposed:

$$\begin{aligned} (1+|p^3|+|q^2|)\phi(p,q,0) &\in L^1(dpdq), \\ \phi(0,q,0) &\in L^1(dq), \\ u(x,y,0) &\in L^1(dxdy)\,. \end{aligned} \tag{7.5}$$

This requirement guarantees that $u(\cdot\,,\cdot,t)$ is of class $C^{3,2}$ for $t > 0$, and the relevant derivatives can be carried out under the integral sign and vanish as $r \to \infty$. The formal time derivative of this expression brings down a factor of p^{-1}:

$$\frac{\partial u}{\partial t} = \fint_{-\infty}^{\infty} dp \int dq\, i\left(p^3 - \frac{3\epsilon^2 q^2}{p}\right) e^{i\left[px+qy+\left(p^3-\frac{3\epsilon^2 q^2}{p}\right)t\right]}\,\phi(p,q,0)\,, \tag{7.6}$$

thereby producing a a singularity rendering the existence of $\partial u/\partial t$ unclear. Existence is guaranteed requiring the condition

$$\phi(0,q,0) = 0\,. \tag{7.7}$$

Note. Eq. (7.7) implies both (7.2) and

$$\phi(p=0,q,t) = 0\,, \quad \text{where} \quad \phi(p,q,t) = \phi(p,q,0)e^{i(p^3-3\epsilon^2 q^2/p)t}\,. \tag{7.8}$$

Further, u is C^1 with respect to t and (7.6) holds with $\frac{\partial u}{\partial t} \to 0$ as $r \to \infty$.

If the condition (7.7) is not satisfied, the issue of whether (7.4) solves (7.1) is much more delicate. Essentially, it amounts to proving that the integral (7.6) exists, that all derivatives $\partial_{yy}, \partial_{xxx}, \partial_x^{-1}$ and ∂_t can be taken within the integral sign and that the integrals involved in (7.6) commute. The difficulty can be put down to the fact that the integral is neither absolutely convergent, nor well defined, in principle, as a Cauchy integral; hence the standard analysis theorems do not apply. In spite of this difficulty in [41] it was proven that for all $t > 0$ one has the following (note that it is sufficient to consider positive times, since we are considering an evolution equation):

7.1.1. *Result I*

Assume that the conditions (7.5) hold, and that in the vicinity of the origin $\phi(0, q, 0) \neq 0$ and

$$|\phi(p, q, 0) - \phi(0, q, 0)| \leq g(q)|p|^\delta \quad \text{for some} \quad \delta > 0,\ g \in L_1 . \tag{7.9}$$

Then one has:

a) The iterated integrals of (7.6) exist and are equal, i.e.,

$$\left(\fint_{-\infty}^{\infty} dp \int_{-\infty}^{\infty} dq - \int_{-\infty}^{\infty} dq \fint_{-\infty}^{\infty} dp\right) \epsilon^{i[px+qy+(p^3-3\epsilon^2 q^2/p)t]} \frac{\phi(p,q,0)}{p} = 0 . \tag{7.10}$$

The integral appearing in the right hand side of (7.6) exists for all t as a principal value integral.

b) Formula (7.6) holds, i.e., differentiation with respect to t under the integral sign in (7.4) is permitted.

c) For all time $t > 0$ the following formula holds:

$$-\partial_x^{-1} u_{yy}(x, y, t) = \fint_{-\infty}^{\infty} dp \int dq\, q^2 \frac{e^{i\left[px+qy+\left(p^3-\frac{3\epsilon^2 q^2}{p}\right)t\right]}}{ip} \phi(p, q, 0) . \tag{7.11}$$

That is, interchange of the x'-integral with those over p and q is permitted. Here

$$\partial_x^{-1} = a \int_{-\infty}^{x} - b \int_{x}^{\infty}, \qquad a + b = 1.$$

d) At time $t = 0$ one has that

$$-\partial_x^{-1} u_{yy}(x, y, 0) = \fint_{-\infty}^{\infty} dp \int dq \frac{e^{i(px+qy)}}{ip} \phi(p, q, 0) + \pi(2a-1)D(y), \tag{7.12}$$

where

$$D(y) \equiv \int dq e^{iqy} q^2 \phi(0,q,0) = -\frac{1}{2\pi}\chi_1(y,0)\,. \tag{7.13}$$

Notes.

1) It follows that only in the symmetric case $a = b = \frac{1}{2}$ does (7.11) at the initial time equal (7.12). On these grounds, it was suggested in [41] that $a = b = \frac{1}{2}$ is the natural choice for ∂_x^{-1}.

2) The proof of the above statements is given in [41]. Some technical considerations are proven in the appendix at the end of this chapter. Important implications that follow from the proof are the following:

7.1.2. *Result II*
Assume that the initial data satisfy the conditions (7.5), (7.9), but not (7.2), i.e., $\phi(0,q,0) \neq 0$. Then:

a) At any time $t > 0$ (7.2) is attained, although nonuniformly, and moreover

$$\lim_{x\to\infty} \int_{-x}^{x} u(x',y,t)dx' = 0 \qquad \text{pointwise} \quad (\forall t > 0), \tag{7.14}$$

b) For large $|x|$, $u(x,y,t)$ decays weakly as $|t/x^3|^{\frac{1}{4}}$ and hence, even with strongly decaying initial data, the solution does not belong to $L^1(\mathbb{R}^2)$ for any $t > 0$. If, besides, $u(x,y,0)$ belongs to $L^2(\mathbb{R}^2)$, then $u(x,y,t)$ belongs too to $L^2(\mathbb{R}^2)$, for all $t \geq 0$, and the solution is unique in an L^2 setting.

The proof is based in establishing (see the appendix) that the latter integral behaves as

$$\begin{aligned}\lim_{x\to\infty} \int_{-x}^{x} u(x',y,t)dx' &= \lim_{x\to\infty} \int dq e^{iqy} J_0(2\sqrt{3q^2xt})\phi(0,q,0) \\ &= \int dq\, e^{iqy}\phi(0,q,0) \lim_{x\to\infty} J_0(2\sqrt{3q^2xt}) = 0\,,\end{aligned} \tag{7.15}$$

since $J_0(2\sqrt{3q^2xt}) \propto (xt)^{-1/4}$ as $x \to \infty$ and

$$|J_0(2\sqrt{3q^2xt})\phi(0,q,0)| \leq |\phi(0,q,0)|$$

(J_0 is the standard Bessel function). This also implies by differentiation that for large x

$$u(x,y,t) - u(-x,y,t) = \int dq\, \phi(0,q,0)\, e^{iqy} \sqrt{\frac{q^2 t}{x}}\, J_1(2\sqrt{3q^2 xt})$$

$$= \left|\frac{t}{x^3}\right|^{\frac{1}{4}} \int dq\, \phi(0,q,0) e^{iqy} |q|^{\frac{1}{2}} + o\Big(\left|\frac{t}{x^3}\right|^{\frac{1}{4}}\Big) \text{ as } x \to \infty\,. \tag{7.16}$$

Square integrability follows from (7.4) and Parseval's relationship.

7.1.3. *Result III*

Assume that the initial data $f(x,y) \equiv u(x,y,0)$ satisfy

$$(1+x)f \in L_1(dxdy)\,,$$

and that they are of class $C^{5,4}$ with the derivatives

$$f_{yy}, f_{xxx}, f_{yyyy}, f_{xxxxx} \in L_1(dxdy).$$

Then (7.9), (7.5) hold, and hence so too results I and II.

This is all one needs to establish that the linear KP evolution with initial data not satisfying (7.2) exists, along with the properties of such an evolution for $t > 0$.

Several other researchers have revisited this problem. Properties of the solution in a weak or distributional sense with quickly decaying initial data (concretely data belonging to the Schwartz space) were considered in [43]. The issue of whether a time reversal evolution ($t < 0$) is possible along with its properties was also considered in [43]. It was pointed out that (for KPI) the time derivative of the solution suffers a jump at $t = 0$:

$$\lim_{t\to 0^+} \frac{\partial u}{\partial t}(x,y,t) - \lim_{t\to 0^-} \frac{\partial u}{\partial t}(x,y,t) \neq 0\,. \tag{7.17}$$

Likewise ∂_x^{-1} is discontinuous as $t \to 0^+$ and

$$\lim_{t\to 0^+} \int_{-\infty}^{x} u_{yy}(x,y,t) = \int_{-\infty}^{x} u_{yy}(x,y,0)\,,$$

or, in view of (7.14),

$$\lim_{t\to 0^+} \partial_x^{-1} u_{yy}(x,y,t) = \int_{-\infty}^{x} u_{yy}(x,y,0)\,. \tag{7.18}$$

We finish this part by making a few remarks in this regard. The function $\partial u/\partial t$ is only defined (via (7.6)) for $t > 0$. If this formula is used to extend its definition to $t \leq 0$ then, assuming that the conditions of Result I hold, one can establish the following:

7.1.4. *Result IV*
Define $\frac{\partial u}{\partial t}(x, y, 0)$ as the right-hand side of (7.6) at $t = 0$. Then

i)

$$\frac{\partial u}{\partial t}(x, y, 0^+) - \frac{\partial u}{\partial t}(x, y, 0^-) = 2\left(\frac{\partial u}{\partial t}(x, y, 0^+) - \frac{\partial u}{\partial t}(x, y, 0)\right) = -6\pi D(y) . \tag{7.19}$$

ii)

$$\lim_{t \to 0+} \partial_x^{-1} u_{yy}(x, y, t) - \partial_x^{-1} u_{yy}(x, y, 0) = [2a - 1) + \epsilon^2]\pi D(y). \tag{7.20}$$

iii) Higher-order time derivatives $\frac{\partial^n u}{\partial t^n}$ exist for all $n \geq 2$ and $t > 0$, but become singular at the initial time.

Proof. Use (7.12) and Lemma 3 of the appendix.

Consequence. $\frac{\partial u}{\partial t}(x, y, t)$ is discontinuous at $t = 0$ no matter what choice of ∂_x^{-1} is taken. The choice

$$a = \frac{1 - \epsilon^2}{2} = \begin{cases} 1, & \text{for KPI} \\ 0, & \text{for KPII} \end{cases}$$

renders $\partial_x^{-1} u_{yy}(x, y, t)$ continuous for $t \geq 0$ but $u(x, y, t)$ with $\frac{\partial u}{\partial t}(x, y, 0)$ as defined above does not solve KP at $t = 0$.

If the symmetric choice $a = \frac{1}{2}$ is taken, then $\partial_x^{-1} u_{yy}$ is discontinuous at the initial time but

$$\frac{\partial u}{\partial t} + 3\epsilon^2 \partial_x^{-1} u_{yy}$$

is continuous and hence $u(x, y, t)$ solves KP for $t \geq 0$.

Remark. From a physical perspective one does not expect that different choices $\partial_x^{-1} u$ arise. Besides, as it has been pointed out, initial time discontinuities can be expected from physical requirements [42]. All this suggests again that the symmetric choice $a = \frac{1}{2}$ is preferred.

7.2. THE NONLINEAR KPII THEORY

Following [41], in this section we show how the above results generalize to deal with the full nonlinear eq. (7.1); for convenience, we take $\epsilon^2 = 1$, i.e., KPII. We shall assume that

$$(1 + |p|^3 + |q|^2)F(k)\mu(x, y, k) \in L^1,$$

which implies that u_{xxx} and u_{yy} tend to zero as $r \to \infty$, and that differentiation can be performed under the integral sign in (2.20). With small norm the results of chapter 2 corresponding to the direct side still hold. Insight into the behavior of (2.20) is obtained when we transform to coordinates $p \equiv -2k_R$, $q \equiv 4k_R k_I$ defined in (2.10):

$$\begin{aligned} u(x, y, t) &= \frac{1}{4\pi}\frac{\partial}{\partial x}\int\!\!\int dp\, dq\, e^{i(px+qy)}\frac{f(p, q)}{p}\mu\left(x, y, -\frac{1}{2}\left(p + i\frac{q}{p}\right)\right) \\ &= \frac{1}{4\pi}\int\!\!\int dp\, dq\, e^{i(px+qy)} f(p, q)\zeta\,, \end{aligned} \tag{7.21}$$

where $\psi = \mu_x/p, \zeta \equiv i\mu + \psi$ and $f(p, q, t) \equiv F(-p/2, -q/2p, t)$. For all time there exists an eigenfunction μ satisfying $\mu \to 1$ as $|k| \to \infty$ whose temporal evolution corresponds still to formula (2.22), or

$$f(p, q, t) = f(p, q, 0)e^{i(p^3 - 3q^2/p)t}\,. \tag{7.22}$$

We note that $p \to 0$, q fixed, corresponds to $k_R \to 0$, $k_I \to \infty$, whereupon (2.17) yields

$$\mu_x \sim \frac{\mu_{1x}}{ik_I} = -\frac{2p}{iq}\mu_{1x}(x, y, t)\,, \tag{7.23}$$

and consequently ψ is regular as $p \to 0$. The analogue of (7.7) — that we do not assume in this section — is $F(k_R, k_I) = o(k_R)$. Consequently the potential $u(x, y, t)$ takes the form

$$u(x, y, t) = \int\!\!\int dp\, dq\, e^{i(px+qy+(p^3-3q^2/p)t)}\tilde{\phi}(x, y, p, q, t)\,, \tag{7.24}$$

where

$$\tilde{\phi}(x, y, p, q, t) = \frac{1}{4\pi}f(p, q, 0)\zeta\Big(x, y, t, -\frac{1}{2}\Big(p + i\frac{q}{p}\Big)\Big). \tag{7.25}$$

The analogy with the linear problem is clear. Moreover we can formally evaluate the integral of $u(x, y, t)$. Using (2.17) and (7.15) and the Appendix we get

$$\lim_{x\to\infty}\int_{-x}^{x} u(x', y, t)dx'$$

$$= -\lim_{x\to\infty} \frac{1}{4\pi} \fint_{-\infty}^{\infty} dp \int dq\, f(p,q,t) e^{iqy} \left(\frac{e^{ipx} - e^{-ipx}}{p} \right)$$

$$= \frac{1}{4\pi} \lim_{x\to\infty} \int dq\, q^2 e^{iqy} J_0(2\sqrt{3q^2 xt}) f(0,q,0) = 0\,. \tag{7.26}$$

We have just seen, at least in a formal way, that, as happens for the linear problem, KPII evolution without the condition $F(k_R, k_I) = o(k_R)$, i.e., when the constraints (7.2) are not required, exists. Besides, for $t > 0$ we have $\int_{-\infty}^{\infty} u(x,y,t)dx = 0$. We expect that all that is required for a rigorous proof is that the small norm condition (6.10), or (6.12), of the previous section is satisfied.

Next we shall briefly comment on the Hamiltonian of the problem:

$$H = \fint_{-\infty}^{\infty} dp \int dq \left\{ p^2 |f|^2(p,q,t) - \frac{3q^2}{p^2} |f|^2(p,q,t) \right\}, \tag{7.27}$$

which is divergent as it stands. It is regularized as follows;

$$H = \fint_{-\infty}^{\infty} dp \int dq \left\{ p^2 |f|^2(p,q,t) - \frac{3q^2}{p^2} \left(|f|^2(p,q,t) - |f|^2(0,q,t) \right) \right\} \tag{7.28}$$

where H is conserved and reduces to the "standard" one if $f(0,q,0) = 0$. In order to check that it is the Hamiltonian of the problem, it suffices to check that the equations of motion follow from $u_t = \frac{\partial}{\partial x}\frac{\delta H}{\delta u}$, which it does.

Other regularizations can be used to give the equations of motion. Since the regularizing term is a Casimir, i.e., it commutes with any integral of motion, the regularization is not unique.

7.3. NONLINEAR KPI THEORY

The rigorous aspects of KPI evolution without initial value constraints $\chi_1(y,0) = 0$ have been studied in [44], assuming that the initial data are in the Schwartz space. The basic result is the following:

7.3.1. *Result V*

Assume that the initial data are in the Schwartz space and that

$$\int (1+p^2)|\hat{u}(p,y)|dy \ll 1\,, \tag{7.29}$$

where

$$\hat{u}(p,y) \equiv \frac{1}{2\pi} \int dx e^{ipx} u(x,y,0)\,. \tag{7.30}$$

Then a solution $u(x,y,t)$ to KPI (7.1) exists that satisfies

i)

$$u(x,y,t) \in C_\infty\{(x,y,t); t>0\}\,.$$

ii) $u(x,y,t)$ satisfies (7.26).

iii) $u(x,y,t)$, with $\partial_t u(x,y,0)$ properly chosen, solves KP for $t \geq 0$.

We refer to [44] for the proof.

8. Appendix

In the sequel we assume that

$$g(p,q), \frac{\partial g}{\partial p} \in L_1(dpdq) \quad \text{and} \quad g(0,q) \in L_1(dq)\,. \tag{8.1}$$

Then one has:

Lemma 1

$$\lim_{|x|\to\infty} \fint_{-\infty}^{\infty} dp \int dq\, e^{ipx}\, \frac{g(p,q)}{ip} = \pi \int dq\, g(0,q)\,.$$

Proof. Use the decomposition

$$\frac{g(p,q)}{p} = \frac{g(0,q)}{p(1+p^2)} + \tilde{g}\,,$$

where

$$\tilde{g}(p,q) \equiv \frac{1}{p}(g(p,q) - \frac{g(0,q)}{1+p^2}) \in L_1(dpdq)\,.$$

Lemma 2

Let

$$N(x,t) \equiv \fint_{-\infty}^{\infty} dp \int dq e^{i\left(px - \frac{\epsilon^2 q^2 t}{p}\right)}\, \frac{g(p,q)}{ip}\,.$$

Then, as $|x| \to \infty$,

$$N(x,t) = -2\pi \int dq J_0(2\sqrt{|\epsilon^2 q^2 x|t}) g(0,q)\, \mathrm{sign}(-\epsilon^2 x) + o(1) \underset{|x|\to\infty}{\longrightarrow} 0.$$

where J_0 is the standard Bessel function.

Proof.

With $\epsilon^2 = 1$, we can write the integral as $N(x,t) = (1) + (2) + (3)$, where

$$(1) \equiv \int_{|p|\geq 1} dp \int dq\, e^{i\left(px - \frac{q^2 t}{p}\right)} \frac{g(p,q)}{ip}$$
$$(2) \equiv \int_{-1}^{1} dp \int dq\, e^{i\left(px - \frac{q^2 t}{p}\right)} \frac{\tilde{g}(p,q)}{ip}$$

tend to zero as $x \to \infty$ by the Riemann–Lebesgue lemma. Here $\tilde{g} \equiv g(p,q) - g(0,q)$. Finally we consider

$$(3) \equiv \int_{-1}^{1} dp \int dq g(0,q) \frac{e^{i\left(px - \frac{q^2 t}{p}\right)}}{ip}.$$

We close the latter contour considering an indentation above $p = 0$ and a semicircular arc with radius one in the upper/lower (for $x >$, $x < 0$) half plane. The integrals over these arcs go to zero exponentially. Hence

$$\begin{aligned}(3) &= -2\pi \int dq\, g(0,q) \operatorname{Res}\left(\frac{e^{i\left(px - \frac{q^2 t}{p}\right)}}{p}\right)_{p=0} \theta(-x) + O(e^{-|x|}) \\ &= -2\pi \int dq\, g(0,q) J_0(2\sqrt{q^2|x|t})\, \theta(-x) + O(e^{-|x|}).\end{aligned}$$

By Lebesgue's theorem (since $J_0(2\sqrt{3q^2|x|t}) \propto |xt|^{-1/4}$ as $|x| \to \infty$) the claim is obtained.

Lemma 3

As $t \to 0^+$ we have

$$\lim_{t \to 0+} N(x,t) = N(x,0) - \pi \operatorname{sign}(\epsilon^2) \int dq\, g(0,q).$$

Proof. One has

$$N(x,t) = \int_{-\infty}^{\infty} dp \int dq \frac{e^{-i\frac{\epsilon^2 q^2 t}{p}}}{ip} \left(e^{ipx} g(p,q) - \frac{g(0,q)}{1+p^2}\right) + H(t).$$

Here

$$\begin{aligned}H(t) &\equiv \int_{-\infty}^{\infty} dp \int dq\; e^{-i\frac{\epsilon^2 q^2 t}{p}} \frac{g(0,q)}{ip(1+p^2)} \\ &= \int_{-\infty}^{\infty} dp \int dq\; e^{-i\epsilon^2 q^2 tp} \frac{pg(0,q)}{i(1+p^2)} \\ &= -\pi \operatorname{sign}(\epsilon^2) \int dq\, g(0,q) e^{-|\epsilon^2|t}.\end{aligned}$$

An application of dominated convergence shows that

$$\lim_{t\to 0^+} H(t) = -\pi\ \mathrm{sign}(\epsilon^2) \int dq\, g(0,q)\,,$$

$$\lim_{t\to 0^+} \fint_{-\infty}^{\infty} dp \int dq \frac{e^{-i\frac{\epsilon^2 q^2 t}{p}}}{ip} \left(e^{ipx} g(p,q) - \frac{g(0,q)}{i(1+p^2)} \right)$$
$$= N(x,0) = \fint_{-\infty}^{\infty} dp \int dq \frac{e^{ipx} g(p,q)}{ip}\,.$$

On the proof of (7.11)–(7.14)

Here we finally sketch the proof of eqs. (7.11)–(7.14). Set

$$\partial_x^{-1} u_{yy}(x',y,t)\, dx' = \lim_{L\to\infty} \left(a \int_{-L}^{x} - b \int_{x}^{L} \right) u_{yy}(x',y,t)\, dx'\,.$$

Using (7.4) and the fact that $\phi(p,q,t) \in L^1(I\!R^2)$, and taking into account that the domain of integration is finite, we obtain

$$\left(a \int_{-L}^{x} - b \int_{x}^{L} \right) dx'\, u_{yy}(x',y,t)$$
$$- \fint_{-\infty}^{\infty} dp\, dq \left(a \int_{-L}^{x} - b \int_{x}^{L} \right) dx'\, e^{i\left(px'+qy+p^3 t - \frac{\epsilon^2 q^2 t}{p}\right)} \phi(p,q,t) q^2$$
$$= - \fint_{-\infty}^{\infty} dp \int dq\, e^{i\left(px+qy+p^3 t - \frac{\epsilon^2 q^2 t}{p}\right)} \frac{\phi(p,q,t)}{ip} q^2$$
$$+ \fint_{-\infty}^{\infty} dp \int dq \left(b\, e^{ipL} + a e^{-ipL} \right) \frac{\phi(p,q,t)}{ip} q^2 e^{iqy+p^3 t - i\frac{\epsilon^2 q^2 t}{p}}\,.$$

Thus the proof reduces to showing that the last integral goes to zero as $L\to\infty$. There are two cases, $t=0$ and $t>0$. With $t=0$ Lemma 1 yields (7.12), while if $t>0$ Lemma (2) yields eqs. (7.11) and (7.14).

Acknowledgments

This work was partially supported by the NSF under grant number DMS-0070792 and and by Junta de Castilla-Leon JADZ, SA43/00B in Spain.

References

1. M.J. Ablowitz, H. Segur, *J. Fluid Mech.*, **92**, (1979), 6791.
2. B.B. Kadomtsev, V.I. Petviashvili *Sov. Phys. Doklady*, **15**, (1970) 539.
3. A.S. Fokas and M.J. Ablowitz, *Stud. Appl. Math.*, **69**, (1983) 211.

4. M.J. Ablowitz, D. Bar Yaacov, A.S. Fokas, *Stud. Appl. Math.*, **69**, (1983) 135.
5. S.V. Manakov, *Physica D*, **3**, (1981) 420.
6. M.J. Ablowitz and A.S. Fokas, *Lecture Notes in Physics* **189**, Springer Verlag (Berlin), (1983). See also the contribution of A.S. Fokas and M.J. Ablowitz in the same Lecture Notes.
7. M.J. Ablowitz, H. Segur, *Solitons and the Inverse Scattering Transform*, SIAM, Phila., PA (1981).
8. M.J. Ablowitz and P.A. Clarkson, *Solitons, Nonlinear Evolution Equations and Inverse Scattering,* Cambridge University Press, Cambridge, UK, (1991).
9. V. Dryuma, *Sov. Phys. JETP*, **19**, (1974) 381.
10. J. Satsuma, *J. Phys. Soc. Japan*, **40** (1976) 276.
11. S.V. Manakov, V.E. Zakharov, L.A. Bordag, A.R. Its and V.B. Matveev, *Phys. Lett. A*, **63** (1977) 205.
12. R.S. Johnson and S. Thompson, *Phys. Lett. A*, **66** (1978) 279.
13. K.A. Gorshov, D.E. Pelinovskii and Yu. A. Stepahyants, *JETP*, **77** (1993) 237.
14. V.M. Galkin, D.E. Pelinovsky, and Yu A. Stepanyants, *Physica D*, **80** (1995) 246.
15. R.S. Ward, *Phys. Lett. A*, **208** (1995) 203.
16. M.J. Ablowitz and J. Villarroel, *Physical Review Letters*, **78** (1997) 570.
17. J. Villarroel and M.J. Ablowitz, *Commun. Math. Phys.*, **207** (1999) 1.
18. M.J. Ablowitz and J. Villarroel, *Scattering in Pure and Applied Science*, Chapter 6.3, R. Pike and P. Sabatier eds., Academic Press (2002) 1792.
19. M. Mañas and P. Santini, *Phys. Lett. A*, **227** (1997) 325.
20. M.J. Ablowitz, S. Chakravarty, A.D. Trubatch and J. Villarroel, *Phys.Lett.A*, **267** (2000) 132.
21. J. Villarroel and M.J. Ablowitz, *SIAM J. Math Anal.* (2003, in press).
22. Q.P. Liu and M. Mañas, *J. Nonlinear Sci.*, **9**(1999) 213.
23. M. Wickerhauser,*Commun. Math. Phys.*, **108** (1987) 67.
24. J. Villarroel and M.J. Ablowitz, *Stud. Appl. Math.*, **109** (2002) 151.
25. M. Boiti, F. Pempinelli, A. Progrebkov, and M. Polivanov, *Inverse problems*, **8** (1992) 331.
26. M. Boiti, F. Pempinelli, A. Progrebkov, *Inverse problems*, **13** (1997) L7.
27. M. Boiti, F. Pempinelli, A. Progrebkov, and B. Prinari, *Inverse problems*, **4** (2001) 937.
28. M: Boiti, F. Pempinelli, A. Progrebkov, and B. Prinari, *Phys. Lett. A*, **285** (2001) 307
29. A. Fokas, D. Pelinovsky, C. Sulem, *Physica D*, **152** (2001) 189.
30. A. Fokas, A. Progrebkov, *Inverse scattering transform for KPI on the background of one soliton*, preprint.
31. L. Faddeev, *J. Math.Phys.*, **4** (1963) 72.
32. P. Deift and E. Trubowitz. *Comm. Pure and Appl. Math.*, **32** (1979) 121.
33. H. Segur, *Mathematical methods in Hydrodynamics and integrability in dynamical systems* (M. Tabor and Y.M. Treve, eds.), AIP conference proceedings 88 (1982) 211.
34. X. Zhou, *Commun. Math. Phys.*, **128** (1990) 551.
35. A. Degasperis, *Nonlinear wave eqs. solvable by the spectral transform*, Lectures at the International School of Physics E. Fermi, (1988), Varenna, Italy.
36. P. Grinevich, *Lett. Math. Phys.*, **40** (1997) 59.
37. A.S. Fokas, L.Y. Sung, *Inverse problems*, **8** (1992) 673.
38. V. E. Zakharov, E. Schulman *Physica D*, **1** (1980) 192.
39. Z. Jiang, R. K. Bullough, S. V. Manakov, *Phys. D*, **18** (1986) 305.

40. J. Lin, H. Chen, *Phys. Lett. A*, **89** (1982) 163.
41. M.J. Ablowitz and J. Villarroel, *Stud. Appl. Math.*, **85** (1991) 195.
42. M.J. Ablowitz and X-P. Wang, *Stud. Appl. Math.*, **98** (1997) 121.
43. M. Boiti, F. Pempinelli and A. Pogrebkov, *Inverse Problems* **10** (1994) 505.
44. A.S. Fokas, L.-Y. Sung, *Math. Proc. Camb. Philos. Soc.*, **125** (1999) 113.
45. E. Olmedilla, *Physica D*, **25** (1987) 330.

PARTIALLY SUPERINTEGRABLE (INDEED ISOCHRONOUS) SYSTEMS ARE NOT RARE

F. CALOGERO (francesco.calogero@roma1.infn.it)
Dipartimento di Fisica, Università di Roma "La Sapienza", 00185 Roma, Italy, and Istituto Nazionale di Fisica Nucleare, Sezione di Roma

Abstract. We call *partially superintegrable (indeed isochronous)* those dynamical systems *all* solutions of which are *completely periodic* with a fixed period ("isochronous") in a *part* of their phase space, and we review a recently introduced trick that allows to manufacture many such systems. Several examples are discussed.

1. Introduction

Recently a trick — amounting essentially to a change of independent, and often as well of dependent, variables — has been introduced [1], that allows to modify a dynamical system so that the new (modified) system thereby obtained features a lot of *completely periodic (indeed isochronous)* solutions. This approach has been applied in a number of contexts [2]–[16]. In this paper we review these results, we report some additional ones, but especially we introduce the notion of *partially superintegrable (indeed isochronous)* dynamical systems and we emphasize the fact, already advertised in the title of this paper, that *these systems are not rare* – indeed, given a generic dynamical system, one can generally associate to it (via the trick) one or more somewhat analogous systems that are *partially superintegrable (indeed isochronous)*.

We shall now explain this fact, but firstly we must introduce the notion of *partially superintegrable (indeed isochronous)* system, as employed in this paper.

Integrable, and *superintegrable*, systems have a standard definition in the context of finite-dimensional Hamiltonian dynamics. Let N be the number of degrees of freedom of a Hamiltonian system. Then this system is *integrable* if there exist $N-1$ nontrivial functionally independent univalent functions of the N canonical variables and of the N conjugate momenta

A.B. Shabat et al. (eds.), New Trends in Integrability and Partial Solvability, 49–77.

that Poisson-commute (namely, their Poisson brackets vanish) among themselves and with the Hamiltonian, so that, including the Hamiltonian itself, the system features N constants of motion in involution. The Hamiltonian system is said to be *superintegrable*(or perhaps *maximally superintegrable* – but we will refrain from using such hyperinflated terminology) if there exist $N-1$ *additional* functionally independent univalent functions of the N canonical variables and of the N conjugate momenta that Poisson-commute with the Hamiltonian, so that, including the Hamiltonian itself, the system features $2N-1$ nontrivial constants of motion. It is then easily seen that, for systems that only feature *confined* motions — and in this paper we restrict attention to such systems — the property of *superintegrability* entails that *all* motions are *completely periodic.* Because of this fact, in this paper any system (including non Hamiltonian systems, and systems with an infinite number of degrees of freedom) is termed *superintegrable* if its generic motions (namely, *all*, or *almost all*, its motions) are *completely periodic*; indeed the type of complete periodicity we encounter is moreover characterized by a *fixed* period, independent of the initial data (although some motions may also, or only, be *completely periodic* with a period which is an integer submultiple, or an integer multiple, of that basic period), hence these motions should perhaps more properly be characterized as *isochronous.*

We call a system *partially superintegrable (indeed isochronous)* if it features this property, but *only* in a domain of phase space (rather than in the entire phase space), a domain which must however be *completely open* (namely, open in all dimensions). In particular the *partially superintegrable (indeed isochronous)* systems considered in this paper are characterized by the property to possess an open domain in phase space such that *all* the motions evolving from a set of initial data in it are *completely periodic* with the same period (*isochronous*). The measure of such an open domain does not vanish; it might, or it might not, be infinite when the measure of the entire phase space is itself infinite (for instance, if the entire phase space is the two-dimensional Euclidian plane, such a domain might be the exterior, or the interior, of a circle of finite radius).

For instance a well-known *superintegrable* system is the one-dimensional many-body problem characterized by the Hamiltonian [2]

$$H(\underline{p},\underline{q})=\frac{1}{2}\sum_{n=1}^{N}\left(p_n^2+\omega^2 q_n^2\right)+\frac{1}{4}g^2\sum_{n,m=1,m\neq n}^{N}\left(q_n-q_m\right)^{-2}, \qquad (1.1)$$

and correspondingly by the Newtonian equations of motion

$$\ddot{q}_n + \omega^2 q_n = g^2 \sum_{m=1, m \neq n}^{N} (q_n - q_m)^{-3} . \qquad (1.2)$$

Here and always below ω is a *positive* constant, $\omega > 0$, and the rest of the notation is, we trust, self-evident. Indeed, in the *real* domain, *all* the solutions of these equations of motion are *completely periodic* ("*isochronous*") with period

$$T = 2\pi/\omega, \qquad (1.3)$$

$$\underline{q}(t+T) = \underline{q}(t). \qquad (1.4)$$

This is not quite true in the *complex* domain, namely if we consider the Newtonian equations of motion

$$\ddot{z}_n + \omega^2 z_n = g^2 \sum_{m=1, m \neq n}^{N} (z_n - z_m)^{-3} , \qquad (1.5)$$

which are identical to (1.2) except that we now consider the dependent variables $z_n \equiv z_n(t)$ to be *complex* rather than *real* (and we also allow the "coupling constant" g to be *complex*, while we always consider the constant ω to be *real*, indeed, without loss of generality, *positive*, $\omega > 0$; and of course we always assume the independent "time" variable, t, to be *real*). Then *all* motions, which take of course place in the complex plane, are again *completely periodic*, but the period may be an integer multiple of T, see (1.3): indeed also in this case the particle configuration does repeat itself with period T, but the individual particles might exchange their roles through the motion, so that the period of the motion of each individual particle might end up being an integer multiple of T (this cannot happen in the real case, when the motion takes place on the real line and the ordering of the particles cannot change throughout the motion due to the singular character of the repulsive two-body interaction, see (1.2)) [2].

Hence the many-body problem characterized by the Hamiltonian (1.1) is *superintegrable*, both in the *real* and in the *complex* domains. The more general many-body problem characterized by the following Hamiltonian,

$$H(\underline{p}, \underline{z}) = \frac{1}{2} \sum_{n=1}^{N} \left(p_n^2 + \omega^2 z_n^2 \right) + \frac{1}{4} \sum_{n,m=1, m \neq n}^{N} g_{nm}^2 (z_n - z_m)^{-2} , \qquad (1.6)$$

is instead only *partially superintegrable*, and only if it is considered in the *complex* domain. Indeed it has recently been shown [6] that the Newtonian equations of motion

$$\ddot{z}_n + \omega^2 z_n = \sum_{m=1, m\neq n}^{N} g_{nm}^2 \left(z_n - z_m\right)^{-3}, \tag{1.7}$$

that clearly obtain from the Hamiltonian (1.6) (provided of course $g_{nm} = g_{mn}$, as we hereafter assume), yield a *completely periodic* motion provided the initial data fall in an appropriate (open) domain, which however generally does not include only real data. (Incidentally: a remarkable generalization of this result is reported and proved below, see Sections 2 and 4 and the Appendix).

These examples present rather special dynamical systems. Let us now justify the statement made above (and in the title of this paper), by showing that to *any* given dynamical system (with the only qualification that its equations of motion be *analytic*, see below) one can generally associate one or more somewhat analogous ones that are *partially superintegrable (indeed isochronous)* — implying that such *partially superintegrable (indeed isochronous) systems are not rare.*

Let us use the following notation to denote a generic (but *analytic*) dynamical system:

$$\zeta_\tau(\underline{x}, \tau) = F\left[\zeta(\underline{x}, \tau); \underline{x}, \tau\right]. \tag{1.8}$$

Here $\zeta \equiv \zeta(\underline{x}, \tau)$ is the dependent variable, which might be a scalar, a vector, a matrix, a tensor, you name it, and that depends of course on the independent variable τ and might moreover depend on a number of additional variables or parameters, here denoted collectively with the vector notation $\underline{x}$; the notation $\zeta_\tau(\underline{x}, \tau)$ denotes the (partial) derivative of $\zeta(\underline{x}, \tau)$ with respect to the independent variable τ; and the quantity in the right-hand side, $F\left[\zeta(\underline{x}, \tau); \underline{x}, \tau\right]$, has of course the same character (scalar, vector,...) as $\zeta(\underline{x}, \tau)$, it might depend explicitly on the independent variable τ as well as on the additional variables and parameters $\underline{x}$, and it is generally a (nonlinear) functional of the dependent variable $\zeta(\underline{x}, \tau)$, say a (nonlinear) function of $\zeta(\underline{x}, \tau)$ and possibly of its (partial) derivatives with respect to some of the parameters subsumed in the notation $\underline{x}$ (or perhaps also of some integrals of $\zeta(\underline{x}, \tau)$ over these parameters). We are therefore considering at this stage a fairly general evolutionary system, although in the following our consideration will be generally restricted to (systems of) Ordinary Differential Equations (ODEs) and to Partial Differential Equations (PDEs), namely to dynamical systems (especially to those interpretable as many-body problems, such as the examples discussed above) and to evolution equations of standard type.

We shall however generally assume that the evolutionary equation (1.8) is defined in the *complex*, namely that not only the dependent variable

$\zeta(\underline{x}, \tau)$ take its values in the complex, but that as well the independent variable τ is *complex*; and this of course requires that the derivative $\zeta_\tau(\underline{x}, \tau)$ with respect to this variable be well defined, namely that the general context be that of *analytic* (of course, not necessarily singularity-free, namely not necessarily *entire*) functions of the complex variable τ. In this context it is then a general consequence of the existence/uniqueness/analyticity theorem for analytic evolution equations that the solution $\zeta(\underline{x}, \tau)$ of the initial-value problem (namely, the solution $\zeta(\underline{x}, \tau)$ that evolves out of a given "initial datum" $\zeta(\underline{x}, 0)$) for the evolutionary system (1.8) is a *holomorphic* function of the *complex* variable τ in a disk of the complex τ-plane centered at $\tau = 0$ and having a nonvanishing radius, say $\rho > 0$ — provided, as we hereafter assume, the right-hand side of (1.8) is *nonsingular* initially, namely for $\tau = 0$ and $\zeta = \zeta(\underline{x}, 0)$. The value of ρ is determined by the properties of the right-hand side of (1.8) in the neighborhood of $\tau = 0$ and $\zeta = \zeta(\underline{x}, 0)$, and it is generally larger the smaller (in modulus) that right-hand side is in that neighborhood; and one can always identify a sufficiently small value of this radius ρ, so that there exist an open set of initial data $\zeta(\underline{x}, 0)$ such that *all* the corresponding solutions $\zeta(\underline{x}, \tau)$ be *holomorphic* functions of the independent (complex) variable τ in the disk of radius ρ centered, in the complex τ-plane, at the origin, $\tau = 0$.

Let us now introduce the following change of independent variable:

$$z(\underline{x}, t) = \zeta(\underline{x}, \tau) \tag{1.9}$$

with

$$\tau \equiv \tau(t) = \eta \left[\exp(i\omega t) - 1\right] / (2i), \tag{1.10}$$

where η and ω are two constants, with the essential requirement that ω be *real* (indeed, without loss of generality, *positive*, $\omega > 0$; while η is at this stage an arbitrary complex number, but see below).

Now we consider $z \equiv z(\underline{x}, t)$ to be the new dependent variable, and t the new independent variable, with the important restriction to consider this latter *real*, indeed to interpret it hereafter as the "time" variable. Of course the new dynamical problem reads now, in self-evident notation,

$$\dot{z}(\underline{x}, t) = (\eta\omega/2) \exp(i\omega t) F\left[z(\underline{x}, t); \underline{x}, \tau(t)\right]. \tag{1.11}$$

Before proceeding further, let us note that the change of variable (1.9) with (1.10) entails that the "initial" data for the evolutionary systems (1.8) and (1.11) are very simply related:

$$z(\underline{x}, 0) = \zeta(\underline{x}, 0). \tag{1.12}$$

And now our conclusion: we saw above that *all* the solutions $\zeta(\underline{x}, \tau)$ of (1.8) evolving from an (appropriately chosen, but certainly) open set of initial data are *holomorphic* functions of the independent (complex) variable τ in a disk of (sufficiently small but certainly) *positive* radius ρ, $\rho > 0$, centered, in the complex τ-plane, at the origin, $\tau = 0$. But, via (1.9) with (1.10), this immediately implies that *all* the corresponding solutions $z(\underline{x}, t)$ of the new dynamical problem (1.11) are *completely periodic* functions with period T, see (1.3), of the new, *real*, "time" variable t,

$$z(\underline{x}, t + T) = z(\underline{x}, t), \tag{1.13}$$

provided there holds the simple inequality

$$|\eta| < \rho. \tag{1.14}$$

Indeed the relation (1.10) clearly entails that τ is a periodic function of t with period T, see (1.3), and moreover that as t varies over one period, say from 0 to T, τ travels counterclockwise, from $\tau = 0$ back to $\tau = 0$, on the circle in the complex τ-plane the diameter of which, of length $|\eta|$, has one end at $\tau = 0$ and the other at $\tau = i\eta$ (hence its center at $\tau = i\eta/2$); this, via the condition (1.14), guarantees that it remains within the disk of radius ρ centered at $\tau = 0$, namely that it stays in the region in which the solution $\zeta(\underline{x}, \tau)$ is holomorphic in τ; and this of course entails, via (1.9), that $z(\underline{x}, t)$ is *completely periodic* in t, see (1.13).

Clearly, once this argument is understood, its conclusion becomes essentially trivial. It is moreover clear that such a conclusion might also obtain if a different change of independent variable had been chosen instead of (1.10), provided it also implied that τ, as a function of the *real* variable t, be periodic and confined to a neighborhood of the origin, $\tau = 0$. But what makes this argument interesting in spite of its triviality is the possibility that the new dynamical system (1.11) obtained in this manner be itself interesting. Generally a necessary, if not sufficient, condition for this to be the case is the requirement that the new evolutionary equation, see (1.11), be *autonomous* (namely, *not* feature an explicit time dependence). As we will see, a virtue of the particular change of dependent variable (1.10) is to greatly facilitate the attainment of this goal, although to achieve this it is often required to perform an additional change of dependent variable, in the guise of the introduction of a prefactor depending (generally exponentially) on t in the relation among the new and old dependent variable, see (1.9). The prefactor should of course have itself the property to be periodic in t with the appropriate period. For instance the change of (dependent and independent) variable

$$z_n(t) = \exp(-i\omega t)\zeta_n(\tau) \tag{1.15}$$

with

$$\tau = [\exp(2i\omega t) - 1] / (2i\omega) \tag{1.16}$$

entails that the Newtonian equations of motion (1.7) correspond to the following (quite analogous) set of ODEs:

$$\zeta_n'' = \sum_{m=1, m \neq n}^{N} g_{nm}^2 \left(\zeta_n - \zeta_m\right)^{-3}, \tag{1.17}$$

where of course the appended primes denote differentiations with respect to the independent variable τ. And this fact was key, utilizing the argument we just described, to proving [6] the *partial superintegrability (indeed isochronicity)* of the Newtonian equations of motion (1.7), as defined above.

This concludes our presentation of the main idea of this paper. In the following Sections, by reviewing the results it has yielded thus far and by presenting some additional findings, we hope to convince the reader that this approach, in spite (or perhaps because) of the trivial character of the observation that underlies its applicability, is remarkably effective inasmuch as it leads to the identification of many interesting evolutionary systems that possess lots of *completely periodic (indeed isochronous)* solutions, and of simple prescriptions to manufacture many more such systems. The potential applicative interest of such an approach is, we trust, self-evident, although of course its actual relevance shall have to wait for future developments in order to be fully demonstrated. Below, in Section 2 we review some results on many-body problems, in Section 3 we report on the existence of "nonlinear harmonic oscillators", in Section 4 we report some results on other *isochronous* oscillators, in Section 5 we exhibit twelve examples of single nonlinear ODEs with *periodic (indeed isochronous)* solutions, in Section 6 we tersely show, via a single example, how this same approach can be used to "deform" nonlinear evolution PDEs so that they possess many *periodic (indeed isochronous)* solutions, and finally in Section 7 we tersely offer some final remarks. Except when we present new findings the treatment is generally limited to a terse description of results, with references to the literature for their derivation as well as for more detailed discussions of them. The paper is completed by an Appendix where some computations are reported, the presentation of which in the body of the paper would have excessively interfered with the flow of the discourse there.

2. Many-body problems

In this Section we review instances of *partially superintegrable (indeed isochronous)* many-body problems. Several other such systems can be found in [2], although the full implications of the trick were understood only as that book [2] was being completed and are therefore only partially reflected there. Of course many other models can be easily manufactured via the technology described above and exemplified below.

We need not repeat here the analysis [6] of the many-body problem characterized by the Newtonian equations of motion (1.7), except to note that a detailed study of that model *outside* of the region of phase space in which it behaves as a *superintegrable (indeed isochronous)* system is also quite interesting [10], although it exceeds the scope of this paper. Let us moreover note that, in Section 4, a nontrivial extension of this system is also shown to be *partially superintegrable* (again, provided it is considered in the *complex*). It is the many problem characterized by the Hamiltonian

$$H(\underline{p},\underline{z}) = \frac{1}{2}\sum_{n=1}^{N}\left(p_n^2+\omega^2 z_n^2\right) + \frac{1}{4}\sum_{n,m=1,m\neq n}^{N}\left[g_{nm}^2\left(z_n-z_m\right)^{-2}+\frac{1}{2}\gamma_{nm}\left(z_n-z_m\right)^{-4}\right] \quad (2.1)$$

correspondingly by the Newtonian equations of motion

$$\ddot{z}_n+\omega^2 z_n = \sum_{m=1,m\neq n}^{N}\left[g_{nm}^2\left(z_n-z_m\right)^{-3}+\gamma_{nm}\left(z_n-z_m\right)^{-5}\right]. \quad (2.2)$$

Indeed the treatment given in the following section entails that the same result obtains even for the more general dynamical systems characterized by equations of motion

$$\ddot{z}_n+\omega^2 z_n = \sum_{k=0}^{K} f_{k,n}(\underline{z}), \quad (2.3a)$$

provided one retains the freedom to assign a sufficiently large value of the real constant ω and the functions $f_{k,n}(\underline{z})$ of the N variables z_n satisfy the same scaling property as featured by the right-hand sides of (2.2),

$$f_{k,n}(\lambda\underline{z}) = \lambda^{-3-2k} f_{k,n}(\underline{z}), \quad (2.3b)$$

and are otherwise essentially arbitrary (but analytic in the N variables z_n).

Next we report the (Hamiltonian [2]) N-body problem in the plane characterized by the Newtonian equations of motion

$$\ddot{\vec{r}}_n = \omega\hat{k}\wedge\dot{\vec{r}}_n + 2\sum_{m=1,m\neq n}^{N} r_{nm}^{-2}(\alpha_{nm}+\alpha'_{nm}\hat{k}\wedge) \cdot\left[\dot{\vec{r}}_n(\dot{\vec{r}}_m\cdot\vec{r}_{nm})+\dot{\vec{r}}_m(\dot{\vec{r}}_n\cdot\vec{r}_{nm})-\vec{r}_{nm}(\dot{\vec{r}}_m\cdot\dot{\vec{r}}_m)\right]. \quad (2.4)$$

Here the N two-vectors $\vec{r}_n\equiv\vec{r}_n(t)$ identify the positions of the moving point-particles in a plane which for notational convenience is immersed in three-dimensional space, so that $\vec{r}_n \equiv (x_n, y_n, 0)$; $\widehat{k}$ is the unit three-vector orthogonal to that plane, $\widehat{k} \equiv (0,0,1)$, so that $\widehat{k} \wedge \vec{r}_n \equiv (-y_n, x_n, 0)$; $\vec{r}_{nm} \equiv \vec{r}_n - \vec{r}_m$, hence $r_{nm}^2 \equiv \vec{r}_{nm} \cdot \vec{r}_{nm} \equiv (x_n - x_m)^2 + (y_n - y_m)^2$; superimposed dots denote of course time derivatives; ω is as usual a *positive* constant; and the "coupling constants" $\alpha_{nm}, \alpha'_{nm}$ are *a priori* arbitrary (of course *real*; a sufficient condition for this system to be Hamiltonian is the requirement that these constants be symmetrical in their two indices, $\alpha_{nm} = \alpha_{mn}, \alpha'_{nm} = \alpha'_{mn}$ [2]).

To treat this many-body problem it is generally convenient to identify the real "physical" plane in which the points $\vec{r}_n \equiv (x_n, y_n, 0)$ move with the complex plane in which the complex numbers $z_n \equiv x_n + iy_n$ move. Indeed via this correspondence these equations of motion take the following simpler form:

$$\ddot{z}_n - i\omega \dot{z}_n = 2 \sum_{m=1, m\neq n}^{N} a_{nm} \dot{z}_n \dot{z}_m / (z_n - z_m) \tag{2.5}$$

with

$$a_{nm} = \alpha_{nm} + i\alpha'_{nm}. \tag{2.6}$$

Then one notes that, via the change of independent variable

$$z_n(t) = \zeta_n(\tau), \tag{2.7}$$

$$\tau = [\exp(i\omega t) - 1] / (i\omega), \tag{2.8}$$

which is the simple version of the trick appropriate to this case, and entails the following very simple relations among the initial data for z_n and ζ_n,

$$z_n(0) = \zeta_n(0), \dot{z}_n(0) = \zeta'_n(0), \tag{2.9}$$

the equations of motion (2.5) become

$$\zeta''_n = 2 \sum_{m=1, m\neq n}^{N} a_{nm} \zeta'_n \zeta'_m / (\zeta_n - \zeta_m), \tag{2.10}$$

where of course primes denote differentiations with respect to the (complex) independent variable τ. And thereby one easily concludes [3] that the N-body problem (2.5) is *partially superintegrable (indeed isochronous)*, featuring motions that are *completely periodic (indeed isochronous)* with period $T = 2\pi/\omega$ provided they evolve from an appropriate open set of initial data, which, as can be easily guessed by looking at the right-hand side of (2.10) and by recalling the simple relation (2.9) among the initial data, are characterized by small (in modulus) values of the initial velocities

$\dot{z}_n(0)$ and by large (in modulus) values of the initial interparticle distances $z_n(0) - z_m(0)$. Moreover, let us again note that the detailed analysis [11] of this N-body problem *outside* of this region of phase space is as well quite interesting, although it exceeds the scope of this paper. And let us also point out that, in special cases, the system is actually *superintegrable* indeed *solvable*: for instance if all the coupling constants are unity, $a_{nm} = 1$ (in which case it was suggested [17] that the model deserves to be considered a "goldfish"; although extension of this honorary title to the general system (2.5) was also later suggested [11]), and in some other cases as well [9] [16].

The next example we report in this section is a deformed version of the classical N-body gravitational problem [12], characterized by the equations of motion

$$\ddot{\vec{r}}_n + i\omega\dot{\vec{r}}_n + 2\omega^2\,\vec{r}_n = \sum_{m=1, m\neq n}^{N} M_m(\vec{r}_m - \vec{r}_n)r_{nm}^{-3}, \tag{2.11}$$

where the N three vectors $\vec{r}_n$ denote the positions in three-dimensional (ordinary) space of N point particles of masses M_n (actually, the restriction to three-dimensional space is merely for convenience, to make contact with physical reality, see below; the results hold independently of the number of dimensions of ambient space). Of course here r_{nm} denotes the standard Euclidian distance among the n-th and m-th particle, $r_{nm}^2 = [(\vec{r}_n - \vec{r}_m)\cdot(\vec{r}_n - \vec{r}_m)]$.

For $\omega = 0$ clearly this N-body problem, (2.11), coincides with the classical (of course Hamiltonian) N-body gravitational problem. We consider instead the case with *positive* ω, $\omega > 0$, when the equations of motion (2.11) define a *deformed* version of the classical N-body gravitational problem; note that in this *deformed* case the motion takes necessarily place in the *complexified* three-dimensional space, due to the second term in the left-hand side of (2.11). It is then convenient to introduce [12] the following change of dependent and independent variables (the version of the trick appropriate to this case):

$$\vec{r}_n(t) = \exp(-2i\omega t)\vec{\rho}_n(\tau), \tag{2.12}$$

$$\tau = [\exp(3i\omega t) - 1]/(3i\omega). \tag{2.13}$$

It is then plain that the equations of motion (2.11) take the new form

$$\vec{\rho}_n'' = \sum_{m=1, m\neq n}^{N} M_m(\vec{\rho}_m - \vec{\rho}_n)\rho_{nm}^{-3}, \tag{2.14}$$

that only differs from (2.11) due to the absence of the "ω-terms" in the left-hand side, and of course because the superimposed dots are now replaced by appended primes, which obviously signify differentiations with respect to the new (complex) independent variable τ, see (2.13). And it is then easy to prove [12], via the approach outlined above, that the system (2.11) is *partially superintegrable (indeed isochronous)*, namely that there is an open set of (complex) initial data for this problem that yield motions *completely periodic (indeed isochronous)* with period $T = 2\pi/\omega$. The behavior of this system for initial data that are *outside* this domain is an interesting open problem.

The last example we report in this section is characterized by the following equations of motion (see [2, Exercise 5.6.20]; note the change of notation from, and the correction of the misprint in, eq. (5.6.20-41a) there):

$$\ddot{z}_n - i(4+p)\omega\dot{z}_n - 2(2+p)\omega^2 z_n = F_n(\underline{z}), \tag{2.15}$$

where we assume p to be a *positive integer*, $p = 1, 2, 3, \ldots$, the constant ω to be as usual *real* (indeed, without loss of generality, *positive*, $\omega > 0$), we indicate with $\underline{z}$ the N-vector the components of which are the N dependent variables $z_n \equiv z_n(t)$, and we assume the N functions $F_n(\underline{z})$ to be *holomorphic* at $z_n = 0$ and to satisfy the scaling property

$$F_n(\lambda\underline{z}) = \lambda^{1+p} F_n(\underline{z}). \tag{2.16}$$

These equations of motion (that include of course the Newtonian equations of motion of an M-body problem in S-dimensional space, with $N = MS$) define a system that is *partially superintegrable (indeed isochronous)*, inasmuch as there is, in the neighborhood of the equilibrium configuration $\dot{\underline{z}} = \underline{z} = 0$, a ball of initial data, of nonvanishing volume in the phase space (having $4N$ real dimensions) of this system, such that all the motions evolving from it are completely periodic, with period $T = \pi/\omega$ if p is *even*, with period $T = 2\pi/\omega$ if p is *odd*. This conclusion is easily proved via the trick consisting of the following change of dependent and independent variables:

$$z_n(t) = (\exp(2i\omega t)\, \zeta_n(\tau), \tag{2.17}$$

$$\tau = [\exp(ip\omega t) - 1] / (ip\omega), \tag{2.18}$$

whereby the equations of motion (2.15) with (2.16) become (in self-evident notation)

$$\zeta_n'' = F_n(\underline{\zeta}). \tag{2.19}$$

Note incidentally that the gravitational case treated above, see (2.11), may be considered a special case of (2.15), by relinquishing the restriction to *positive* values of the parameter p; indeed (2.11) is clearly a special case of

(2.15) with (2.16) and $p = -3$; and indeed, for this value $p = -3$, there is as well correspondence among (2.12,2.13) and (2.17,2.18), up to a (merely notational) change of the sign of ω.

3. Nonlinear harmonic oscillators

In this section we report tersely a result obtained with V. Inozemtsev [14] via the approach outlined above: a rather remarkable finding, as indicated by the title of this section, that might appear an oxymoron since most physicists consider, in association with the substantive "oscillator", the adjectives "linear" and "harmonic", and conversely "nonlinear" and "unharmonic", as synonymous. Let us emphasize that the systems considered in this section must be considered *superintegrable*, not just *partially superintegrable.*

The starting point is the fact that the matrix evolution equation

$$Y''(\tau) = c\,[Y(\tau)]^3\,, \tag{3.1}$$

is *integrable*, and more importantly, that *all* its solutions are *meromorphic* functions of the independent variable τ, for *all* (finite) values of this complex variable [18]. Here the dependent variable $Y \equiv Y(\tau)$ is a *matrix* of *arbitrary* rank, c is an arbitrary *scalar* constant, and of course primes denote differentiations with respect to the independent variable τ.

We now use the following version of the trick:

$$M(t) = \exp(i\omega t)Y(\tau), \tag{3.2}$$

$$\tau = [\exp(i\omega t) - 1]\,/(i\omega), \tag{3.3}$$

where as usual ω is a *real* (say, without loss of generality, *positive,* $\omega > 0$) constant. It is then plain that the matrix $M(t)$ satisfies the matrix ODE

$$\ddot{M}(t) - 3i\omega\dot{M}(t) - 2\omega^2 M(t) = c\,[M(t)]^3\,, \tag{3.4}$$

while the fact that *all* the solutions of (3.1) are *meromorphic* functions of the independent variable τ clearly entails, via (3.2) with (3.3), that *all* the *nonsingular* solutions of this matrix ODE are *completely periodic* with period $T = 2\pi/\omega$,

$$M(t+T) = M(t). \tag{3.5}$$

But clearly the matrix evolution equation (3.4) represents a collection of *nonlinear* (linear plus cubic) oscillators, while the result we just mentioned justifies calling them *harmonic* (indeed, for arbitrary initial conditions, they *all* oscillate with a *single period*, except for the exceptional cases in which they diverge: nothing could be less cacophonous, namely more harmonic, featuring a single note, of course with all its harmonics). So, the title of this section is justified, in spite of its ossimoronic appearance.

This finding can be given a more remarkable look by using appropriate parametrizations [20] [2] [14] of the matrix $M(t)$, which are preserved by the evolution equation (3.4) and transform this matrix ODE into a system of coupled oscillator equations that may take various interesting forms, for instance, as systems of coupled *rotation-invariant* evolution equations satisfied by an arbitrary number of vectors in a space with an arbitrary number of dimensions, say (in self-evident notation) [14]

$$\ddot{\vec{z}}_{nm} - 3i\omega\dot{\vec{z}}_{nm} - 2\omega^2 z_{nm} = c \sum_{n'=1}^{N} \sum_{m'=1}^{M} \vec{z}_{nm'}(\vec{z}_{n'm'} \cdot \vec{z}_{n'm}) \tag{3.6}$$

with $n = 1, \ldots, N$, $m = 1, \ldots, m$, or

$$\ddot{\vec{z}}_{nm} - 3i\omega\dot{\vec{z}}_{nm} - 2\omega^2 z_{nm} = c \sum_{n'=1}^{N} \sum_{m'=1}^{M} \vec{z}_{n'm'}(\vec{z}_{n'm'} \cdot \vec{z}_{nm}), \tag{3.7}$$

with $n = 1, \ldots, N$, $m = 1, \ldots, m$. These vectors $\vec{z}_{nm} \equiv \vec{z}_{nm}(t)$ are of course *complex*, as entailed by the evolution equations they satisfy, see the second term in the left hand side of (3.6) and (3.7); but *real* equations can of course be easily obtained by introducing the *real* and *imaginary* parts of these vectors, say $\vec{z}_{nm}(t) = \vec{x}_{nm}(t) + i\,\vec{y}_{nm}(t)$.

Let us also note that modified versions of these Newtonian equations of motion can also be exhibited [14] which are not only *rotation-invariant*, but as well *translation-invariant*, while retaining the property of *complete isochronicity* of *all* their nonsingular solutions.

In these examples of applications of the trick an important underlying idea was to restrict attention to *autonomous* (i.e., time-independent) evolution equations, and more specifically to evolution equations that are *autonomous* in both *avatars*, before and after application of the trick, namely both in the version with the *real* independent variable t ("time") as well as in the related version with the *complex* independent ("time-like") variable τ; indeed the main focus of some recent papers has been the identification (and/or investigation) of certain ODEs [7] [8], or of classes of ODEs [5] [13], or of (classes of) PDEs [4] [5], possessing such a property. But this restriction is not necessary for the applicability of this approach. Indeed in the following section we indicate how information can be fairly easily evinced, via this approach, on the behavior — in particular, the existence of *completely periodic* ("isochronous") solutions emerging out of certain open domains of initial data — of certain interesting, *autonomous*, systems of ODEs describing a "physical" evolution as it unfolds over the *real* time t, even though the versions of these systems of ODEs obtained via the trick and characterizing the evolution as function of the complex variable τ are themselves *no more autonomous.*

4. Isochronous oscillators

The main new idea — described at the end of the preceding section — that underlies the results reported in this section was developed in collaboration with Jean-Pierre Françoise [15]. We report here the main findings, referring to the recent joint paper with Françoise for proofs and comments, including a discussion of how such results could (or rather could not) be obtained by more standard techniques [15].

Let us consider the system of N nonlinear ODEs

$$\dot{z}_n - i r_n \omega z_n = f_n(\underline{z}), \tag{4.1}$$

where $z_n \equiv z_n(t)$ are the N dependent variables, t is of course the *real* independent variable ("time"), ω is a *positive* number, $\omega > 0$, that sets the time scale and to which we associate as usual the period $T = 2\pi/\omega$, the N numbers r_n are *real* and *rational,*

$$r_n = \frac{p_n}{q_n} \geqslant 0 \tag{4.2}$$

with p_n, q_n two coprime integers and $q_n > 0$, and each of the N functions $f_n(\underline{z})$ of the N components z_n of the N-vector $\underline{z}$ is arbitrary except for the requirements to be *holomorphic* in each of the complex variables z_n near $\underline{z} = \underline{0}$ and to actually vanish at $\underline{z} = \underline{0}$ faster than linearly,

$$\lim_{\delta \to 0} \left[\delta^{-1} f_n(\delta z_1, \ldots, \delta z_m, \ldots, \delta z_N) \right] = 0, \tag{4.3}$$

and moreover to satisfy there the N conditions

$$\lim_{\epsilon \to 0} \left[\epsilon^{-1-r_n} f_n(\epsilon^{r_1} z_1, \ldots, \epsilon^{r_m} z_m, \ldots, \epsilon^{r_N} z_N) \right] = \text{finite}. \tag{4.4}$$

If the N rational numbers r_n are either *all nonnegative* or *all nonpositive,* no additional conditions are required; if instead both *positive* and *negative* numbers r_n are featured by the system (4.1), then one must additionally require that *all* the N functions $f_n(\underline{z})$ such that the corresponding rational numbers r_m have one of the two signs (namely, are either *positive* or *negative)* depend *polynomially* on *all* the coordinates z_m.

Provided the functions $f_n(\underline{z})$ are so restricted, one can then prove [15] that there exists a *positive* number $Z > 0$ such that the N conditions on the "initial data" $z_n(0)$,

$$|z_n(0)| < Z, \tag{4.5}$$

entail that *all* the corresponding solutions $z_n(t)$ of (4.1) are *completely periodic* with period (at least) $\widetilde{T} = qT = 2\pi q/\omega$, where q is the minimum common multiple of the N integers q_n. One therefore concludes that, for small enough initial data, the system (4.1) with (4.2) and the conditions on the functions $f_n(\underline{z})$ specified above (see (4.3) and (4.4), as well as the sentence following (4.4)), retains the *isochronicity* property of its linear part, namely of the system that obtains from (4.1) by equating its left-hand side to zero, the solutions of which read obviously

$$z_n(t) = z_n(0)\exp(ir_n\omega t) \tag{4.6}$$

and are therefore *completely periodic (indeed isochronous)* with period $\widetilde{T}$. .Let us emphasize that, as discussed in [15], this result includes the *Hamiltonian case*, which is characterized by an even N, $N = 2M$, by "oppositely paired" r_n's, say $r_{M+m} = -r_m, m = 1, \ldots, M$, and by functions $f_n(\underline{z})$ obtainable by differentiation from a single function $h(\underline{z})$ as follows:

$$\begin{aligned} f_m(\underline{z}) &= \frac{\partial h(\underline{z})}{\partial z_{M+m}}, \\ f_{M+m}(\underline{z}) &= -\frac{\partial h(\underline{z})}{\partial z_m}, \qquad m = 1, \ldots, M. \end{aligned} \tag{4.7}$$

As a second example of application of the same type of approach — and there are of course many others, as made clear by these examples — let us now consider the Hamiltonian many-body problem characterized by the Newtonian evolution equations (see (2.2))

$$\ddot{z}_n + \omega^2 z_n = \sum_{m=1, m\neq n}^{N} \left[g_{nm}^2 (z_n - z_m)^{-3} + \gamma_{nm} (z_n - z_m)^{-5} \right]. \tag{4.8}$$

The version of the trick we now apply reads

$$z_n(t) = \exp(-i\omega t)\zeta_n(\tau), \tag{4.9}$$

$$\tau = [\exp(2i\omega t) - 1]/(2i\omega). \tag{4.10}$$

It is easily seen that this change of dependent and independent variables entails the following relations among the "initial data":

$$z_n(0) = \zeta_n(0), \dot{z}_n(0) = \zeta_n'(0) - i\omega\zeta_n(0), \tag{4.11}$$

as well as the following evolution equations for the new dependent variable $\zeta_n(\tau)$:

$$\zeta_n'' = \sum_{m=1, m\neq n}^{N} \left[g_{nm}^2 (\zeta_n - \zeta_m)^{-3} + \gamma_{nm} (1 + 2i\omega\tau) (\zeta_n - \zeta_m)^{-5} \right], \tag{4.12}$$

where the primes denote of course differentiations with respect to the new independent variable τ.

Although these evolution equations, (4.12), are *not* autonomous (unless *all* the "coupling constants" γ_{nm} vanish), it is nevertheless still fairly obvious, and indeed easy to prove (see Appendix) that, provided the initial data $\zeta_n(0), \zeta_n'(0)$ satisfy the following conditions,

$$|\zeta_n(0) - \zeta_m(0)| > R, \tag{4.13}$$

$$|\zeta_n'(0)| < V, \tag{4.14}$$

with R chosen sufficiently *large* and V chosen sufficiently *small* (see (8.18) with (8.13), and (8.19), in the Appendix for an explicit version of such conditions), then the solutions $\zeta_n(\tau)$ of (4.12) are *holomorphic* in τ in the disk $|\tau| \leq 1/\omega$, hence these quantities, considered via (4.10) as functions of the real variable t, are all periodic with period $\widehat{T} = \pi/\omega$, hence the corresponding (via (4.9) and (4.10)) solutions $z_n(t)$ of (4.8) are *completely antiperiodic* with the same period $\widehat{T}$, $z_n(t+\widehat{T}) = -z_n(t)$, hence *completely periodic (indeed isochronous)* with period $T = 2\pi/\omega$, $z_n(t+T) = z_n(t)$. Note that the conditions we just mentioned, (4.13) and (4.14), define an *open* set of initial data $\zeta_n(0), \zeta_n'(0)$ for the problem (4.12), to which there corresponds via (4.11) an *open* set of initial data $z_n(0), \dot{z}_n(0)$ for the original problem (4.8). However generally this *open* set of initial data $z_n(0), \dot{z}_n(0)$ need not include only *real* data.

Before concluding this section let us also note that the Newtonian equations of motion (4.8) obviously admit the (exact, explicit) solution

$$z_n(t) = a\cos(\omega t) + b\sin(\omega t) + c_n \tag{4.15}$$

with a and b arbitrary constants and the N constants c_n required to satisfy the algebraic system

$$\omega^2 c_n = \sum_{m=1, m\neq n}^{N} \left[g_{nm}^2 (c_n - c_m)^{-3} + \gamma_{nm} (c_n - c_m)^{-1} \right]. \tag{4.16}$$

The investigation of the behavior of this model, see (4.8), for initial data that are *outside* the domain identified above, see (4.13) and (4.14), is an interesting open problem.

5. Nonlinear ODEs

In this section we exhibit, with minimal commentary, *twelve* single nonlinear ODEs and their solutions, which are generally *isochronous* (or quasi isochronous, see below). These results, elementary as they are, are perhaps interesting thanks to their pedagogical value, as examples useful to test numerical integration routines, as tools to investigate the behavior of more general nonlinear systems which possess *periodic* solutions but are not amenable to exact treatment, and as additional illustrations of the efficacy of a technology (to identify evolution equations possessing *lots of isochronous solutions*) which is likely to prove fruitful in various applicative contexts, in particular in the mathematical modeling of cyclic phenomena, which are a common feature of the world we live in. Here, however, we do not elaborate on how this technique works, since we feel this has been sufficiently illustrated above; we merely exhibit *twelve* examples which have indeed all been arrived at via such a technique (although they could of course also be obtained otherwise).

Notation: t is the independent variable (hereafter assumed to be real: "time"), and superposed dots denote differentiations with respect to this variable t. The ODEs are generally written below in their neater version; additional arbitrary constants can of course be introduced via trivial transformations, for instance by multiplying by constant factors the independent and dependent variables or by adding a constant to the latter. The evolution equations written below are generally *complex*, hence the dependent variables are as well generally *complex*; this entails the possibility (left as an option for the diligent reader) to reformulate these equations as two coupled *real* ODEs by introducing the standard representations of a complex number via its real and its imaginary parts, or via its amplitude and its phase. But in some cases the equations are *real*, hence their solutions may be *real* as well. Whenever a square (or cubic) root is written, it is understood that the results are valid for anyone of its two (or three) determinations; this fact justifies our usage sometimes below of the plural to denote some of the displayed solutions, or equations, since more than one is entailed by the possibility to choose different determinations of the square (or cubic) roots.

Example 1. The *real* ODE

$$2\ddot{x}x - 3\dot{x}^2 = x^2(1 - x^2) \tag{5.1}$$

has the *general solution*

$$x(t) = \frac{\sinh(a)}{\cosh(a) - \cos(t - t_0)}. \tag{5.2}$$

Here a and t_0 are two arbitrary constants (*real* if the solution is *real*). Note that this solution is periodic with period 2π (and it is *nonsingular* in the *real* case). Also note that this ODE, (5.1), is the Newtonian equation of motion yielded, in the standard manner [2], by the Hamiltonian

$$H(x,p)=\frac{1}{2}\left[cp^2x^3+\frac{1+x^2}{cx}\right], \tag{5.3}$$

where c is an arbitrary constant ($c\neq 0$) that does not appear in the equation of motion (5.1) and p is the "canonical momentum" conjugated to the "canonical variable" x.

Example 2. The *real* ODE

$$\ddot{x}\,\sin(x)\,\sin(2x)=\dot{x}^2\,\sin(3x)+\sin^3(x)\,\cos(2x) \tag{5.4}$$

has the *general solution*

$$x(t)=\arctan\left[\frac{\sinh(a)}{\cosh(a)-\cos(t-t_0)}\right]. \tag{5.5}$$

Here a and t_0 are two arbitrary constants (*real* if the solution is *real*). Note that, in the *real* case, this solution is periodic with period 2π (and it is of course *nonsingular*). Also note that this ODE, (5.4), is the Newtonian equation of motion yielded, in the standard manner [2], by the Hamiltonian

$$H(x,p)=\frac{1}{2}\left[cp^2\,\sin^2(x)\,\sin(2x)+\frac{1}{c\,\sin(2x)}\right], \tag{5.6}$$

where c is an arbitrary constant ($c\neq 0$) that does not appear in the equation of motion (5.4) and p is the "canonical momentum" conjugated to the "canonical variable" x.

Example 3. The *real* ODE

$$\ddot{x}\sin(x)\,\sin(2x)=\dot{x}^2\sin(3x)+\sin^3(x) \tag{5.7}$$

has the *general solution*

$$x(t)=\arctan\left[\frac{\sin(a)}{\cos(a)-\cos(t-t_0)}\right]. \tag{5.8}$$

Here a and t_0 are two arbitrary constants (*real* if the solution is *real*). Note that, in the *real* case, this solution is *nonsingular*, and it is periodic with period 2π. [Indeed this formula, (5.8), entails that $\dot{q}(t)$ vanishes at $t=t_0+n\pi$, $\dot{q}(t_0+n\pi)=0$, and that $[\ddot{q}(t_0+n\pi)]\,(-)^{n+1}>0$, hence the maxima and minima of $q(t)$ are alternatively attained at $t=t_0+n\pi$, $n=0,1,2,3,\ldots$ and they have the same values, namely $q_{\rm min}=q(t_0+2m\pi)=-\arctan\left[\frac{\sin(a)}{1-\cos(a)}\right]<0$, $q_{\rm max}=q(t_0+\pi+2m\pi)=\arctan\left[\frac{\sin}{1+\cos(a)}\right]>0$, $m=0,1,2,\ldots$; here we take of course always the principal value of the arctan function, namely the value in the interval between 0 and π, and we assumed for simplicity (without significant loss of generality) that $0 < a < \frac{\pi}{2}$ namely $0 < \sin(a) < 1$].

Also note that this ODE, (5.7), is the Newtonian equation of motion yielded, in the standard manner [2], by the Hamiltonian

$$H(x,p) = \frac{1}{2}\left[c\, p^2 \sin^2(x)\sin(2x) + \frac{\cot(2x)}{c}\right], \tag{5.9}$$

where c is an arbitrary constant ($c \neq 0$) that does not appear in the equation of motion (5.7) and p is the "canonical momentum" conjugated to the "canonical variable" x.

Example 4. The *complex* ODE

$$\ddot{z} + z = (\dot{z} + iz)^3 \tag{5.10}$$

has the *general solutions*

$$z(t) = a\exp(-it) + [i + b\exp(-2it)]^{1/2}, \tag{5.11a}$$
$$z(t) = \tilde{a}\exp[-i(t-t_0)] + \left\{i + \tilde{b}\exp[-2i(t-t_0)]\right\}^{1/2}. \tag{5.11b}$$

In (5.11a) a and b are two arbitrary *complex* constants; the expression (5.11b) corresponds to (5.11a) with $\tilde{a} = a\exp(-it_0), \tilde{b} = b\exp(-2it_0)$ and displays a feature of the solution clearly entailed by the *autonomous* character of the ODE (5.10). Note that these solutions, (5.11a), are *periodic* with period 2π unless they are *singular*; a necessary and sufficient condition to exclude this (for real t) is validity of the inequality $|b| \neq 1$; moreover, if and only if $|b| > 1$ these solutions are *antiperiodic* with period π, $z(t+\pi) = -z(t)$.

This ODE, (5.10), possesses moreover the *special solution*

$$z(t) = c\exp(-it), \tag{5.12}$$

with c an arbitrary *complex* constant. This solution is clearly *nonsingular* and *periodic* with period 2π (*antiperiodic* with period π). It is not a special case of the general solution (5.11).

Example 5. The *complex* ODE

$$\ddot{z} + 3i\dot{z} + z = -4(\dot{z} - iz/2)^3 + 3\dot{z}^2/z \tag{5.13}$$

has the *general solutions*

$$z(t) = \left\{i + a\exp(it) + \left[\{i + a\exp(it)\}^2 + b^2\exp(2it)\right]^{1/2}\right\}^{1/2}, \tag{5.14}$$

as well as the *special solution*

$$z(t) = c\exp(it/2). \tag{5.15}$$

Here a, b and c are arbitrary *complex* constants. In (5.14) one may of course replace t with $t - t_0$ provided one also replaces a, b with $\tilde{a} = a\exp(it_0), \tilde{b} = b\exp(it_0)$, and likewise in (5.15) one may replace t with $t - t_0$ provided one also replaces c with $\tilde{c} = c\exp(it_0/2)$. Clearly the *special solution* (5.15) in *nonsingular* and *periodic* with period 4π (*antiperiodic* with period 2π). The *general solutions* (5.14) are as well *periodic*: if $b = 0$ the solution (5.14) vanishes or becomes simply $z(t) = \pm\{2[i + a\exp(it)]\}^{1/2}$, in which case it is *periodic* with period 2π if $|a| < 1$, *periodic* with period 4π (*antiperiodic* with period 2π) if $|a| > 1$, *singular* if $|a| = 1$. If $b \neq 0$ the *general solutions* (5.14) are *singular* for real t if and only if one of the two quantities

$$F_{\pm}(a,b) = \left|a^2 + b^2\right|^2 - \left[|a|^2 + |b|^2 \pm 2\,\mathrm{Im}(ab^*)\right] \tag{5.16}$$

vanishes, $F_+(a,b) = 0$ or $F_-(a,b) = 0$; otherwise they are *periodic*, with period 2π or 4π depending on the signs of the two quantities $F_{\pm}(a,b)$, see (5.16).

Example 6. The *complex* ODE

$$\ddot{z} - 3i\dot{z} - 2z = (\dot{z} - iz)\ z \tag{5.17}$$

has the *general solution*

$$z(t) = 2a\ \exp(it)\ \tan\left[-ia\ \exp(it) + b\right] \tag{5.18}$$

and the *special solution*

$$z(t) = c\ \exp(it). \tag{5.19}$$

Here a, b and c are three arbitrary complex constants ($a \neq 0$). Clearly the *special solution* (5.19) is periodic with period 2π (and entire), and the *general solution* (5.18) is as well periodic with period 2π unless it is singular (and conditions on the constants a, b sufficient to exclude this eventuality are obvious). It is also clear how to display the property of the *general solution* (5.18), respectively of the *special solution* (5.19), that corresponds to the *autonomous* character of the ODE (5.17): replace t with $t - t_0$ and a with $\tilde{a} = a\ \exp(it_0)$ respectively c with $\tilde{c} = c\ \exp(it_0)$.

Example 7. The *complex* ODE

$$\ddot{z} + i\dot{z} + 2z = (z + i\dot{z})^3 z \tag{5.20}$$

has three *general solutions*, which coincide with the three roots of the following cubic equation in z:

$$z^3 + az\exp(2it) + b\exp(3it) - 2 = 0 \tag{5.21}$$

where a, b are two arbitrary *complex* constants. The coefficients of this cubic equation are clearly periodic functions of t with period 2π, hence the set of its three roots is as well periodic in t with period 2π, while each of the three roots is *periodic* with period $2\pi, 4\pi$ or 6π due to the possibility that, through the time evolution, different roots exchange, as it were, their roles. It is left for the diligent reader, using if need be the explicit expressions of the solutions of cubic equations, to ascertain the limitations on the constants a, b which are necessary and sufficient to guarantee that the solution of (5.20) belong to one of these three periodicity regimes. These limitations identify open domains of the space of values of these *complex* constants (and correspondingly of the initial data, say $z(0), \dot{z}(0)$, which would be assigned in the context of the initial-value problem for (5.20)), domains the boundaries of which correspond to values of the *complex* constants a, b (or correspondingly of the initial data $z(0), \dot{z}(0)$) that yield solutions of (5.20) which become *singular* at some real value of t.

This ODE, (5.20), possesses moreover the *special solution*

$$z(t) = c\exp(it) \tag{5.22}$$

where c is an arbitrary *complex* constant. This solution is clearly *nonsingular* and *periodic* with period 2π (*antiperiodic* with period π).

Note that in (5.21) one can replace t with $t - t_0$, provided one replaces simultaneously a with $\tilde{a} = a\exp(2it_0)$ and b with $\tilde{b} = b\exp(3it_0)$; likewise in (5.22) one can replace t with $t - t_0$ provided one replaces simultaneously c with $\tilde{c} = c\exp(it_0)$.

Example 8. The *complex* ODE

$$\ddot{z} - 5i\dot{z} - 6z = z^2 \tag{5.23}$$

has the *general solution*

$$z(t) = 6\exp(2it)\wp(a - i\exp(it); 0, g_3), \tag{5.24a}$$

$$z(t) = 6\exp\left[2i(t - t_0)\right]\wp(\tilde{a} - i\exp\left[i(t - t_0)\right]; 0, \tilde{g}_3). \tag{5.24b}$$

In (5.24a) a and g_3 are two arbitrary *complex* constants, and $\wp(u; g_2, g_3)$ is the Weierstrass elliptic function; the expression (5.24b) corresponds to (5.24a) with $\tilde{a} = a\exp(-it_0), \tilde{g}_3 = g_3\exp(6it_0)$. Note that this solution, (5.24a), is *periodic* in t with period 2π, unless it is *singular*; a condition necessary and sufficient to exclude that this happen for real t is provided by the inequality $|a - u_p - 2n\omega_1 - 2m\omega_2| \neq 1$, where n, m are two arbitrary integers, ω_1, ω_2 are the two semiperiods of the Weierstrass elliptic function $\wp(u; 0, g_3)$, and $u_p = u_p(g_3)$ is the argument at which the Weierstrass elliptic function $\wp(u; 0, g_3)$ has a (double) pole, $[\wp(u_p; 0, g_3)]^{-1} = 0$.

Example 9. The *complex* ODEs

$$\ddot{z} - 5i\dot{z} - 4z = (z + i\dot{z})\, z \left\{ z^2 + \left[z^4 + 4(z + i\dot{z}) \right]^{1/2} \right\} \tag{5.25}$$

have the *general solution*

$$z(t) = -a \exp(it) \tan \left[(a^3/3) \exp(3it) - b \right] \tag{5.26}$$

and the *special solutions*

$$z(t) = \left[c \exp(-3it) - 1/4 \right]^{-1/3} . \tag{5.27}$$

In (5.26) a, b are two arbitrary *complex* constants; to display the invariance of this solution under time translations one can of course replace t with $t - t_0$ and a with $\tilde{a} = a \exp(it_0)$. This solution, (5.26), is clearly *periodic* with period 2π, unless it is *singular*; this happens for real t only if $|a|^3 = 3\,|b + (2n+1)\pi/2|$ for some *integer* value of n. In the *special solutions* (5.27) c is as well an arbitrary *complex* constant, which should be replaced by $\tilde{c} = c \exp(-3it_0)$ if one wished to replace t with $t - t_0$. These solutions are *singular* if and only if $|c| = 1/4$; otherwise they are *periodic*, with period $2\pi/3$ if $|c| < 1/4$, with period 2π if $|c| > 1/4$.

Example 10. The *complex* ODE.

$$\ddot{z} - 5i\dot{z} - 4z = (z + i\dot{z})\, z^3 - (z + i\dot{z})^2\, z^{-1} \tag{5.28}$$

has the *general solution*

$$z(t) = a\, \exp(it) \left\{ b \exp\left[-a^3 \exp(3it) \right] - 3 \right\}^{-1/3} . \tag{5.29}$$

In this formula a and b are two arbitrary *complex* constants; to display the property of this solution that corresponds to the translation invariance of the ODE (5.28) one can of course replace t with $t - t_0$ and simultaneously a with $\tilde{a} = a\, \exp(it_0)$. This solution, (5.29), is periodic with period 2π unless it is singular, and clearly conditions sufficient to exclude that this happen are either $|b| > 3 \exp\left(|a|^3\right)$ or $|b| < 3 \exp\left(-\,|a|^3\right)$.

Example 11. The *complex* ODEs

$$\ddot{z} - i\dot{z} = \dot{z}^2 \left[1 + \alpha / \log(z) \right] / z \tag{5.30}$$

have the *general solutions*

$$z(t) = \exp\left\{ a \exp\left[\exp(it) + b \right]^{1/(1-\alpha)} \right\}, \quad \alpha \neq 1 \tag{5.31a}$$

$$z(t) = \exp\left\{ a \exp\left[b \exp(it) \right] \right\}, \quad \alpha = 1. \tag{5.31b}$$

Our usage of the plural in characterizing the ODEs (5.30) as well as the solutions (5.31a) is motivated by the possibility to choose different determinations of $\log(z)$ in (5.30) and of the power $1/(1-\alpha)$ in (5.31a). In these formulas, (5.30) and (5.31a), α is an arbitrary *complex* constant (of course the same!); in (5.31) a, b are two arbitrary *complex* constants. In (5.31a) one may of course replace t with $t-t_0$ provided one also replaces a with $\tilde{a} = a \exp\left[it_0/(1-\alpha)\right]$ and b with $\tilde{b} = b\exp(-it_0)$, and likewise in (5.31b) one may replace t with $t-t_0$ provided one also replaces b with $\tilde{b} = b\exp(it_0)$. Clearly the solution (5.31b) is *nonsingular* and *periodic* with period 2π. The solution (5.31a) is *nonsingular*, indeed *entire*, if $\alpha = 1 - 1/p$ with p an arbitrary *positive integer*; otherwise these solutions, (5.31a), are *singular*, at a real value of t, if and only if $|b| = 1$. If $|b| > 1$ these solutions, (5.31a), are *periodic* with period 2π (irrespective of the value of the constant α), while if instead $|b| < 1$ they are *periodic* only if α is a *real rational* number, say $\alpha = m/n$ with m, n two coprime integers, with period $2\,|m-n|\,\pi$.

Example 12. The *complex* third-order ODE

$$\dddot{z} - 9i\ddot{z} - 26\dot{z} + 24iz = z(\dot{z} - 2iz) \tag{5.32}$$

has the *general solution*

$$z(t) = 12\exp(2it)\wp(a - i\exp(it); g_2, g_3), \tag{5.33a}$$
$$z(t) = 12\exp\left[2i(t-t_0)\right]\wp(\tilde{a} - i\exp\left[i(t-t_0)\right]; g_2, g_3), \tag{5.33b}$$

and the *special solution*

$$z(t) = c\exp(2it). \tag{5.34}$$

In (5.33a) a, g_2 and g_3 are three arbitrary *complex* constants, and $\wp(u\,; g_2, g_3)$ is the Weierstrass elliptic function; the expression (5.33b) corresponds to (5.33a) with $\tilde{a} = a\exp(-it_0)$, $\tilde{g}_2 = g_2\exp(4it_0)$, $\tilde{g}_3 = g_3\exp(6it_0)$. Note that this solution, (5.33a), is *periodic* with period 2π, unless it is *singular*; a condition necessary and sufficient to exclude that this happen for real t is provided by the inequality $|a - u_p - 2n\omega_1 - 2m\omega_2| \neq 1$, where n, m are two arbitrary integers, ω_1, ω_2 are the two semiperiods of the Weierstrass elliptic function $\wp(u; g_2, g_3)$, and $u_p = u_p(g_2, g_3)$ is the argument at which the Weierstrass elliptic function $\wp(u; g_2, g_3)$ has a (double) pole, $\left[\wp(u_p; g_2, g_3)\right]^{-1} = 0$. The special solution (5.34) (where c is of course an arbitrary *complex* constant) is clearly *nonsingular* and *periodic* with period π (*antiperiodic* with period $\pi/2$).

6. Nonlinear PDEs

In the previous sections various applications of the trick to manufacture evolution equations featuring many *periodic (indeed isochronous)* solutions have been reported in the context of (single, or systems of) ODEs. The fact that this approach is as well applicable to PDEs was pointed out in [4] and elaborated in [5]. In this section we merely outline the main idea, on the basis of a single example.

Consider the hierarchy of (C-integrable) nonlinear evolution PDEs

$$\varphi_\tau = \{R\,[\varphi;\xi]\}^m\,(\varphi^2\varphi_{\xi\xi}), \quad m = 0, 1, 2, \ldots, \tag{6.1}$$

where $\varphi \equiv \varphi(\xi, \tau)$ is the dependent variable, ξ and τ are the independent variables, subscripted variables denote partial differentiations, for instance $\varphi_\tau \equiv \partial\varphi(\xi,\tau)/\partial\tau$, and the "recursion operator" $R\,[\varphi;\xi]$ is defined by the formula

$$R\,[\varphi;\xi] = \varphi^2(\partial/\partial\xi)^2\varphi(\partial/\partial\xi)^{-1}\varphi^{-2}, \tag{6.2}$$

so that the first three PDEs of this hierarchy, (6.1), read as follows:

$$\varphi_\tau = \varphi^2\varphi_{\xi\xi}, \tag{6.3}$$

$$\varphi_\tau = \frac{1}{2}\varphi^2(\varphi^2)_{\xi\xi}, \tag{6.4}$$

$$\varphi_\tau = \varphi_{\xi\xi\xi\xi}\varphi^4 + 6\varphi_{\xi\xi\xi}\varphi_\xi\varphi + 4\varphi_{\xi\xi}^2\varphi^3 + 7\varphi_{\xi\xi}\varphi_\xi^2\varphi^2. \tag{6.5}$$

The fact that (6.1) is a hierarchy of *C-integrable* nonlinear evolution PDEs linearizable via an appropriate hodograph transformation is well-known [21].

Now introduce the following change of dependent and independent variables (the "trick"):

$$\begin{aligned} w(x,t) &= \exp(i\omega t)\varphi(\xi,\tau), && (6.6a)\\ \xi &= x\exp(ir\omega t), && (6.6b)\\ \tau &= \Big[\exp\{i\,[m+1+r(m+2)]\,\omega t\} - 1\Big] \\ &\quad \times \Big\{i\,[m+1+r(m+2)]\,\omega\Big\}^{-1}, && (6.6c) \end{aligned}$$

where ω is an arbitrary *positive* constant and r is an arbitrary *real* and *rational* number (possibly zero). Note that this change of variables, (6.6), depends on the positive integer m that identifies the particular equation of the hierarchy (6.1) to which it shall be applied, that it reduces to the identity (up to a notational change) if ω vanishes, and that generally it entails that old and new dependent variables coincide "initially",

$$w(x,0) = \varphi(x,0). \tag{6.7}$$

It is now easy to verify that the new dependent variable $w(x,t)$ satisfies the evolution PDE

$$w_t - i\omega w - ir\omega x w_x = \{R\,[w;x]\}^m\,(w^2 w_{xx}). \tag{6.8}$$

This PDE, (6.8), is *autonomous* (i.e., there is no explicit t-dependence), nor does it feature any explicit x-dependence if $r=0$. It is a "deformation" of (6.1), characterized by the two parameters ω and r: when ω vanishes, $\omega=0$, it reduces to (6.1) (up to trivial notational changes); when ω does *not* vanish (recall we assumed it is *positive*), it presumably features *lots of periodic solutions* (or rather, solutions that feature interesting periodicity properties in the real "time" variable t), since the solution $w(x,t)$ of this PDE, (6.8), can be obtained via ((6.6) from the solution $\varphi(\xi,\tau)$ (having the same initial datum, see (6.7)) of (6.1), and clearly to every solution $\varphi(\xi,\tau)$ of (6.1) that is *meromorphic* in τ there corresponds a solution $w(x,t)$ of (6.8) that is *periodic* in the (*real*) "time" variable t (unless it is *singular* — a nongeneric happening). The periodicity to be expected is of course characterized by the basic period $T=2\pi/\omega$, but integer multiples of it might come into play if r is not an integer, see (6.6c). Similar, but more rich, periodicity properties are moreover entailed for solutions $w(x,t)$ of (6.8) that correspond to solutions $\varphi(\xi,\tau)$ of (6.1) featuring *rational* branch points as functions of τ.

7. Outlook

The observation that *partially integrable (indeed isochronous) systems* — as defined in this paper — *are not rare* might be considered surprising by some, and instead trivial by others. Be it as it may, it leads to the identification of many dynamical systems and evolution equations that feature *lots of completely periodic (indeed isochronous) solutions.* Such systems are likely to be of applicative interest in many contexts, and, at least in some cases [10] [11], their investigation turned out to be remarkably rewarding also from the theoretical point of view. Much, clearly, remains to be done to take full advantage of the potentialities of this approach.

A final remark, on the distinction among *partially superintegrable* and *superintegrable* systems. Obviously a *partially superintegrable* system can be promoted to become a *superintegrable* one by appropriately restricting the phase space in which the system evolves — or equivalently, in the context of the initial-value problem, by restricting the domain within which its initial conditions are to be assigned. We prefer to keep the distinction among the two types of systems, and always imagine that every system is allowed to live in its *natural* phase space.

8. Appendix

The result we prove in this Appendix is the existence of two *positive* numbers R and V, such that all N components of the solution $\underline{\zeta}(\tau)$ of the (second-order, nonautonomous) evolution equations (4.12) are *holomorphic* functions of the complex variable τ in the disk $|\tau| \leq 1/\omega$ provided the "initial data" $\underline{\zeta}(0), \underline{\zeta}'(0)$ are restricted by the conditions (4.13) and (4.14). Here the underlined symbol $\underline{\zeta}$ denotes of course an N-vector, $\underline{\zeta} \equiv (\zeta_1, \ldots, \zeta_N)$.

To prove this result we take advantage of the following standard Theorem (see for instance Section 12.21 of [22]): let a system of analytic evolution equations read

$$w_j' = F_j(\underline{w}, \tau), \quad j = 1, \ldots, J, \tag{8.1}$$

with the solution $\underline{w} \equiv \underline{w}(\tau)$ (where $\underline{w} \equiv (w_1, \ldots, w_J)$ is of course a J-vector) characterized by the simple initial datum

$$\underline{w}(0) = \underline{0}. \tag{8.2}$$

The functions $F_j(\underline{w}, \tau)$ are *analytic* in all their $J + 1$ arguments, and *holomorphic* at $\underline{w} = \underline{0}$ and $\tau = 0$ (namely where the initial data are assigned, see (8.2)). Then the J components $w_j(\tau)$ of the solution $\underline{w}(\tau)$ are *holomorphic* functions of the complex variable τ (at least) in the disk $|\tau| \leq \rho$, with the positive quantity ρ bounded below by the formula

$$\rho > a\left[1 - \exp\left\{-b/\left[(J + 1)aF\right]\right\}\right], \tag{8.3}$$

where the *positive* quantities a and b are characterized by the requirement that *all* the functions $F_j(\underline{w}, \tau)$ be *holomorphic* in *all* their $J + 1$ arguments provided

$$|\tau| \leq a \tag{8.4}$$

and

$$|w_j| \leq b, \quad j = 1, \ldots, J, \tag{8.5}$$

while the quantity F is defined by the formula

$$F = \max_{\substack{j,k=1,\ldots,J \\ |\tau| \leq a;\ |w_k| \leq b}} |F_j(\underline{w}, \tau)|. \tag{8.6}$$

Note that the right-hand side of (8.3) is of course an increasing function of a and of b, and a decreasing function of F, hence the lower bound (8.3) is valid *a fortiori* if we underestimate a and b, and we overestimate F, as we shall do below.

To apply this Theorem to our case we set

$$J = 2N \tag{8.7}$$

and

$$w_n(\tau) = \zeta_n(\tau) - \zeta_n(0), w_{N+n}(\tau) = \alpha\left[\zeta_n'(\tau) - \zeta_n'(0)\right], n = 1, \ldots, N, \quad (8.8)$$

with α a *positive* constant, $\alpha > 0$, the value of which will be conveniently assigned below. The condition (8.2) is then automatically satisfied, while our evolution equations, (4.12), can now be identified with the standard evolution equations (8.1) by setting

$$F_n(\underline{w}, \tau) = \zeta_n'(0) + \alpha^{-1} w_{N+n} \quad (8.9a)$$

$$F_{N+n}(\underline{w}, \tau) = \alpha \sum_{m=1, m\neq n}^{N} \Big(g_{nm}^2 \left[\zeta_n(0) - \zeta_m(0) + w_n - w_m\right]^{-3}$$

$$+ \gamma_{nm}(1 + 2i\omega\tau)\left[\zeta_n(0) - \zeta_m(0) + w_n - w_m\right]^{-5} \Big), \quad (8.9b)$$

$$n = 1, \ldots, N.$$

We then note, first of all, that we can make the assignment $a = \infty$ (since clearly these functions $F_j(\underline{w}, \tau)$ are *nonsingular* for *all* values of τ), hence we replace (8.3) with

$$\rho > b / \left[(2N+1)F\right], \quad (8.10)$$

where we also used (8.7).

Next we set, say,

$$b = R/4, \quad (8.11)$$

an assignment that is clearly adequate, via (4.13), to guarantee that *all* the functions $F_j(\underline{w}, \tau)$, see (8.9), be *holomorphic* in *all* their arguments w_j in the ball characterized by the restrictions (8.5).

Now we must find an upper bound to F, see (8.6). Clearly via (8.9), (4.14), and (4.13) with (8.5) and (8.11), we get

$$F < \max\left\{V + R/(4\alpha), \alpha R\Omega^2 / \left[2(2N+1)\right]^2\right\}, \quad (8.12)$$

with

$$\Omega = \frac{4(2N+1)}{R^3}\left\{(N-1) \max_{\substack{n,m=1,\ldots,N \\ n\neq m}} \left[2\left|g_{nm}^2\right| + \frac{|\gamma_{nm}|}{R^2}\right]\right\}^{1/2}. \quad (8.13)$$

To get the last formula, (8.13), we used the fact that, since we only need to consider values of τ in the disk $|\tau| \leq 1/\omega$, we can use the maximization $|1 + i\omega\tau| \leq 2$.

The maximum in (8.12) must be taken by choosing one or the other argument in the square bracket; we now choose α so that these two terms coincide, namely we set

$$\alpha = \frac{2(2N+1)^2\, V}{R\Omega^2} \left\{ 1 + \left[1 + \left(\frac{R\Omega}{2(2N+1)\, V} \right)^2 \right]^{1/2} \right\}, \tag{8.14}$$

hence we get

$$F < \frac{V}{2} \left\{ 1 + \left[1 + \left(\frac{R\Omega}{2(2N+1)\, V} \right)^2 \right]^{1/2} \right\}. \tag{8.15}$$

Insertion of this expression and (8.11) in (8.10) yields

$$\rho > \frac{R}{2(2N+1)V} \left\{ 1 + \left[1 + \left(\frac{R\Omega}{2(2N+1)\, V} \right)^2 \right]^{1/2} \right\}, \tag{8.16}$$

and it is then plain that the condition we must prove,

$$\rho \geq 1/\omega, \tag{8.17}$$

is entailed by the two inequalities

$$\Omega < \omega, \tag{8.18}$$

$$V < \frac{R\,(\omega^2 - \Omega^2)}{4\,(2N+1)\,\omega}. \tag{8.19}$$

Clearly the first, (8.18), of these two inequalities can always be satisfied, see (8.13), by choosing a sufficiently large value of R, and then the second, (8.19), of these two inequalities can as well be satisfied by choosing a sufficiently small value of V. Q. E. D.

References

1. F. Calogero, "A class of integrable hamiltonian systems whose solutions are (perhaps) all completely periodic", J. Math. Phys. **38**, 5711-5719 (1997).
2. F. Calogero, *Classical many-body problems amenable to exact treatments*, Lecture Notes in Physics Monograph **m 66**, Springer, 2001.
3. F. Calogero and J.-P. Françoise, "Periodic solutions of a many-rotator problem in the plane", Inverse Problems **17**, 1-8 (2001).
4. F. Calogero and J.-P. Françoise, "Periodic motions galore: how to modify nonlinear evolution equations so that they feature a lot of periodic solutions", J. Nonlinear Math. Phys. **9**, 99-125 (2002).

5. F. Calogero, "Differential equations featuring many periodic solutions", in: *Geometry and integrability* (edited by L. Mason and Y. Nutku), London Mathematical Society Lecture Notes, vol. 295, Cambridge University Press, 2002 (in press).
6. F. Calogero, "Periodic solutions of a system of complex ODEs", Phys. Lett. **A293**, 146-150 (2002).
7. F. Calogero, "On a modified version of a solvable ODE due to Painlevé", J. Phys. A: Math. Gen. **35**, 985-992 (2002).
8. F. Calogero, "On modified versions of some solvable ODEs due to Chazy", J. Phys. A: Math. Gen. **35**, 4249-4256 (2002).
9. F. Calogero, "Solvable three-body problem and Painlevé conjectures", Theor. Math. Phys. **133**, 1443-1452 (2002).
10. F. Calogero and M. Sommacal, "Periodic solutions of a system of complex ODEs. II. Higher periods", J. Nonlinear Math. Phys. **9**, 1-33 (2002).
11. F. Calogero, J.-P. Françoise and M. Sommacal, "Periodic solutions of a many-rotator problem in the plane. II. Analysis of various motions", J. Nonlinear Math. Phys. (in press).
12. F. Calogero, "A complex deformation of the classical gravitational many-body problem that features a lot of completely periodic motions", J. Phys. A: Math. Gen. **35**, 3619-3627 (2002).
13. F. Calogero and J.-P.- Françoise, "Nonlinear evolution ODEs featuring many periodic solutions", Theor. Math. Phys. (in press).
14. F. Calogero and V. I. Inozemtsev, "Nonlinear harmonic oscillators", J. Phys. A: Math. Gen. **35**, 10365–10375 (2002).
15. F. Calogero and J.-P. Françoise, "Isochronous motions galore: nonlinearly coupled oscillators with lots of isochronous solutions", *Proceedings* of the Workshop on Superintegrability in Classical and Quantum Systems, Centre de Recherches Mathématiques (CRM), Université de Montréal, September 16-21, 2002 (in press).
16. F. Calogero, "General solution of a three-body problem in the plane", J. Phys. A: Math. Gen. **36**, 7291–7299 (2003).
17. F. Calogero, "The "neatest" many-body problem amenable to exact treatments (a "goldfish"?)", Physica **D 152-153**, 78-84 (2001).
18. V. I. Inozemtsev, "Matrix analogues of elliptic functions ", Funct. Anal. Appl. **23**, 323-325 (1990) [Russian original : Funct. Anal. Pril. **23**, 81-82 (1989)].
19. M. Bruschi and F. Calogero, "On the integrability of certain matrix evolution equations", Physics Lett. **A273**, 167-172 (2000).
20. M. Bruschi and F. Calogero, "Integrable systems of quartic oscillators", Physics Lett. **A273**, 173-182 (2000).
21. N. Euler, M. L. Gandarias, M. Euler and O. Lindblom, "Auto-Hodograph Transformations for a Hierarchy of Nonlinear Evolution Equations", J. Math. Anal. Appl. **257**, 21-28 (2001).
22. E. L. Ince, *Ordinary Differential Equations*, Dover, New York, 1956.

INITIAL-BOUNDARY VALUE PROBLEMS FOR LINEAR PDES: THE ANALYTICITY APPROACH

A. DEGASPERIS (antonio.degasperis@roma1.infn.it)
Dipartimento di Fisica, Università di Roma "La Sapienza", Roma, Italy
Istituto Nazionale di Fisica Nucleare, Sezione di Roma

S. V. MANAKOV (manakov@itp.ac.ru)
Landau Institute for Theoretical Physics, Moscow, Russia

P. M. SANTINI (paolo.santini@roma1.infn.it)
Dipartimento di Fisica, Università di Roma "La Sapienza", Roma, Italy
Istituto Nazionale di Fisica Nucleare, Sezione di Roma

Abstract. It is well-known that one of the main difficulties associated with any method of solution of initial-boundary value problems for linear PDEs is due to the presence of boundary values which cannot be arbitrarily assigned. To deal efficiently with this difficulty, we have recently proposed two alternative (but interrelated) methods in Fourier space: the Analyticity approach and the Elimination by Restriction approach. In this work we present the Analyticity approach and we illustrate its power by studying the well-posedness of initial-boundary value problems for second and third order evolutionary PDEs, and by constructing their solution. We also show the connection between the Analyticity approach and the Elimination by Restriction approach in the particular case of the Dirichlet and Neumann problems for the Schrödinger equation in the n-dimensional quadrant.

1. Introduction

It is well-known that one of the main difficulties associated with Initial-Boundary Value (IBV) problems for linear PDEs of the type

$$\mathcal{L}(\nabla, \frac{\partial}{\partial t})u(\mathbf{x}, t) = f(\mathbf{x}, t), \qquad \mathbf{x} \in V \subset I\!R^n, \quad t > 0, \tag{1.1}$$

where $\nabla = (\frac{\partial}{\partial x_1}, \dots, \frac{\partial}{\partial x_n})$, $\mathcal{L}$ is a constant coefficients partial differential operator, $f(\mathbf{x}, t)$ is a given forcing and $u(\mathbf{x}, t)$ is the unknown field, which

A.B. Shabat et al. (eds.), New Trends in Integrability and Partial Solvability, 79–103.

may be asked to satisfy Dirichlet, or Neumann, or Robin, or mixed, or periodic boundary conditions on ∂V, is due to the presence of unknown Boundary Values (BVs) in any method of solution. To deal efficiently with this difficulty, we have recently proposed two alternative (but interrelated) methods in Fourier space: the *Analyticity* approach and the *Elimination by Restriction* (EbR) approach.

The first step, common to both methods, consists in rewriting the PDE (1.1), defined in a space-time domain $\mathcal{D}$, in the corresponding Fourier space, using the Green's formula. The PDE in Fourier space takes the form of a linear relation among the Fourier Transforms (FTs) of the solution, of the initial condition and of a set of BVs, only a subset of which is given a priori. The important observation here is that this relation is always supplemented by analyticity requirements on all the FTs involved, as a consequence of the geometric properties of the space-time domain $\mathcal{D}$.

The second step is where the two methods differ from each other; once the problem is formulated in Fourier space, we propose the following two alternative strategies.

i) The *Analyticity* approach consists in using systematically the analyticity properties of all the FTs involved in the above relation to derive a system of linear equations which allows one to express the unknown BVs in terms of the known ones, and therefore to solve the problem.
ii) *The Elimination by Restriction* approach consists, instead, in applying to the above linear relation in Fourier space a suitable annihilation operator, which eliminates all the unknown BVs; a new transform is therefore generated, which is well-suited to the specific IBV problem under scrutiny. The inversion of this new transform (if it exists) leads to the solution.

The Analyticity approach is inspired by Fokas's recent discovery of the global relation, obtained first within the $x - t$ transform approach [1] and more recently using differential forms [2]. The use of the global relation to study the well-posedness and to solve IBV problems is illustrated, for instance, in [3], [4], [5]. In [5], in particular, general results on the well-posedness of IBV problems for dispersive $1+1$ dimensional equations of arbitrary order are discussed.

Our main contribution to the method consists, after formulating the IBV problem in Fourier space, in imposing systematically the analyticity properties of all the Fourier transforms involved in the problem, to derive a cascade of analyticity constraints which allow one to express the unknown BVs in terms of the known ones, and therefore to solve the problem. In particular, Fokas' global relation appears, in the methodology we propose, as a "zero residue condition" for the FT of the solution.

The Analyticity approach we propose is very elementary and, above all, has the great conceptual advantage to originate itself from a single guiding principle: satisfying the analyticity properties of all the Fourier transforms involved in the problem. It is the type of approach that can be easily taught in University courses, combining nicely standard PDE theory tools, like the Green's formula and the Fourier transform, with elementary notions in Complex Functions theory.

The essential aspects of the Analyticity approach were first presented by the authors at the Workshop "Boundary value problems" in Cambridge, December 2001, inside the programme: "Integrable Systems". The method is systematically presented for the first time in this paper, with illustrations of IBV problems of various type (Dirichlet, Neumann, mixed, periodic) for some second and third order classical PDEs of the Mathematical Physics: the Schrödinger, the heat and the linear Korteweg-de Vries equations. Also its connections with the EbR approach are illustrated here, in the particular example of the Schrödinger equation in the n-dimensional quadrant. A general account of the EbR approach is given in [6]. A different approach, valid for semibounded domains, has been recently presented in [7]. A general review of the basic spectral methods of solution of IBV problems for linear and soliton PDEs is presented in [8].

Here below, §2 is devoted to the presentation of the Analyticity approach, while §3 is dedicated to its application to some IBV problems for second and third order evolutionary PDEs in $1+1$ and in $n+1$ dimensions. In §4 we finally discuss the connections between the Analyticity approach and the EbR approach.

2. The Analyticity Approach

2.1. THE FOURIER TRANSFORM AND ITS ANALYTICITY PROPERTIES

The natural FT associated with the space-time domain $\mathcal{D} = V \otimes (0, \infty)$ (in short: $FT_{\mathcal{D}}$) is defined by

$$\hat{F}(\mathbf{k}, q) = \int_{\mathcal{D}} d\mathbf{x} dt e^{-i(\mathbf{k}\cdot\mathbf{x}+qt)} F(\mathbf{x}, t) \qquad (2.1)$$

for any smooth function $F(\mathbf{x}, t)$, $(\mathbf{x}, t) \in \mathcal{D}$, assuming that $F(\mathbf{x}, t) \to 0$ fast enough if $t \to \infty$; where $\mathbf{k} = (k_1, \cdots, k_n) \in I\!R^n$, $q \in I\!R$ and $\mathbf{k} \cdot \mathbf{x} = \sum_{j=1}^{n} k_j x_j$. Its inverse:

$$F(\mathbf{x}, t)\chi_{\mathcal{D}}(\mathbf{x}, t) = \int_{I\!R^{n+1}} \frac{d\mathbf{k} dq}{(2\pi)^{n+1}} e^{i(\mathbf{k}\cdot\mathbf{x}+qt)} \hat{F}(\mathbf{k}, q) \qquad (2.2)$$

reconstructs $F(\mathbf{x}, t)$ in $\mathcal{D}$ and zero outside, where $\chi_{\mathcal{D}}(\mathbf{x}, t)$ is the characteristic function of the domain $\mathcal{D}$: $\chi_{\mathcal{D}}(\mathbf{x}, t) = 1$, $(\mathbf{x}, t) \in \mathcal{D}$, $\chi_{\mathcal{D}}(\mathbf{x}, t) =$

0, $(\mathbf{x}, t) \notin \mathcal{D}$ (therefore: $\chi_{\mathcal{D}}(\mathbf{x}, t) = \chi_V(\mathbf{x})H(t)$, where $H(t)$ is the usual Heaviside (step) function).

If the space domain is the whole space: $V = \mathbb{R}^n$, the $FT_{\mathcal{D}}$ (2.1) is defined in $\mathcal{A} = \mathbb{R}^n \otimes \bar{\mathcal{I}}_q$, where $\bar{\mathcal{I}}_q$ is the closure of the lower half q-plane $\mathcal{I}_q$, analytic in $q \in \mathcal{I}_q$, $\forall \mathbf{k} \in \mathbb{R}^n$ and exhibits a proper asymptotic behavior for large q in the analyticity region. If the space domain V is bounded, the $FT_{\mathcal{D}}$ acquires strong analyticity properties in all the Fourier variables: it is defined in $\mathcal{A} = \mathbb{C}^n \otimes \bar{\mathcal{I}}_q$, analytic in $q \in \mathcal{I}_q$, $\forall \mathbf{k} \in \mathbb{C}^n$, entire in every complex k_j, $j = 1, \ldots, n$, $\forall q \in \bar{\mathcal{I}}_q$, and exhibits a proper asymptotic behavior, for large $(\mathbf{k}, q)$, in the analyticity regions. If the space domain is semi-bounded, then the analyticity in the Fourier variables k_j, $j = 1, \ldots, n$ is limited to open regions of the complex plane, depending on the geometric properties of the domain V. We are therefore led to the following definition:

Definition of admissibility. Given a space-time domain $\mathcal{D}$, a function of $(\mathbf{k}, q)$ is an *admissible Fourier transform for the domain* $\mathcal{D}$ (an admissible $FT_{\mathcal{D}}$) iff it possesses the analyticity properties and the asymptotic behavior corresponding to that domain.

2.2. THE IBV PROBLEM IN FOURIER SPACE

We find it convenient to rewrite the IBV problem (1.1) in Fourier space. This goal is conveniently achieved using the well-known *Green's formula (identity)*:

$$b\mathcal{L}a - a\tilde{\mathcal{L}}b = \text{div } J(\mathbf{x}, t), \tag{2.3}$$

and its integral consequence, the celebrated *Green's integral identity*:

$$\int_{\mathcal{D}} (b\mathcal{L}a - a\tilde{\mathcal{L}}b) d\mathbf{x}dt = \int_{\partial\mathcal{D}} J(\mathbf{x}, t) \cdot \nu d\sigma, \tag{2.4}$$

obtained by integrating (2.3) over the domain $\mathcal{D}$ and by using the divergence theorem. In equation (2.3), $\tilde{\mathcal{L}}$ is the formal adjoint of $\mathcal{L}$: $\tilde{\mathcal{L}} = \mathcal{L}(-\nabla, -\frac{\partial}{\partial t})$, $J(\mathbf{x}, t)$ is an $(n+1)$-dimensional vector field, div is the $(n+1)$-dimensional divergence operator and $a(\mathbf{x}, t)$ and $b(\mathbf{x}, t)$ are arbitrary functions. In equation (2.4), $d\sigma$ is the hypersurface element of the boundary and ν is its outward unit normal. We remark that, given $\mathcal{L}$, its formal adjoint $\tilde{\mathcal{L}}$ and two arbitrary functions a and b, an $(n+1)$-dimensional vector field $J(\mathbf{x}, t)$ satisfying the Green's formula (2.3) always exists and can be algorithmically found to be a linear expression of a, b and their partial derivatives of order up to $N-1$, if $\mathcal{L}$ is of order N.

The arbitrariness of a and b allows one to extract from (2.3) and (2.4) several important informations on the IBV problem; with the particular choice

$$a = u(\mathbf{x}, t), \qquad b = e^{-i(\mathbf{k}\cdot\mathbf{x}+qt)}/\mathcal{L}(i\mathbf{k}, iq), \tag{2.5}$$

where $\mathcal{L}(i\mathbf{k}, iq)$ is the eigenvalue of the operator $\mathcal{L}$, corresponding to the eigenfunction $e^{i(\mathbf{k}\cdot\mathbf{x}+qt)}$, the vector field J takes the following form: $J = e^{-i(\mathbf{k}\cdot\mathbf{x}+qt)}J'(\mathbf{x}, t; \mathbf{k}, q)/\mathcal{L}(i\mathbf{k}, iq)$ and the Green's integral identity (2.4) gives the $FT_{\mathcal{D}}$ of the solution in terms of the $FT_{\mathcal{D}}$ of the forcing and (of appropriate FTs) of all the IBVs:

$$\begin{aligned} \hat{u}(\mathbf{k}, q) &= \frac{1}{\mathcal{L}(i\mathbf{k}, iq)}(\hat{f}(\mathbf{k}, q) - \int_{\partial\mathcal{D}} e^{-i(\mathbf{k}\cdot\mathbf{x}+qt)}J'(\mathbf{x}, t; \mathbf{k}, q)\cdot\nu d\sigma) \\ &=: \frac{\hat{\mathcal{N}}(\mathbf{k}, q)}{\mathcal{L}(i\mathbf{k}, iq)}, \qquad (\mathbf{k}, q) \in \mathcal{A}. \end{aligned} \tag{2.6}$$

If the PDE has the following evolutionary form:

$$\mathcal{L}(\nabla, \frac{\partial}{\partial t}) = \frac{\partial}{\partial t} - \mathcal{K}(\nabla), \tag{2.7}$$

which we assume from now on just for the sake of simplicity, then

$$\hat{u}(\mathbf{k}, q) = \frac{\hat{f}(\mathbf{k}, q) + \hat{u}_0(\mathbf{k}) + \hat{\mathcal{B}}(\mathbf{k}, q)}{\mathcal{L}(i\mathbf{k}, iq)} =: \frac{\hat{\mathcal{N}}(\mathbf{k}, q)}{\mathcal{L}(i\mathbf{k}, iq)}, \qquad (\mathbf{k}, q) \in \mathcal{A} \tag{2.8}$$

and the linear relation (2.8) makes clear how the different contributions coming from the equation (the denominator $\mathcal{L}$), from the forcing $\hat{f}$, from the initial condition $\hat{u}_0$,

$$\hat{u}_0(\mathbf{k}) = \int_V d\mathbf{x} e^{-i\mathbf{k}\cdot\mathbf{x}}u_0(\mathbf{x}), \qquad u_0(\mathbf{x}) := u(\mathbf{x}, 0), \tag{2.9}$$

and from the set of boundary values $\hat{\mathcal{B}}$ separate in Fourier space.

Its inverse transform (2.2) gives the corresponding *Fourier representation* of the solution:

$$U(\mathbf{x}, t) = u(\mathbf{x}, t)\chi_{\mathcal{D}}(\mathbf{x}, t) = \int_{\mathbb{R}^{n+1}} \frac{d\mathbf{k}dq e^{i(\mathbf{k}\cdot\mathbf{x}+qt)}}{(2\pi)^{n+1}}\frac{\hat{\mathcal{N}}(\mathbf{k}, q)}{\mathcal{L}(i\mathbf{k}, iq)},$$
$$(\mathbf{x}, t) \in \mathbb{R}^{n+1}. \tag{2.10}$$

Two sources of problems arise at a first glance of equation (2.6):

i) the RHS of the equation depends on known and unknown BVs;
ii) apparently the RHS of the equation is not an admissible $FT_{\mathcal{D}}$.

It is very satisfactory that the analyticity constraints which make the r. h. s. of (2.6) an admissible $FT_{\mathcal{D}}$ provide also a number of relations among the IBVs which are sufficient to express the unknown BVs in terms of known data.

2.3. THE ANALYTICITY CONSTRAINTS AND THEIR RESOLUTION

In general, $\mathcal{L}(i\mathbf{k}, iq)$, the denominator of equation (2.8), is an entire and, most frequently, polynomial function of all its complex variables. Let $\mathcal{S}$ be the manifold in which this entire function is zero:

$$\mathcal{S} = \{(\mathbf{k}, q) \in \mathcal{C}^{n+1} : \ \mathcal{L}(i\mathbf{k}, iq) = 0\}. \tag{2.11}$$

Then the r. h. s. of equation (2.8) provides an admissible $FT_{\mathcal{D}}$ of the solution of the IBV problem under investigation if the numerator $\hat{\mathcal{N}}(\mathbf{k}, q)$ of $\hat{u}$ in (2.8) satisfies in $\mathcal{A} \cap \mathcal{S}$, hereafter called the *Singularity Manifold (SM) of the IBV problem*, the following *Zero Residue Condition* (ZRC):

$$\hat{\mathcal{N}}(\mathbf{k}, q) = 0, \qquad (\mathbf{k}, q) \in \mathcal{A} \cap \mathcal{S}. \tag{2.12}$$

If the singularity manifold $\mathcal{A} \cap \mathcal{S}$ contains the real axis (which is usually a part of the boundary of $\mathcal{A}$) and if this singularity is not already taken care of by the ZRC (2.12), then the FT (2.8) of $u(\mathbf{x}, t)$ is still admissible, even if it is singular on the real axis, provided one applies the *Denominator Regularization* (DR):

$$\mathcal{L}(i\mathbf{k}, iq) \to \mathcal{L}_{reg}(i\mathbf{k}, iq). \tag{2.13}$$

In fact, for the PDEs considered here, see (2.7), the zeroes of the denominator $\mathcal{L}(i\mathbf{k}, iq) = iq - \mathcal{K}(i\mathbf{k})$ are simple and then one regularizes the denominator, see (2.13), moving the zeroes of $\mathcal{L}$ a bit off the real axis, outside the domain $\mathcal{A}$.

The ZRC plus the DR constitute the main set of *Analyticity Constraints* (ACs) that must be imposed to the r. h. s. of (2.8) in order to obtain an admissible $FT_{\mathcal{D}}$ of the solution of the IBV problem under investigation. The ZRC (2.12) provides a (linear) relation among the FTs of the forcing, of the initial condition and of all the BVs. Moreover, it is important to realize that the analyticity properties of all these FTs generate, through the admissibility argument, a cascade of further analyticity constraints, so that the analysis goes on until all these conditions are finally met. This procedure defines, in principle, a set of relations (a system of equations) among the IBVs. We then conclude that:

a) *The unique solvability of such a system, together with the admissibility of the obtained solution, are equivalent to the study of the unique solvability of all the IBV problems associated with (1.1).*

b) *By solving this system for a set of BVs in terms of the remaining ones, one expresses all quantities in terms of known data and, from equation (2.10), one obtains the Fourier representation of the solution.*

In most of the examples considered in this paper, this system of equations is algebraic, with entire coefficients. Therefore, if M is the squared matrix of the coefficients of the unknown BVs, the admissibility argument imposes that the countable set of zeroes of *det* M:

$$\{q_m\}_{m\in\mathbb{Z}}, \qquad \det\ M(q_m) = 0 \tag{2.14}$$

lies outside the analyticity domain of an admissible $FT_{\mathcal{D}}$:

$$q_m \notin \mathcal{I}_q, \qquad m \in \mathbb{Z}. \tag{2.15}$$

It turns out that the set (2.14) coincides with the spectrum arising in the eigenfunction expansion approach [9] and coincides also with the restricted domain in which the EbR method works. These deep connections justify for (2.14) the name of *spectrum of the IBV problem.*

The admissibility argument imposes also that the constructed solution of such system exhibit the proper asymptotic behavior in the analyticity domain. It is actually convenient to impose first this asymptotic admissibility, the easiest to be checked, which enables one to disregard without effort all the IBV problems which are ill-posed because incompatible with asymptotics.

2.4. GENERAL REMARKS

Remark 1. *Analyticity vs Causality.* It is well-known that there are definite connections between the analyticity properties of the FT of the solution of evolution equations and the causality principle. In our general setting it is straightforward to show that:
The analyticity properties of the $FT_{\mathcal{D}}$ of the solution of the IBV problem (1.1) imply the causality principle.
Indeed, using the convolution theorem, the inverse FT (2.10) of the r. h. s. of equation (2.8) (in which all the analyticity constraints have been preliminary imposed) is equivalent to the following Green's representation of the solution:

$$u(\mathbf{x},t) = \int_0^t dt' \int_V d\mathbf{x}' G_{RF}(\mathbf{x}-\mathbf{x}';t-t')\mathcal{N}(\mathbf{x}',t'), \quad (\mathbf{x},t) \in \mathcal{D}, \tag{2.16}$$

where $\mathcal{N}(\mathbf{x},t)\chi_{\mathcal{D}}(\mathbf{x},t)$ is the inverse FT (2.2) of $\hat{\mathcal{N}}(\mathbf{k},q)$ and G_{RF} is the celebrated retarded-fundamental Green's function of the operator $\mathcal{L}$:

$$G_{RF}(\mathbf{x},t) = \int_{I\!R^{n+1}} \frac{d\mathbf{k}dq}{(2\pi)^n} \frac{e^{i(\mathbf{k}\cdot\mathbf{x}+qt)}}{\mathcal{L}_{reg}(i\mathbf{k},iq)}, \tag{2.17}$$

which satisfies the important property: $G_{RF}(\mathbf{x},t) = 0,\ t < 0$, due to the regularization of $\mathcal{L}(i\mathbf{k},iq)$. Equation (2.16) is the usual way in which the causality principle becomes transparent.

Remark 2. *Regularization and Fourier representation.* As we have already written, even if the zeroes of the denominator on the real axis are all cured by the ZRC, some regularization must be introduced in the calculation of the Fourier representation (2.10), before splitting $\hat{\mathcal{N}}$ in the sum of the different contributions (each one singular on the real axis) in (2.8) coming from the forcing, from the initial condition and from the BVs. The most convenient regularization is obviously that in (2.13) and it leads to the following Fourier representation:

$$\begin{aligned} U(\mathbf{x},t) &= u(\mathbf{x},t)\chi_{\mathcal{D}}(\mathbf{x},t) = \int_{I\!R^{n+1}} \frac{d\mathbf{k}dqe^{i(\mathbf{k}\cdot\mathbf{x}+qt)}}{(2\pi)^{n+1}} \frac{\hat{f}(\mathbf{k},q)}{\mathcal{L}_{reg}(i\mathbf{k},iq)} \\ &+ \int_{I\!R^{n+1}} \frac{d\mathbf{k}dqe^{i(\mathbf{k}\cdot\mathbf{x}+qt)}}{(2\pi)^{n+1}} \frac{\hat{u}_0(\mathbf{k})}{\mathcal{L}_{reg}(i\mathbf{k},iq)} \\ &+ \int_{I\!R^{n+1}} \frac{d\mathbf{k}dqe^{i(\mathbf{k}\cdot\mathbf{x}+qt)}}{(2\pi)^{n+1}} \frac{\hat{\mathcal{B}}(\mathbf{k},q)}{\mathcal{L}_{reg}(i\mathbf{k},iq)}, \end{aligned} \tag{2.18}$$

where $(\mathbf{x},t) \in I\!R^{n+1}$.

3. Illustrative Examples

In this section we apply the Analyticity approach to the following classical equations of the Mathematical Physics in 1+1 dimensions, the Schrödinger, the heat and the linear Korteweg-de Vries (KdV) equations:

$$\frac{\partial u}{\partial t} - \alpha\frac{\partial^2 u}{\partial x^2} = f, \qquad \alpha = i, 1, \qquad x \in V,\ t > 0, \tag{3.1}$$

$$\frac{\partial u}{\partial t} - \eta\frac{\partial^3 u}{\partial x^3} = f, \qquad \eta = \pm 1, \qquad x \in V,\ t > 0, \tag{3.2}$$

prototype examples respectively of second and third order evolutionary PDEs and basic universal models for the description of dispersive and diffusive phenomena, where the space domain V is either the segment $(0, L)$ or the semiline $(0, \infty)$. Hereafter the BVs will be indicated by

$$v_0^{(j)}(t) := \frac{\partial^j u}{\partial x^j}(x,t)|_{x=0}, \qquad v_L^{(j)}(t) := \frac{\partial^j u}{\partial x^j}(x,t)|_{x=L}, \qquad j \in I\!N \tag{3.3}$$

and their Fourier transforms by $\hat{v}_0^{(j)}(q)$, $\hat{v}_L^{(j)}(q)$:

$$\hat{v}_0^{(j)}(q) := \int_0^\infty dt e^{-iqt} v_0^{(j)}(t), \qquad \hat{v}_L^{(j)}(q) := \int_0^\infty dt e^{-iqt} v_L^{(j)}(t). \tag{3.4}$$

We also apply the method to the study of IBV problems for the multi-dimensional analogue of equation (3.1), for $\alpha = i$:

$$\frac{\partial u}{\partial t} - i \triangle u = f, \qquad x \in V,\ t > 0, \qquad \triangle := \sum_{j=1}^n \frac{\partial^2}{\partial {x_j}^2} \tag{3.5}$$

in the n-dimensional quadrant

$$V = \{\mathbf{x} :\ x_j \geq 0,\ j = 1, \dots, n\}. \tag{3.6}$$

The corresponding BVs will be indicated by:

$$v_{0j}^{(0)}(\mathbf{x}_j, t) = u(\mathbf{x}, t)|_{x_j=0}, \quad v_{0j}^{(1)}(\mathbf{x}_j, t) = \frac{\partial u}{\partial x_j}(\mathbf{x}, t)|_{x_j=0} \tag{3.7}$$

and their FTs by:

$$\hat{v}_{0j}^{(m)}(\mathbf{k}_j, q) = \int_0^\infty dt \int_{V_j} d\mathbf{x}_j e^{-i(\mathbf{k}_j \cdot \mathbf{x}_j + qt)} v_{0j}^{(m)}(\mathbf{x}_j, t), \qquad m = 0, 1. \tag{3.8}$$

In equations (3.7)–(3.8) $\mathbf{x}_j = (x_1, \dots, \underline{x_j}, \dots, x_n) \in I\!R^{n-1}$, $\mathbf{k}_j = (k_1, \dots, \underline{k_j}, \dots, k_n) \in I\!R^{n-1}$, $\int_{V_j} d\mathbf{x}_j = \int_0^{L_1} dx_1 \cdot \cdot (\underline{\int_0^{L_j} dx_j}) \cdot \cdot \int_0^{L_n} dx_n$, and a line under a variable indicates that that variable is removed.

The application of the Analyticity approach to higher order problems and to other relevant examples will be presented in [8].

3.1. THE SECOND ORDER PDES

In the case of the PDEs (3.1), equations (2.3) and (2.5) imply:

$$\tilde{\mathcal{L}} = -\tfrac{\partial}{\partial t} - \alpha \tfrac{\partial^2}{\partial x^2}, \qquad J = \left(ab, \alpha \left[a \tfrac{\partial b}{\partial x} - b \tfrac{\partial a}{\partial x}\right]\right), \\ \mathcal{L}(ik, iq) = i(q - i\alpha k^2). \tag{3.9}$$

In addition, if V is the segment $(0, L)$, equation (2.8) yields

$$\begin{aligned} \hat{u}(k,q) &= \frac{\hat{\mathcal{N}}(k,q)}{i(q - i\alpha k^2)}, \\ \hat{\mathcal{N}}(k,q) &= \hat{f}(k,q) + \hat{u}_0(k) - \alpha\Big([\hat{v}_0^{(1)}(q) + ik\hat{v}_0^{(0)}(q)] \\ &\qquad - e^{-ikL}[\hat{v}_L^{(1)}(q) + ik\hat{v}_L^{(0)}(q)]\Big) \end{aligned} \tag{3.10}$$

and the Fourier representation (2.18) of the solution takes the following form:

$$\begin{aligned} u(x,t) &= \int_{I\!R^2} \frac{dqdk}{(2\pi)^2 i} e^{i(kx+qt)} \frac{\hat{f}(k,q)}{q - i\alpha k^2 - i0} + \int_{I\!R} \frac{dk}{2\pi} e^{ikx - \alpha k^2 t} \hat{u}_0(k) \\ &\quad - \alpha \int_{\gamma(\alpha)} \frac{dk}{2\pi} e^{ikx - \alpha k^2 t} [\hat{v}_0^{(1)}(i\alpha k^2) + ik\hat{v}_0^{(0)}(i\alpha k^2)] \\ &\quad + \alpha \int_{\gamma(\alpha)} \frac{dk}{2\pi} e^{-ik(x-L) - \alpha k^2 t} [\hat{v}_L^{(1)}(i\alpha k^2) - ik\hat{v}_L^{(0)}(i\alpha k^2)]\,; \\ x &\in (0,L), \quad t > 0\,, \end{aligned} \tag{3.11}$$

where the integration path $\gamma^{(\alpha)} = \partial\mathcal{K}_1^{(\alpha)}$ is the counterclockwise oriented boundary of $\mathcal{K}_1^{(\alpha)}$, and $\mathcal{K}_1^{(i)}$ and $\mathcal{K}_0^{(i)}$ are respectively the first and third quadrant of the complex k-plane, $\mathcal{K}_m^{(1)} = \hat{\rho}_{\frac{\pi}{4}}\mathcal{K}_m^{(i)}$, $m = 0, 1$, where $\hat{\rho}_{\frac{\pi}{4}}$ is the $\pi/4$ rotation operator: $\hat{\rho}_{\frac{\pi}{4}} : k \to e^{\frac{i\pi}{4}} k$ (see Figs 1a, b).

The corresponding expressions for the semiline or for the infinite line cases, with rapidly decreasing conditions at ∞, follow immediately from the ones above, setting $\hat{v}_L^{(0)} = \hat{v}_L^{(1)} = 0$ in the semiline case, or setting $\hat{v}_0^{(0)} = \hat{v}_0^{(1)} = \hat{v}_L^{(0)} = \hat{v}_L^{(1)} = 0$ in the infinite line case.

It is instructive to first apply the Analyticity approach to the simplest case in which the space domain is the whole space, with rapidly decreasing BVs at $x = \pm\infty$.

3.1.1. *The whole line $V = (-\infty, \infty)$*
Equation (3.10b) reduces to

$$\hat{\mathcal{N}}(k,q) = \hat{f}(k,q) + \hat{u}_0(k) \tag{3.12}$$

and the admissibility argument imposes that $\hat{\mathcal{N}}(k,q)/\mathcal{L}$ be defined in $(k,q) \in \mathcal{A} = I\!R \otimes \bar{\mathcal{I}}_q$ and be analytic in $q \in \mathcal{I}_q$, $\forall k \in I\!R$. If $\alpha = 1$, the denominator is singular for $q = ik^2$, $k \in I\!R$, outside the definition domain, and no regularization is needed. If, instead, $\alpha = i$, the denominator is singular for $q = -k^2 < 0$, on the real negative axis, at the boundary of the analyticity domain, and the only analyticity constraint to be fulfilled is the Denominator Regularization (2.13):

$$\mathcal{L}(ik, iq) = i(q - i\alpha k^2) \to \mathcal{L}_{reg}(ik, iq) = i(q - i\alpha k^2 - i0). \tag{3.13}$$

The regularization (3.13) is sufficient to make the r. h. s. of the first equation in (3.10) an admissible $FT_{\mathcal{D}}$, from which we recover the well-known Fourier representation of the solution of equations (3.1):

$$u(x,t) = \int_{I\!R^2} \frac{dqdke^{i(kx+qt)}}{(2\pi)^2 i} \frac{\hat{f}(k,q)}{q - i\alpha k^2 - i0} + \int_{I\!R} \frac{dk}{2\pi} e^{ikx - \alpha k^2 t} \hat{u}_0(k) \tag{3.14}$$

for $x \in I\!R,\ t > 0$.

3.1.2. *The semiline* $V = (0, \infty)$

In this case:

$$\hat{\mathcal{N}}(k,q) = \hat{f}(k,q) + \hat{u}_0(k) - \alpha[\hat{v}_0^{(1)}(q) + ik\hat{v}_0^{(0)}(q)] \tag{3.15}$$

and admissibility imposes that $\hat{\mathcal{N}}/\mathcal{L}$ be defined in $\mathcal{A} = \bar{\mathcal{I}}_k \otimes \bar{\mathcal{I}}_q$, be analytic in $q \in \mathcal{I}_q$, $\forall k \in \bar{\mathcal{I}}_k$ and be analytic in $k \in \mathcal{I}_k$, $\forall q \in \bar{\mathcal{I}}_q$. Therefore the singularity manifolds $\mathcal{A} \cap \mathcal{S}^{(\alpha)}$, corresponding to $\alpha = i, 1$, are parametrizable either in terms of k or in terms of q in the following way:

$$\begin{aligned} \mathcal{A} \cap \mathcal{S}^{(\alpha)} &= \{q = i\alpha k^2,\ k \in \overline{\mathcal{K}_0^{(\alpha)}}\} \\ &= \{k = k_0^{(\alpha)}(q),\ \pi \le \arg q \le 2\pi\}, \end{aligned} \tag{3.16}$$

where

$$k_0^{(\alpha)}(q) = \begin{cases} iq^{\frac{1}{2}}, & \alpha = i \\ e^{\frac{3}{4}\pi i} q^{\frac{1}{2}}, & \alpha = 1. \end{cases} \tag{3.17}$$

If $\alpha = 1$, there is no singularity on the real axis and no regularization is needed (the singularity at $k = q = 0$ is eliminated by the ZRC). If $\alpha = i$, there are two singularities for $k \in I\!R$; that corresponding to $k < 0$ is cured by the ZRC (2.12), while that corresponding to $k > 0$ is cured instead by the regularization (3.13).

The ZRC (2.12) is conveniently parametrized in terms of q in the following way:

$$\hat{\mathcal{N}}(k_0^{(\alpha)}(q), q) = \hat{f}(k_0^{(\alpha)}(q), q) + \hat{u}_0(k_0^{(\alpha)}(q)) - \alpha[\hat{v}_0^{(1)}(q) + ik_0^{(\alpha)}(q)\hat{v}_0^{(0)}(q)] = 0, \tag{3.18}$$

for $\pi \leq \arg q \leq 2\pi$. It is one equation involving 4 $FT's$ which are therefore dependent. If we are interested in solving the Dirichlet and Neumann problems, we use this ZRC to express the unknown BVs in terms of the known ones:

$$\begin{aligned} Dirichlet: \ \alpha\hat{v}_0^{(1)}(q) &= \hat{f}(k_0^{(\alpha)}(q), q) + \hat{u}_0(k_0^{(\alpha)}(q)) - \alpha i k_0^{(\alpha)}(q)\hat{v}_0^{(0)}(q) \\ Neumann: \ i\alpha k_0^{(\alpha)}(q)\hat{v}_0^{(0)}(q) &= \hat{f}(k_0^{(\alpha)}(q), q) + \hat{u}_0(k_0^{(\alpha)}(q)) - \alpha\hat{v}_0^{(1)}(q), \end{aligned} \tag{3.19}$$

for $\pi \leq \arg q \leq 2\pi$. It is easy to see that the unknown BVs are expressed by (3.19) as admissible FTs which, inserted in (3.11), give the wanted solution of the Dirichlet and Neumann problems. The Robin problem can be similarly solved.

We remark that the ZRC (3.18) could also be solved for $\hat{u}_0$ (using now, for convenience, the variable k):

$$\hat{u}_0(k) = -\hat{f}(k, i\alpha k^2) + \alpha[\hat{v}_0^{(1)}(i\alpha k^2) + ik\hat{v}_0^{(0)}(i\alpha k^2)], \qquad k \in \overline{\mathcal{K}_0^{(\alpha)}} \tag{3.20}$$

but, in this case, the solution would not be, in general, an admissible FT, since the r. h. s. of (3.20) cannot be extended to the rest of the lower half k-plane. Even in the special case in which the forcing and the assigned BVs were on a compact support in t, corresponding to entire FTs, the solution $\hat{u}_0(k)$ would not be admissible, because it would not possess, in general, the proper asymptotics. This means that the (unphysical) problem in which we assign arbitrarily u and its space derivative at $x = 0$ cannot be treated by the FT method, unless the above BVs are suitably constrained.

3.1.3. *The segment $V = (0, L)$*

Now admissibility implies that $\hat{\mathcal{N}}/\mathcal{L}$ be defined in $\mathcal{A} = C \otimes \bar{\mathcal{I}}_q$, be analytic in $q \in \mathcal{I}_q$, $\forall k \in C$ and be analytic in $k \in C$, $\forall q \in \bar{\mathcal{I}}_q$, with proper asymptotics for large $|k|$ and/or $|q|$ in the analyticity regions. Therefore the singularity manifold on which the ZRC (2.12) is defined is the union of two sectors:

$$\begin{aligned} \mathcal{A} \cap \mathcal{S}^{(\alpha)} &= \bigcup_{m=0}^{1} \{q = i\alpha k^2, \ k \in \overline{\mathcal{K}_m^{(\alpha)}}\} \\ &= \bigcup_{m=0}^{1} \{k = k_m^{(\alpha)}(q) = (-)^m k_0^{(\alpha)}(q), \ \pi \leq \arg q \leq 2\pi\}. \end{aligned} \tag{3.21}$$

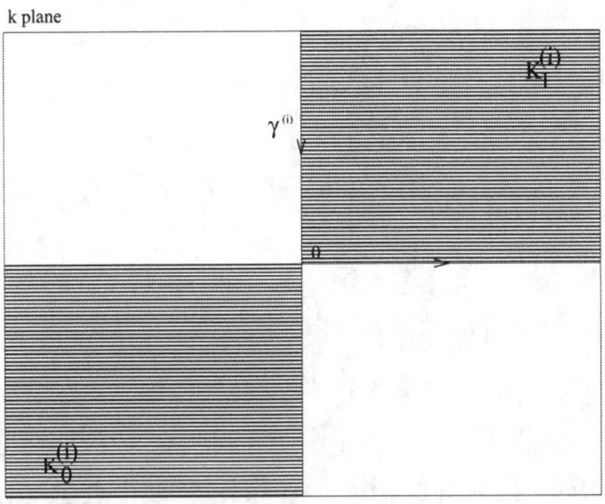

Fig.1a The SM $\mathcal{A} \cap \mathcal{S}^{(\alpha)}$ $(\alpha = i)$

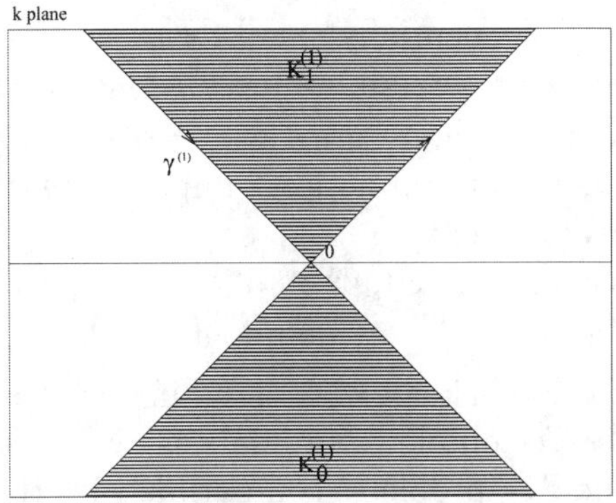

Fig.1b The SM $\mathcal{A} \cap \mathcal{S}^{(\alpha)}$ $(\alpha = 1)$

Both singularities on the real axis are cured by the ZRC and no regularization is needed. The regularization (3.13), however, is still introduced, according to the Remark 2 of §2.4, in computing the Fourier representation (2.18) of the solution, and it leads to the r. h. s. of (3.11). The ZRC (2.12), conveniently parametrized using q, consists of the following system of two linear algebraic equations:

$$\hat{\mathcal{N}}(k_m^{(\alpha)}(q), q) = 0, \qquad m = 0, 1 \qquad \pi \le \arg\, q \le 2\pi, \tag{3.22}$$

where $\hat{\mathcal{N}}$ is given by (3.10). Since this system contains four BVs, we expect that two out of four BVs can be arbitrarily assigned. To establish which pairs of BVs can be arbitrarily given, one should impose that the corresponding solutions of the algebraic system (3.22) define admissible FTs; i.e., the following two conditions must be satisfied.

i) The system must be uniquely solvable for the unknown pair of BVs in its definition domain. More precisely, indicating by M the 2×2 matrix of the coefficients of the unknown BVs, the admissibility condition imposes that the countable set $\{q_j\}_{j \in I\!N}$ of zeroes of det M, *the spectrum of the IBV problem*, lie outside the analyticity domain:

$$q_j \notin \mathcal{I}_q, \qquad j \in I\!N. \tag{3.23}$$

ii) The solution of the system must define admissible Fourier Transforms; in particular, it must exhibit the proper asymptotics in the analyticity domain.

Studying first the asymptotics of (3.22), one infers without any effort which pairs of BVs *cannot* be assigned arbitrarily. The asymptotics of (3.22) imply immediately that the following expressions:

$$\hat{f}(k_0^{(\alpha)}(q), q) + \hat{u}_0(k_0^{(\alpha)}(q)) - \alpha[\hat{v}_0^{(1)}(q) + ik_0^{(\alpha)}(q)\hat{v}_0^{(0)}(q)], \tag{3.24}$$

$$e^{-ik_0^{(\alpha)}(q)L}[\hat{f}(-k_0^{(\alpha)}(q), q) + \hat{u}_0(-k_0^{(\alpha)}(q))] + \alpha[\hat{v}_L^{(1)}(q) - ik_0^{(\alpha)}(q)\hat{v}_L^{(0)}(q)]$$

are exponentially small for $q \sim \infty$ in $\pi \leq \arg q \leq 2\pi$. Since the asymptotic series of the admissible FTs appearing in the LHS of equations (3.24) are inverse power series of $q^{\frac{1}{2}}$, equations (3.24) impose severe constraints on the involved functions, implying that, if the forcing and the initial data are given, then:
asymptotic admissibility is compatible with assigning at $x = 0$ any BV between $(v_0^{(0)}, v_0^{(1)})$ and, at $x = L$, any BV between $(v_L^{(0)}, v_L^{(1)})$. It is not compatible instead with assigning arbitrarily the pairs $(v_0^{(1)}, v_0^{(0)})$ or $(v_L^{(1)}, v_L^{(0)})$.

To complete our analysis, we must check if the spectrum associated with the IBV problems compatible with the asymptotics lie outside the definition domain. The analysis is straightforward and produces the following results.
Proposition (the spectrum). *Assigning arbitrarily* $(v_0^{(0)}, v_L^{(0)})$ *(the Dirichlet problem) or* $(v_0^{(1)}, v_L^{(1)})$ *(the Neumann problem), the spectrum is characterized by the equation* $\sin(kL) = 0 \Leftrightarrow k_m = \frac{\pi m}{L}$, $m \in \mathbb{Z}$ *and is given by the negative eigenvalues* $\{q_m\}_{n \in \mathbb{N}}$, $q_m = -k_m^2 = -(\frac{\pi m}{L})^2$, $m \in \mathbb{N}$, *if* $\alpha = i$, *and by the purely imaginary eigenvalues* $q_m = ik_m^2 = i(\frac{\pi m}{L})^2$, $m \in \mathbb{N}$, *if* $\alpha = 1$. *Assigning instead* $(v_0^{(0)}, v_L^{(1)})$ *or* $(v_0^{(1)}, v_L^{(0)})$ *(the mixed problems), the spectrum is characterized by the equation* $\cos(kL) = 0 \Leftrightarrow k_m = \frac{\pi}{L}(2m+1)$, $m \in \mathbb{Z}$ *and is given by the negative eigenvalues* $\{q_m\}_{m \in \mathbb{N}}$, $q_m = -k_m^2 = -(\frac{\pi}{L})^2(2m+1)^2$, $m \in \mathbb{N}$, *if* $\alpha = i$, *and by the purely imaginary eigenvalues* $q_m = ik_m^2 = i(\frac{\pi}{L})^2(2m+1)^2$, $m \in \mathbb{N}$, *if* $\alpha = 1$.

For $\alpha = 1$ the spectrum lies outside the analyticity region and the solutions of the algebraic system (3.22) define directly admissible FTs; if $\alpha = i$ the solutions of the algebraic system (3.22) define admissible FTs after moving these singularities a bit off the real q - axis in the UHP, outside the definition domain (again a regularization!). We conclude that all the IBV problems compatible with admissible asymptotics turn out to be well-posed:
IBV problems for the Schrödinger and heat equations (3.1) are well-posed assigning at $x = 0$ *any BV among* $(v_0^{(0)}, v_0^{(1)})$ *and at* $x = L$ *any BV among* $(v_L^{(0)}, v_L^{(1)})$.

It is interesting to remark that, if one insisted, instead, in solving an IBV problem in which the BVs $(\hat{v}_0^{(1)}(q),\ \hat{v}_0^{(0)}(q))$ are assigned, the corresponding algebraic system would be always uniquely solvable (no point spectrum would arise), but the solution would exhibit an exponential blow up at $q \sim \infty$ in the analyticity region, that cannot be accepted. This undesired blow up could be cured if the assigned BVs were related by the (additional) analyticity constraint:

$$\hat{f}(k_0^{(\alpha)}(q), q) + \hat{u}_0(k_0^{(\alpha)}(q)) - \alpha[\hat{v}_0^{(1)}(q) + ik_0^{(\alpha)}(q)\hat{v}_0^{(0)}(q)] = 0, \qquad (3.25)$$

implying the following admissible solutions of the algebraic system (3.22):

$$\hat{v}_L^{(1)}(q) = -ik_0^{(\alpha)}(q)\hat{v}_L^{(0)}(q) = -\frac{e^{-ik_0^{(\alpha)}(q)L}}{2\alpha}[\hat{f}(k_0^{(\alpha)}(q),q) \tag{3.26}$$

$$+\hat{f}(-k_0^{(\alpha)}(q),q) + \hat{u}_0(k_0^{(\alpha)}(q)) + \hat{u}_0(-k_0^{(\alpha)}(q)) - 2\alpha\hat{v}_0^{(1)}(q)].$$

The additional analyticity constraint (3.25) is not surprising at all, since it is nothing but the ZRC of the semiline problem. Similarly, assigning the right boundary conditions ($v_L^{(0)}$, $v_L^{(1)}$), the unknowns $\hat{v}_0^{(0)}$ and $\hat{v}_0^{(1)}$ would exhibit again an exponential blow up which cannot be accepted; an admissible asymptotics would be guaranteed now by the (additional) analyticity constraint:

$$e^{-ik_0^{(\alpha)}(q)L}[\hat{f}(-k_0^{(\alpha)}(q),q) + \hat{u}_0(-k_0^{(\alpha)}(q))] + \alpha[\hat{v}_L^{(1)}(q) - ik_0^{(\alpha)}(q)\hat{v}_L^{(0)}(q)] = 0, \tag{3.27}$$

implying the following solution of the algebraic system:

$$\hat{v}_0^{(1)}(q) = ik_0^{(\alpha)}(q)\hat{v}_0^{(0)}(q) = \frac{1}{2\alpha}[\hat{f}(k_0^{(\alpha)}(q),q) + \hat{f}(-k_0^{(\alpha)}(q),q) \tag{3.28}$$

$$+\hat{u}_0(k_0^{(\alpha)}(q)) + \hat{u}_0(-k_0^{(\alpha)}(q)) + 2\alpha e^{-ik_0^{(\alpha)}(q)L}\hat{v}_L^{(1)}(q)].$$

3.1.4. *The periodic problem*

If we assume L-periodicity of u and u_x, then $\hat{v}_0^{(1)} = \hat{v}_L^{(1)} =: \hat{v}^{(1)}$, $\hat{v}_0^{(0)} = \hat{v}_L^{(0)} =: \hat{v}^{(0)}$ and the algebraic system (3.22) consists now of two equations for two BVs, which have to be treated therefore as unknowns. The solutions of this system read:

$$\begin{aligned} \hat{v}^{(1)}(q) &= \frac{1}{2\alpha}\left(\frac{\hat{f}(k_0^{(\alpha)}(q),q)+\hat{u}_0(k_0^{(\alpha)}(q))}{1-e^{-ik_0^{(\alpha)}(q)L}} + \frac{\hat{f}(-k_0^{(\alpha)}(q),q)+\hat{u}_0(-k_0^{(\alpha)}(q))}{1-e^{ik_0^{(\alpha)}(q)L}}\right), \\ \hat{v}^{(0)}(q) &= \frac{1}{2i\alpha k_0^{(\alpha)}(q)}\left(\frac{\hat{f}(k_0^{(\alpha)}(q),q)+\hat{u}_0(k_0^{(\alpha)}(q))}{1-e^{-ik_0^{(\alpha)}(q)L}} - \frac{\hat{f}(-k_0^{(\alpha)}(q),q)+\hat{u}_0(-k_0^{(\alpha)}(q))}{1-e^{ik_0^{(\alpha)}(q)L}}\right). \end{aligned} \tag{3.29}$$

They satisfy asymptotic admissibility and the spectrum, characterized by the equation $1-e^{\pm ikL} = 0$, ($\Rightarrow$ $k_n = \frac{2\pi}{L}n$, $n \in \mathbb{Z}$), is given by $q_n = -k_n^2 = -(\frac{2\pi}{L})^2 n^2$, $n \in \mathbb{N}$, for $\alpha = i$, and by $q_n = ik_n^2 = i(\frac{2\pi}{L})^2 n^2$, $n \in \mathbb{N}$, for $\alpha = 1$; therefore the usual regularization is needed again in the Schrödinger case. We conclude that *the periodic problem for equations (3.1), in which one imposes the L- periodicity of u and u_x, is well-posed and no BV can be assigned arbitrarily.*

Remark We remark that the Fourier transforms of the unknown boundary functions exhibit generically a branch point at $q = 0$, due to the well-known slow decay as $t \to \infty$ of the solutions of the dispersive evolution equation under investigation.

The above procedure generalizes with no difficulties to higher order problems. In the following we concentrate on a third order problem only.

3.2. THE LINEAR KDV EQUATION

In this section we investigate IBV problems for 3rd order operators, illustrating the method on the simplest possible example (3.2).

Since the group velocity $v_g = 3\eta k^2$ of the associated wave packet is positive (negative) for η positive (negative), we have the following expectations. In the semiline case, one should be able to assign at $x = 0$ more BVs for positive η than for negative η. In the segment case, for η positive one can assign arbitrarily more BVs at $x = 0$ than at $x = L$ (and vice versa for η negative). The precise indication of "how many" and "which" BVs can be assigned in order to have a well-posed IBV problem follows again in a straightforward way from the Analyticity approach.

Equations (2.3) and (2.5) imply:

$$\tilde{\mathcal{L}} = -\mathcal{L}, \qquad J = \left(ab, -\eta\left[b\frac{\partial^2 a}{\partial x^2} - \frac{\partial b}{\partial x}\frac{\partial a}{\partial x} + \frac{\partial^2 b}{\partial x^2}a\right]\right), \qquad (3.30)$$
$$\mathcal{L}(i\mathbf{k}, iq) = i(q + \eta k^3).$$

In addition, if V is the segment $(0, L)$, equation (2.8) yields

$$\begin{aligned} \hat{u}(k,q) &= -i\frac{\hat{\mathcal{N}}(k,q)}{q+\eta k^3}, \\ \hat{\mathcal{N}}(k,q) &= \hat{f}(k,q) + \hat{u}_0(k) - \eta\left([\hat{v}_0^{(2)}(q) + ik\hat{v}_0^{(1)}(q) - k^2\hat{v}_0^{(0)}(q)]\right. \qquad (3.31) \\ &\quad \left. - e^{-ikL}[\hat{v}_L^{(2)}(q) + ik\hat{v}_L^{(1)}(q) - k^2\hat{v}_L^{(0)}(q)]\right) \end{aligned}$$

and the Fourier representation (2.18) of the solution takes the following form:

$$\begin{aligned} u(x,t) &= \int_{I\!R^2} \frac{dqdk}{(2\pi)^2 i} e^{i(kx+qt)} \frac{\hat{f}(k,q)}{q+\eta k^3 - i0} + \int_{I\!R} \frac{dk}{2\pi} e^{i(kx-\eta k^3 t)}\hat{u}_0(k) \\ &+ \eta\int_\gamma \frac{dk}{2\pi} e^{ik^3 t}\sum_{m=0}^{2}(-i\eta k)^m[\hat{v}_0^{(2-m)}(k^3)E^{(m)}(k,\eta x) \qquad (3.32) \\ &- \hat{v}_L^{(2-m)}(k^3)E^{(m)}(k,\eta(x-L))], \qquad x \in (0,L), \quad t > 0, \end{aligned}$$

where γ is the counterclockwise oriented boundary of the sector $\mathcal{K}_0^{(-)}$, $\gamma = \partial\mathcal{K}_0^{(-)}$ (see Fig.2a), and

$$E^{(m)}(k,x) := H(x)(\rho_{m+1}e^{-i\rho_1 kx} + \rho_{2-m}e^{-i\rho_2 kx}) - H(-x)e^{-ikx}, \qquad (3.33)$$

ρ_m being the cubic roots of unity:

$$\rho_m = e^{\frac{2\pi i}{3}m}. \tag{3.34}$$

3.2.1. *The segment* $V = (0, L)$

Now $\hat{\mathcal{N}}/\mathcal{L}$ must be defined in $\mathcal{A} = \mathbb{C} \otimes \bar{\mathcal{I}}_q$, analytic in $q \in \mathcal{I}_q$, $\forall k \in \mathbb{C}$ and analytic in $k \in \mathbb{C}$, $\forall q \in \bar{\mathcal{I}}_q$, with proper asymptotics for large $|k|$ and/or $|q|$ in the analyticity regions. Therefore the singularity manifold $\mathcal{A} \cap \mathcal{S}^{(\eta)}$, corresponding to $\eta = \pm 1$, are given by (see Figs 2a,b):

$$\begin{aligned} \mathcal{A} \cap \mathcal{S}^{(\eta)} &= \bigcup_{m=0}^{2} \{q = -\eta k^3, k \in \mathcal{K}_m^{(\eta)}\} \\ &= \bigcup_{m=0}^{2} \{k = k_m^{(\eta)}(q),\ \pi \leq \ \arg q \leq 2\pi\} \end{aligned} \tag{3.35}$$

where $k_m^{(\eta)}(q) = -\eta\rho_m q^{\frac{1}{3}}$ and the three sectors $\mathcal{K}_m^{(\eta)}$ are

$$\mathcal{K}_m^{(-)} = \{k : \ \frac{\pi}{3}(2m+1) \leq \arg k \leq \frac{\pi}{3}(2m+2)\}, \tag{3.36}$$

$\mathcal{K}_m^{(+)} = \hat{\rho}_\pi \mathcal{K}_m^{(-)}$, $\quad m = 0, 1, 2$, and $\hat{\rho}_\pi$ is the involution $\hat{\rho}_\pi :\ k \to -k$.

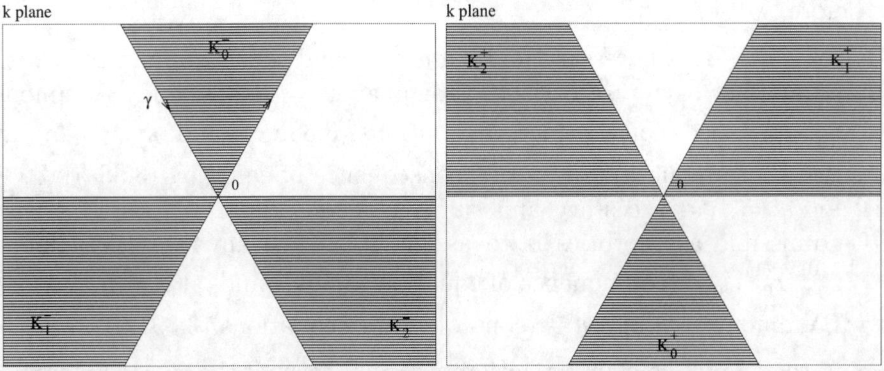

Fig.2a The SM $\mathcal{A} \cap \mathcal{S}^{(\eta)}$ $(\eta = -1)$ Fig.2b The SM $\mathcal{A} \cap \mathcal{S}^{(\eta)}$ $(\eta = 1)$

The ZRC (2.12) consists of the following three equations:

$$\hat{\mathcal{N}}(k_m^{(\eta)}(q), q) = 0, \qquad m = 0, 1, 2, \qquad \pi \leq \arg q \leq 2\pi. \tag{3.37}$$

For $q \in \mathbb{R}$ there is one singularity on the real k - axis, which is cured by one of the three equations (3.37) and no denominator regularization is then needed. The regularization (3.13), however, is still introduced, according to the Remark 2 of §2.2.2, in writing the Fourier representation (3.32) of the solution.

The three3 algebraic equations (3.37) contain six BVs; therefore we expect to be allowed to assign independently only three BVs. As before, a quick asymptotic estimate selects the sets of three BVs which can be assigned independently, compatibly with asymptotic admissibility. The asymptotics of equations (3.37) imply that the following expressions, respectively, for $\eta = -1$:

$$\begin{aligned} & e^{iq^{\frac{1}{3}}L}[\hat{f}(q^{\frac{1}{3}},q)+\hat{u}_0(q^{\frac{1}{3}})]-[\hat{v}_L^{(2)}(q)+iq^{\frac{1}{3}}\hat{v}_L^{(1)}(q)-q^{\frac{2}{3}}\hat{v}_L^{(0)}(q)], \\ & \hat{f}(\rho_1 q^{\frac{1}{3}},q)+\hat{u}_0(\rho_1 q^{\frac{1}{3}})+[\hat{v}_0^{(2)}(q)+i\rho_1 q^{\frac{1}{3}}\hat{v}_0^{(1)}(q)-\rho_2 q^{\frac{2}{3}}\hat{v}_0^{(0)}(q)], \\ & \hat{f}(\rho_2 q^{\frac{1}{3}},q)+\hat{u}_0(\rho_2 q^{\frac{1}{3}})+[\hat{v}_0^{(2)}(q)+i\rho_2 q^{\frac{1}{3}}\hat{v}_0^{(1)}(q)-\rho_1 q^{\frac{2}{3}}\hat{v}_0^{(0)}(q)], \end{aligned} \tag{3.38}$$

and for $\eta = 1$:

$$\begin{aligned} & \hat{f}(-q^{\frac{1}{3}},q)+\hat{u}_0(-q^{\frac{1}{3}})-[\hat{v}_0^{(2)}(q)-iq^{\frac{1}{3}}\hat{v}_0^{(1)}(q)-q^{\frac{2}{3}}\hat{v}_0^{(0)}(q)], \\ & \quad e^{-i\rho_1 q^{\frac{1}{3}}L}[\hat{f}(-\rho_1 q^{\frac{1}{3}},q)+\hat{u}_0(-\rho_1 q^{\frac{1}{3}})] \\ & \quad +[\hat{v}_L^{(2)}(q)-i\rho_1 q^{\frac{1}{3}}\hat{v}_L^{(1)}(q)-\rho_2 q^{\frac{2}{3}}\hat{v}_L^{(0)}(q)], \\ & \quad e^{-i\rho_2 q^{\frac{1}{3}}L}[\hat{f}(-\rho_2 q^{\frac{1}{3}},q)+\hat{u}_0(-\rho_2 q^{\frac{1}{3}})] \\ & \quad +[\hat{v}_L^{(2)}(q)-i\rho_2 q^{\frac{1}{3}}\hat{v}_L^{(1)}(q)-\rho_1 q^{\frac{2}{3}}\hat{v}_L^{(0)}(q)], \end{aligned} \tag{3.39}$$

are exponentially small for $q \sim \infty$ in $\pi \leq \arg q \leq 2\pi$. Therefore, reasoning as before, we see that:

i) for $\eta = -1$, a necessary and sufficient condition to obtain FTs with admissible asymptotics is to assign at $x = 0$ any one BV among $v_0^{(0)}, v_0^{(1)}, v_0^{(2)}$ (consequence of equations (3.38b,c)) and, at $x = L$, any two BVs among $v_L^{(0)}, v_L^{(1)}, v_L^{(2)}$ (consequence of equation (3.38a));

ii) for $\eta = 1$, a necessary and sufficient condition to obtain FTs with admissible asymptotics is to assign at $x = 0$ any two BVs among $v_0^{(0)}, v_0^{(1)}, v_0^{(2)}$ (consequence of equation (3.39a)) and, at $x = L$, any one BV among $v_L^{(0)}, v_L^{(1)}, v_L^{(2)}$ (consequence of equations (3.39a,b)).

Again, to complete our investigation, we must check if the spectrum associated with the above IBV problems selected by the asymptotic admissibility, lie entirely outside the analyticity domain of an admissible FT. It is easy to prove that it is indeed the case.

Proposition (the spectrum of the IBV problem). *Consider any IBV problem on the segment for equation (3.2) compatible with the asymptotic admissibility established above; i.e., in which, for $\eta = -1$, one assigns arbitrarily at $x = 0$ any BV among $v_0^{(0)}, v_0^{(1)}, v_0^{(2)}$ and any two BVs at $x = L$ among $v_L^{(0)}, v_L^{(1)}, v_L^{(2)}$, and in which, for $\eta = 1$, one assigns arbitrarily at $x = 0$ any* *two BVs among $v_0^{(0)}, v_0^{(1)}, v_0^{(2)}$ and any BV at $x = L$ among $v_L^{(0)}, v_L^{(1)}, v_L^{(2)}$. For $\eta = -1$, let $v_0^{(n)}$ be the given BV at $x = 0$ and $v_L^{(m)}$ be the unknown BV at $x = L$ while, for $\eta = 1$, let $v_0^{(n)}$ be the unknown BV at $x = 0$ and $v_L^{(m)}$ be the given BV at $x = L$. Then the corresponding spectrum is characterized by the following equation:*

$$\Delta^{(\eta(m-n))}(k) = 0, \tag{3.40}$$

where:

$$\Delta^{(j)}(k) := e^{-ikL} + \rho_1^j e^{-\rho_1 ikL} + \rho_2^j e^{-\rho_2 ikL}. \tag{3.41}$$

The proof is tedious but straightforward and makes essential use of the well-known algebra of the roots of unity, which implies also that all the above IBV problems lead only to three (similar) purely imaginary discrete spectra $\{k_n^{(j)}\}_{n\in I\!N}$, characterized by the three equations $\Delta^{(j)}(k) = 0,\ j = 0,1,2.$ More precisely:

1) the spectrum characterized by equation $\Delta^{(0)}(k) = 0$ is given by:

$$\begin{aligned} &k_n^{(0)} = -i(\zeta_n^{(0)}/L),\ n \in I\!N^+ : \frac{e^{-\frac{3}{2}\zeta_n^{(0)}}}{2} = -\cos(\tfrac{\sqrt{3}}{2}\zeta_n^{(0)}) \\ &(\zeta_n^{(0)} \sim \tfrac{\pi}{\sqrt{3}}(2n-1),\ n \geq 1). \end{aligned} \tag{3.42}$$

2) The spectrum characterized by equation $\Delta^{(1)}(\zeta) = 0$ is:

$$\begin{aligned} &k_n^{(1)} = -i(\zeta_n^{(1)}/L),\ n \in I\!N : \quad \frac{e^{-\frac{3}{2}\zeta_n^{(1)}}}{2} = \cos(\tfrac{\sqrt{3}}{2}\zeta_n^{(1)} + \tfrac{\pi}{3}), \\ &(\zeta_0^{(1)} = 0, \quad \zeta_n^{(1)} \sim \tfrac{2\pi}{\sqrt{3}}(n - \tfrac{5}{6}),\ n \geq 2). \end{aligned} \tag{3.43}$$

3) The spectrum characterized by equation $\Delta^{(2)}(\zeta) = 0$ is:

$$\begin{aligned} &k_n^{(2)} = -i(\zeta_n^{(2)}/L),\ n \in I\!N : \frac{e^{-\frac{3}{2}\zeta_n^{(2)}}}{2} = \cos(\tfrac{\sqrt{3}}{2}\zeta_n^{(2)} - \tfrac{\pi}{3}), \\ &(\zeta_0^{(2)} = 0, \quad \zeta_n^{(2)} \sim \tfrac{2\pi}{\sqrt{3}}(n - \tfrac{1}{6}),\ n \geq 1). \end{aligned} \tag{3.44}$$

We conclude that all the three discrete spectra

$$\{q_n^{(j)}\}_{n\in I\!N}, \qquad q_n^{(j)} = k_n^{(j)3} = i\left(\frac{\zeta_n^{(j)}}{L}\right)^3, \qquad j = 0,1,2, \tag{3.45}$$

associated with the above IBV problems lie on the positive imaginary axis of the complex q plane, outside the analyticity domain of an admissible FT. Therefore:

IBV problems for equation (3.2) on the segment $(0, L)$ are well-posed iff:

i) for $\eta = -1$*, one assigns at* $x = 0$ *any one BV among* $v_0^{(0)}, v_0^{(1)}, v_0^{(2)}$ *and at* $x = L$ *any two BVs among* $v_L^{(0)}, v_L^{(1)}, v_L^{(2)}$*;*
ii) for $\eta = 1$*, one assigns at* $x = 0$ *any two BVs among* $v_0^{(0)}, v_0^{(1)}, v_0^{(2)}$ *and any one BV at* $x = L$ *among* $v_L^{(0)}, v_L^{(1)}, v_L^{(2)}$.

3.2.2. *The periodic problem*

If we assume L-periodicity of u, u_x and u_{xx}, then $v_0^{(j)} = v_L^{(j)}$, $j = 0, 1, 2$, the algebraic system (3.37) consists now of three equations for three BVs, which have to be treated then as unknowns. The solution of this system satisfy asymptotic admissibility and the spectrum, characterized by the equations $1 - e^{-i\rho_j k} = 0$, $j = 0, 1, 2$ ($\Rightarrow k_n = \frac{2\pi}{L}\rho_j^{-1} n$, $n \in \mathbb{Z}$, $j = 0, 1, 2$), is given by the real numbers $q_n = -\eta k_n^3 = -\eta(\frac{2\pi}{L})^3 n^3$, $n \in \mathbb{Z}$ and must be regularizes in the usual way. We conclude that:
the periodic problem for the linear KdV equation (3.2), in which one imposes L-periodicity to u, u_x *and* u_{xx}*, is well-posed and no BV can be assigned.*

3.2.3. *The semiline* $V = (0, \infty)$

Taking the limit $L \to \infty$ of the results of §3.2.1 we immediately obtain the results on the semiline. In this case, the singularity manifolds are the restrictions of the above ones to the lower half k plane. No spectrum arises and the asymptotic admissibility implies that:
IBV problems for equation (3.2) on the semiline $(0, \infty)$ *are well-posed iff, for* $\eta = -1$*, one assigns at* $x = 0$ *any one BV among* $(v_0^{(0)}, v_0^{(1)}, v_0^{(2)})$ *and, for* $\eta = 1$*, one assigns at* $x = 0$ *any two BVs among* $(v_0^{(0)}, v_0^{(1)}, v_0^{(2)})$.

We remark that, in the cases treated so far, the spectra of all the IBV problems compatible with asymptotic admissibility lie always outside the analyticity domain. We do not have, however, a general argument excluding the situation in which part of the spectrum lie inside. Therefore the complete characterization of the spectrum, the only part of the method in which some technicality is involved, seems to be unavoidable and makes it difficult to prove general results for operators of arbitrary order.

The Analyticity approach applies nicely also to an arbitrary number of dimensions and next section is devoted to an illustration of it. The application of the method to higher order problems and to other relevant examples will be presented in [8].

3.3. MULTIDIMENSIONAL SCHRÖDINGER EQUATION

In this section we study the Dirichlet and Neumann problems for the Schrödinger equation (3.5) in the n-dimensional quadrant (3.6). Then:

$$\begin{aligned}&\tilde{\mathcal{L}} = -\frac{\partial}{\partial t} - i\triangle, \qquad J = (ab, i(a \bigtriangledown b - b \bigtriangledown a), \\ &\mathcal{L}(i\mathbf{k}, iq) = i(q + k^2),\end{aligned} \tag{3.46}$$

where $k^2 = \mathbf{k} \cdot \mathbf{k}$. Equations (2.6) and (3.46) give the following expression of the Fourier transform of the solution in terms of the Fourier transforms of the forcing and of all the IBVs:

$$\begin{aligned}\hat{u}(\mathbf{k}, q) &= \frac{\hat{\mathcal{N}}(\mathbf{k},q)}{i(q+k^2)}, \\ \hat{\mathcal{N}}(\mathbf{k}, q) &:= \hat{f}(\mathbf{k}, q) + \hat{u}_0(\mathbf{k}) - i \sum_{j=1}^{n} [\hat{v}_{0j}^{(1)}(\mathbf{k}_j, q) + ik_j\hat{v}_{0j}^{(0)}(\mathbf{k}_j, q)].\end{aligned} \tag{3.47}$$

The Fourier representation (2.18) of the solution reads:

$$\begin{aligned}u(\mathbf{x}, t) = &\int_{I\!R^{n+1}} \frac{dqd\mathbf{k}}{(2\pi)^{n+1}i} e^{i(\mathbf{k}\cdot\mathbf{x}+qt)} \frac{\hat{f}(\mathbf{k},q)}{q+k^2-i0} + \int_{I\!R^n} \frac{d\mathbf{k}}{(2\pi)^n} e^{i(\mathbf{k}\cdot\mathbf{x}-k^2t)}\hat{u}_0(\mathbf{k}) \\ + &\sum_{j=1}^{n} \int_{I\!R^{n-1}} \frac{d\mathbf{k}_j}{(2\pi)^{n-1}} \int_{\gamma^{(i)}} \frac{dk_j}{2\pi i}\{e^{i(\mathbf{k}\cdot\mathbf{x}-k^2t)}[\hat{v}_{0j}^{(1)}(\mathbf{k}_j, -k^2) + ik_j\hat{v}_{0j}^{(0)}(\mathbf{k}_j, -k^2)],\end{aligned} \tag{3.48}$$

where $d\mathbf{k}_j = dk_1 \cdots \underline{dk_j} \cdots dk_n$.

In view of the distinguished parity properties of the Fourier transforms in (3.47), we shall make an extensive use of the parity operators:

$$\Delta_{\pm} = \prod_{l=1}^{n}(1\pm\hat{\sigma}_l), \qquad \Delta_{\pm}^{(j)} = \prod_{\substack{l=1 \\ l\neq j}}^{n}(1\pm\hat{\sigma}_l), \tag{3.49}$$

where $\hat{\sigma}_j$ is the involution $\hat{\sigma}_j : k_j \rightarrow -k_j$.

In this multidimensional case, the FT of the solution is defined in $\mathcal{A} = \bar{\mathcal{I}}_{k_1} \otimes \cdots \otimes \bar{\mathcal{I}}_{k_n} \otimes \bar{\mathcal{I}}_q$, analytic for $q \in \mathcal{I}_q$, $\forall \mathbf{k} \in \bar{\mathcal{I}}_{k_1} \otimes \cdots \otimes \bar{\mathcal{I}}_{k_n}$ and in $k_j \in \mathcal{I}_{k_j}$, $\forall \mathbf{k}_j \in \bar{\mathcal{I}}_{k_1} \otimes \cdots \otimes \underline{\bar{\mathcal{I}}_{k_j}} \otimes \cdots \otimes \bar{\mathcal{I}}_{k_n}$ and $\forall q \in \bar{\mathcal{I}}_q$. We find it convenient to study the ZRC in the n different regions $\mathcal{Q}_j^- \subset \mathcal{A} \cap \mathcal{S}$, $j = 1, \ldots, n$ defined by:

$$\begin{aligned}\mathcal{Q}_j^- &:= \{(\mathbf{k}, q) \in C\!\!\!\!I^{n+1} : \mathbf{k}_j \in I\!R^{n-1},\ \pi \leq \arg q \leq 2\pi, k_j = \chi_j(\mathbf{k}_j, q)\}, \\ \chi_j(\mathbf{k}_j, q) &:= i(q + \mathbf{k}_j \cdot \mathbf{k}_j)^{\frac{1}{2}} \in \overline{\mathcal{I}_k}, \qquad j = 1, \ldots, n.\end{aligned} \tag{3.50}$$

Therefore the starting point of the analysis is the set of n equations

$$\hat{\mathcal{N}}(\mathbf{k}, q)|_{k_j=\chi_j} = 0, \quad \mathbf{k}_j \in I\!R^{n-1}, \quad \pi \leq \arg q \leq 2\pi, \qquad j = 1, \ldots, n. \tag{3.51}$$

Dirichlet problem. The parity properties in k of the BV terms imply that the application of the parity operator $\Delta_-^{(j)}$ defined in (3.49) to the jth equation (3.51) eliminates all the $\hat{v}_0^{(1)}$'s except $\hat{v}_{0j}^{(1)}$:

$$\begin{aligned} \Delta_-^{(j)}\hat{v}_{0j}^{(1)}(\mathbf{k}_j,q) &= -\Delta_-^{(j)}(W(\mathbf{k},q)|_{k_j=\chi_j}), \qquad j=1,\dots,n, \\ W(\mathbf{k},q) &:= i\hat{f}(\mathbf{k},q)+i\hat{u}_0(\mathbf{k})+i\sum_{j=1}^{n}k_j\hat{v}_{0j}^{(0)}(\mathbf{k}_j,q) \end{aligned} \tag{3.52}$$

and the analyticity properties of the functions $\hat{v}_0^{(1)}$ allow one to express them in terms of known quantities:

$$\hat{v}_{0j}^{(1)}(\mathbf{k}_j,q)=\mathcal{P}^{(j)}\Delta_-^{(j)}\hat{v}_{0j}^{(1)}(\mathbf{k}_j,q)=-\mathcal{P}^{(j)}\Delta_-^{(j)}(W(\mathbf{k},q)|_{k_j=\kappa_j}), \tag{3.53}$$

for $j=1,\dots,n$, applying the lower half plane analyticity projectors in all the k-variables (except k_j):

$$\mathcal{P}^{(j)}=\prod_{\substack{m=1\\ m\neq j}}^{n}\mathcal{P}_m, \qquad \mathcal{P}_m=-\frac{1}{2\pi i}\int_I R\frac{dk'_m}{k'_m-(k_m-i0)}. \tag{3.54}$$

Equations (3.53) summarize all the analyticity informations contained in the ZRC, allow one to express the unknown BVs in terms of given data and, via (3.48), to solve the Dirichlet problem.
Neumann problem. Similar considerations can be made in solving the Neumann BV problem. In this case:

$$\begin{aligned} &i\chi_j(\mathbf{k}_j,q)\Delta_+^{(j)}\hat{v}_{0j}^{(0)}(\mathbf{k}_j,q)=-\Delta_+^{(j)}(V(\mathbf{k},q)|_{k_j=\chi_j}), \qquad j=1,\dots,n, \\ &V(\mathbf{k},q):=i\hat{f}(\mathbf{k},q)+i\hat{u}_0(\mathbf{k})+\sum_{j=1}^{n}\hat{v}_{0j}^{(1)}(\mathbf{k}_j,q) \end{aligned} \tag{3.55}$$

and

$$\hat{v}_{0j}^{(0)}(\mathbf{k}_j,q)=i\mathcal{P}^{(j)}\left(\tfrac{1}{\chi_j(\mathbf{k}_j,q)}\Delta_+^{(j)}(V(\mathbf{k},q)|_{k_j=\chi_j})\right), \quad j=1,\dots,n. \tag{3.56}$$

In this multidimensional context, for the presence of the analyticity projectors, the unknown BVs in Fourier space turn out to be nonlocal expressions of the given data. It is however possible to show that, due to the analyticity properties of the involved FTs, it is not really necessary to apply the above analyticity projectors to construct the unknown BVs and the solution $u(\mathbf{x},t)$ in configuration space. The strategy to avoid unpleasant nonlocalities is outlined in the next section and leads to a Fourier representation of the solution already obtained in [6] using the EbR approach. Therefore this strategy is also the way to establish the connection between the Analyticity and the EbR approaches.

4. Connections between the Analyticity and the EbR approaches

Dirichlet problem. We first remark that the unknown BVs can be constructed directly in terms of known data from the r. h. s. of (3.52b):

$$v_{0j}^{(1)}(\mathbf{x}_j,t) = -\int_{I\!R^n} \frac{d\mathbf{k}_j dq}{(2\pi)^n} e^{i(\mathbf{k}_j\cdot\mathbf{x}_j+qt)} \Delta_-^{(j)}(W(\mathbf{k},q)|_{k_j=\chi_j}), \tag{4.1}$$

for $t>0,\ x_k \geq 0,\ k\neq j$. Indeed, from the analyticity properties of $\hat{v}_{0j}^{(1)}$ we know that its inverse FTs (2.2) is zero outside the domain of definition in configuration space (i.e., for $x_k<0,\ k\neq j$); this implies the formula

$$\int_{I\!R^n} d\mathbf{k}_j dq e^{i(\mathbf{k}_j\cdot\mathbf{x}_j+qt)}[\hat{v}_{0j}^{(1)}(\mathbf{k}_j,q) - \Delta_-^{(j)}\hat{v}_{0j}^{(1)}(\mathbf{k}_j,q)] = 0 \tag{4.2}$$

for $t>0,\ x_j>0,\ j=1,\ldots,n$ and, through (3.52a), equation (4.1).

Also the solution $u(\mathbf{x},t)$ can be reconstructed without going through the nonlocalities associated with the analyticity projectors. Indeed it is possible to show that the following relation holds true:

$$\sum_{j=1}^{n} \hat{v}_{0j}^{(1)}(\mathbf{k}_j,q) \equiv (\Delta_- - 1)W(\mathbf{k},q), \tag{4.3}$$

where the equivalence $\hat{A}(\mathbf{k},q)\equiv\hat{B}(\mathbf{k},q)$ means that the FTs $\hat{A}(\mathbf{k},q)$ and $\hat{B}(\mathbf{k},q)$ are equal under the following Fourier integral projector:

$$\int_{I\!R^{n+1}} d\mathbf{k}dq \frac{e^{i(\mathbf{k}\cdot\mathbf{x}+qt)}}{q+k^2-i0}[\hat{A}(\mathbf{k},q)-\hat{B}(\mathbf{k},q)] = 0, \qquad (\mathbf{x},t)\in\mathcal{D}. \tag{4.4}$$

The equivalence (4.3) and equation (3.52) imply

$$\hat{\mathcal{N}}(\mathbf{k},q) \equiv \Delta_- W(\mathbf{k},q) \tag{4.5}$$

and the following spectral representation of the solution in terms of known data:

$$\begin{aligned} u(\mathbf{x},t) = &\int_{\mathcal{R}^{n+1}} \frac{dqd\mathbf{k}}{(2\pi)^{n+1}} e^{i(\mathbf{k}\cdot\mathbf{x}+qt)} \frac{\Delta_-\hat{f}(\mathbf{k},q)}{q+k^2-i0} \\ &+ \int_{\mathcal{R}^n} \frac{d\mathbf{k}}{(2\pi)^n} e^{i(\mathbf{k}\cdot\mathbf{x}-k^2t)}\Delta_-\hat{u}_0(\mathbf{k}) \\ &+ \sum_{j=1}^{n}\int_{I\!R^{n-1}} \frac{d\mathbf{k}_j}{(2\pi)^{n-1}}\Big(\int_{\gamma^{(i)}} \frac{dk_j}{\pi} e^{i(\mathbf{k}\cdot\mathbf{x}-k^2t)} k_j \Delta_-^{(j)}\hat{v}_{0j}^{(0)}(\mathbf{k}_j,-k^2), \\ &(\mathbf{x},t)\in\mathcal{D}, \end{aligned} \tag{4.6}$$

already obtained in [6] using the EbR approach.
The proof of (4.3) is based on the important fact that all the admissible Fourier transforms $\hat{A}(\mathbf{k},q)$ under consideration satisfy the equivalence

$$\hat{\sigma}_j \hat{A}(\mathbf{k},q) \equiv \hat{A}(\mathbf{k},q)|_{k_j=\chi_j}, \qquad j=1,\ldots,n \tag{4.7}$$

and goes as follows. For $n=2$, the 2 ZRCs (3.51) and their consequences (3.52) yield the 4 equivalence relations:

$$\begin{aligned}
&\hat{v}_{01}^{(1)}(k_2,q)+\hat{\sigma}_1\hat{v}_{02}^{(1)}(k_1,q)\equiv-\hat{\sigma}_1 W(\mathbf{k},q),\\
&\hat{\sigma}_2\hat{v}_{01}^{(1)}(k_2,q)+\hat{v}_{02}^{(1)}(k_1,q)\equiv-\hat{\sigma}_2 W(\mathbf{k},q),\\
&(1-\hat{\sigma}_2)\hat{v}_{01}^{(1)}(k_2,q)\equiv-\hat{\sigma}_1(1-\hat{\sigma}_2)W(\mathbf{k},q),\\
&(1-\hat{\sigma}_1)\hat{v}_{02}^{(1)}(k_1,q)\equiv-\hat{\sigma}_2(1-\hat{\sigma}_1)W(\mathbf{k},q)
\end{aligned} \tag{4.8}$$

and their sum is exactly equation (4.3). To generalize this result to the case of an arbitrary n, consider the n ZRCs (3.51) and all their consequences, obtained applying systematically parity operators characterized by different indexes:

$$\begin{aligned}
&\hat{v}_{0j}^{(1)}(\mathbf{k}_j,q)+\hat{\sigma}_j\sum_{l\neq j}\hat{v}_{0l}^{(1)}(\mathbf{k}_l,q)\equiv -\hat{\sigma}_j W(\mathbf{k},q), \qquad j=1,\ldots,n\\
&(1-\hat{\sigma}_i)\hat{v}_{0j}^{(1)}(\mathbf{k}_j,q)+\hat{\sigma}_j(1-\hat{\sigma}_i)\sum_{l\neq j}\hat{v}_{0l}^{(1)}(\mathbf{k}_l,q)\equiv-\hat{\sigma}_j(1-\hat{\sigma}_i)W(\mathbf{k},q),\\
&i\neq j, \quad j=1,\ldots,n;\\
&\cdots\cdots\cdots\cdots\cdots\cdots\cdots\cdots\cdots\cdots\cdots\cdots\\
&\Delta_-^{(j)}\hat{v}_{0j}^{(1)}(\mathbf{k}_j,q)\equiv-\hat{\sigma}_j\Delta_-^{(j)}W(\mathbf{k},q), \qquad j=1,\ldots,n.
\end{aligned} \tag{4.9}$$

The sum of all these equations with weights $1/\binom{n-1}{m}$ (m is the number of parity operators appearing in the equation) yields the result (4.3).
Neumann problem. Similar considerations can be made in the case of the Neumann IBV problem. Now the unknown BVs are recovered via:

$$v_{0j}^{(0)}(\mathbf{x}_j,t)=-\int_{I\!R^n}\frac{d\mathbf{k}_j dq}{(2\pi)^n}e^{i(\mathbf{k}_j\cdot\mathbf{x}_j+qt)}(\Delta_+^{(j)}V(\mathbf{k},q)|_{k_j=\chi_j}) \tag{4.10}$$

for $t>0$, $x_k\geq 0$, $k\neq j$, and the spectral representation of the solution reads:

$$\begin{aligned}
u(\mathbf{x},t)=&-\int_{I\!R^{n+1}}\frac{dqd\mathbf{k}e^{i(\mathbf{k}\cdot\mathbf{x}+qt)}}{(2\pi)^{n+1}i}\frac{\Delta_+\hat{f}(\mathbf{k},q)}{q+k^2-i0}+\int_{I\!R^n}\frac{d\mathbf{k}e^{i(\mathbf{k}\cdot\mathbf{x}-k^2t)}}{(2\pi)^n}\Delta_+\hat{u}_0(\mathbf{k})\\
&-\sum_{j=1}^{n}\int_{I\!R^{n-1}}\frac{d\mathbf{k}_j}{(2\pi)^{n-1}}\left(\int_{\gamma^{(i)}}\frac{dk_j e^{i(\mathbf{k}\cdot\mathbf{x}-k^2t)}}{\pi i}\Delta_+^{(j)}\hat{v}_{0j}^{(1)}(\mathbf{k}_j,-k^2), \quad (\mathbf{x},t)\in\mathcal{D};\right.
\end{aligned} \tag{4.11}$$

a formula already derived in [6] using the EbR approach.

Acknowledgments

The present work has been carried out during several visits and meetings. We gratefully acknowledge the financial contributions provided by the RFBR Grant 01-01-00929, the INTAS Grant 99-1782 and by the following Institutions: the University of Rome "La Sapienza" (Italy), the Istituto Nazionale di Fisica Nucleare (Sezione di Roma), the Landau Institute for Theoretical Physics, Moscow (Russia) and the Isaac Newton Institute, Cambridge (UK), within the programme "Integrable Systems".

References

1. A. S. Fokas, Proc. Roy. Soc. Lond. A, **53**, 1411 (1997).
2. A. S. Fokas, J. Math. Phys. **41**, 4188 (2000).
3. A. S. Fokas, Proc. Roy. Soc. Lond. A, **457**, 371 (2001).
4. A. S. Fokas and B. Pelloni, Math. Proc. Camb. Phil. Soc., **131**, 521 (2001).
5. B. Pelloni, "On the well-posedness of boundary value problems for integrable evolution equations on a finite interval", in the Proceedings of NEEDS 2001; editors A. Mikhailov and P. M. Santini. To be published in Theor. Math. Phys., 2002.
6. A. Degasperis, S. V. Manakov and P. M. Santini, "Initial-boundary value problems for linear and soliton PDEs", Theor. Math. Phys., **133**, 1472 (2002).
7. A. S. Fokas, "A new transform method for evolution PDEs", preprint 2002.
8. A. Degasperis, S. V. Manakov and P. M. Santini, "Initial-boundary value problems for integrable PDEs: spectral methods of solution"; in preparation.
9. P. M. Morse and H. Feshbach, *Methods of Theoretical Physics*, Mc Graw-Hill, New York (1953).

QUASI-EXACTLY SOLVABLE BOSE SYSTEMS

S.N. DOLYA (dolya@ilt.kharkov.ua)
B. Verkin Institute for Low Temperature Physics and Engineering, 47 Lenin Prospekt, Kharkov 61164, Ukraine

O.B. ZASLAVSKII (aptm@kharkov.ua)
Department of Mechanics and Mathematics, Kharkov V.N. Karazin's National University, Svoboda Sq. 4, Kharkov 61077, Ukraine

Abstract. We extend the notion of quasi-exactly solvable (QES) models to Bose systems. We obtain conditions under which algebraization of the part of the spectrum occurs. In some particular cases simple exact expressions for several energy levels of an anharmonic Bose oscillator are obtained explicitly. The corresponding results do not exploit perturbation theory and include strong coupling regime. A generic Hamiltonian under discussion cannot, in contrast to QES potential models, be expressed as a polynomial in generators of sl_2 algebra. The suggested approach is extendable to many-particle Bose systems with interaction.

1. Introduction

The conception of quasi-exactly solvable (QES) systems, discovered in the 1980s [1]–[12], received in recent years much attention both from the viewpoint of physical applications and their inner mathematical beauty. In turned out that in quantum mechanics there exists a peculiar class of objects that occupy an intermediate place between exactly solvable and non-solvable models in the sense that in an infinite Hilbert space a finite part of a spectrum is singled out within which eigenvectors and eigenvalues can be found from an algebraic equation of a finite degree—in other words, a partial algebraization of the spectrum occurs. For one-dimensional QES models corresponding QES Hamiltonians possess hidden group structure based on $sl(2, R)$ algebra. Thus, they have direct physical meaning, being related to quantum spin systems [13].

Meanwhile, the notion of QES systems is not constrained by potential models and can have nontrivial meaning for any kind of infinite-dimensional systems. In the first place, it concerns Bose Hamiltonians whose physical

A.B. Shabat et al. (eds.), New Trends in Integrability and Partial Solvability, 105–114.

importance is beyond doubt. Here one should distinguish two cases. First, it turns out that some systems of two interacting particles or quasi-particles with Bose operators of creation and annihilation a, a^+ and b, b^+ can be mapped on the problem for a particle moving in a certain type of one-dimensional potentials and, remarkably, these potentials belong just to the QES type [14]–[16]. In particular, such a type of Hamiltonians is widely spread in quantum optics and physics of magnetism [13]. The aforementioned mapping works only for a special class of Bose Hamiltonians which possess an integral of motion. Then the procedure is performed in three steps: (i) the whole Hilbert space splits in a natural way into different pieces with respect to the values of an integral of motion, (ii) in each piece the Schrödinger equation takes a finite-difference form, (iii) it is transformed into the differential equation by means of introducing a generating function. In so doing, the integral of motion under discussion represents a linear combination of numbers of particles a^+a and b^+b.

The second kind of Bose systems looks much more usual—it is simply some polynomial with respect to Bose operators of creation and annihilation of one particle. The fact that only one pair a, a^+ enters Hamiltonian, deprives us, by contrast with the first case, of the possibility to construct a simple integral of motion—in this sense the eigenvalue problem becomes more complicated. In general, the solutions of the Schrödinger equation contain infinite numbers of quasi-particles and only approximate or numerical methods can be applied to such systems. However, it turns out that, if the coefficients at different powers of a, a^+ are selected in a proper way, in some cases a *finite-dimensional* closed subspace is singled out and algebraization of the spectrum occurs similar to what happens in "usual" QES potential models or differential equations. In so doing, the eigenvectors belonging to the subspace under discussion, can be expressed as a finite linear combination of eigenvectors of an harmonic oscillator and, thus, contain a finite number of quasi-particles. The relevant basis functions that compose a finite-dimensional subspace, look very much unlike the wave functions of a harmonic oscillator. As a result, we obtain QES models with an *infinite* number of quasi-particles in this *finite*-dimensional subspace.

2. Anharmonic Bose-oscillator

Consider Hamiltonian

$$H = H_0 + V,\ H_0 = \sum_{p=1}^{p_0} \varepsilon_p (a^+a)^p,$$

$$V = \sum_{s=0}^{s_0} A_s[(a^+a)^s a^2 + (a^+)^2(a^+a)^s]. \tag{2.1}$$

Throughout the paper we assume that all the coefficients of the Hamiltonian are real. For Hamiltonian (2.1) to have a well-defined ground state, one should take $p_0 > s_0 + 2$ independently of the relations between coefficients or $p_0 = s_0 + 2$ provided $\varepsilon_{p_0} \geq 2A_{s_0}$. In the x-representation we obtain

$$H_x = \sum_{p=1}^{p_0} \varepsilon_p \left(x\frac{d}{dx}\right)^p + \sum_{s=0}^{s_0} A_s \left[\left(x\frac{d}{dx}\right)^s \frac{d^2}{dx^2} + x^2\left(x\frac{d}{dx}\right)^s\right]. \tag{2.2}$$

We are interested in the solutions of Schrödinger equation of the type $|\psi\rangle = \sum_{n=0}^{N} b_n |n\rangle$, where $|n\rangle$ is the state with n particles: $a^+a\,|n\rangle = n\,|n\rangle$. For Hamiltonian (2.2) subspaces with even and odd are not mixed. Therefore, it makes sense to consider them separately. In x representation $a \to \frac{d}{dx}$, $a^+ \to x$ the wave function of even states $\Phi = \sum_{l=0} a_l \Phi_l$, $\Phi_l \equiv x^{2l}$.

It follows from (2.2) that

$$\begin{aligned} H_x\Phi_l &= \alpha_l\Phi_{l+1} + \beta_l\Phi_{l-1} + \gamma_l\Phi_l\,,\ l = 0, 1, \ldots, L\,, \qquad (2.3)\\ \alpha_l &= \sum_{s=0}^{s_0} A_s(2l)^s\,,\\ \beta_l &= \sum_{s=0}^{s_0} A_s 2l(2l-1)(2l-2)^s\,,\\ \gamma_l &= \sum_{p=1}^{p_0} \varepsilon_p(2l)^p\,. \end{aligned}$$

We are interested in the possibility of the existence of the invariant basic $F_{2L} = \{1, x^2, x^4, \ldots, x^{2L}\}$. The condition of cut off at $l = L$ reads: $\alpha_L = 0$.

For odd states the invariant basic $F_{2M+1} = \{x, x^3, \ldots, x^{2M+1}\}$, $\tilde{\Phi}_m = x^{2m+1}$ and

$$\begin{aligned} H_x\tilde{\Phi}_m &= \tilde{\alpha}_m\tilde{\Phi}_{m+1} + \tilde{\beta}_m\tilde{\Phi}_{m-1} + \tilde{\gamma}_m\tilde{\Phi}_m\,,\ m = 0, 1, \ldots, M\,, \qquad (2.4)\\ \tilde{\alpha}_m &= \sum_{s=0}^{s_0} A_s(2m+1)^s\,,\\ \tilde{\beta}_m &= \sum_{s=0}^{s_0} A_s(2m+1)2m(2m-1)^s\,,\\ \tilde{\gamma}_m &= \sum_{p=1}^{p_0} \varepsilon_p(2m+1)^p\,. \end{aligned}$$

The subspace with $m \leq M$ is invariant with respect to the action of H provided $\tilde{\alpha}_M = 0$.

The procedure described above is, in fact, nothing else than the Bose version of quasi-exactly solvable (QES) models, applied to an anharmonic oscillator. Now we would like to point out why for the case under consideration Turbiner's theorem, in general, does not hold, so our formulas cannot considered as particular cases of its realization. The point is that Turbiner's theorem implies that the space $F_N = \{1, x, x^2, \ldots, x^N\}$ of *all* polynomials of degree at most N is invariant with respect to J^i. Meanwhile, in our case, only subset F_{2N} (for even states) or F_{2M+1} (for odd ones) is invariant, whereas the set F_N is not. Only in particular cases, when both conditions $\alpha_L = 0$ (for even states) and $\tilde{\alpha}_M = 0$ (for odd states) are satisfied simultaneously, the Hamiltonian becomes an algebraic combination of J^i.

In contrast to [11], where differential equations were the object of research, in our paper the coordinate-momentum representation, in which the operator a becomes differential, is used as an useful device at an intermediate stage only. In principle, one could rely directly on the known formulas of the action of operators a, a^+ on states with a definite number of particles without resorting to the x-representation.

Consider, as an example, the Hamiltonian (2.1) with $s_0 = 2$, whose off-diagonal part reads

$$\begin{aligned} V &= A_0(a^2 + a^{+2}) + A_1[(a^+a)a^2 + a^{+2}(a^+a)] \\ &\quad + A_2[(a^+a)^2a^2 + a^{+2}(a^+a)^2]\,. \end{aligned} \tag{2.5}$$

Now

$$\begin{aligned} \alpha_l &= A_0 + 2lA_1 + (2l)^2A_2\,, \\ \beta_l &= 2l(2l-1)[A_0 + (2l-2)A_1 + (2l-2)^2A_2]\,. \end{aligned} \tag{2.6}$$

$$\begin{aligned} \tilde{\alpha}_m &= A_0 + (2m+1)A_1 + (2m+1)^2A_2\,, \\ \tilde{\beta}_m &= (2m+1)2m[A_0 + (2m-1)A_1 + (2m-1)^2A_2]\,. \end{aligned} \tag{2.7}$$

First, consider even states. In the simplest nontrivial particular case the invariant subspace is two-dimensional, $L = 1$. Then

$$\alpha_1 = A_0 + 2A_1 + 4A_2 = 0\,, \tag{2.8}$$

$\Phi = a_0\Phi_0 + a_1\Phi_1$ and it follows from the Schrödinger equation $H\Phi = E\Phi$ that $-Ea_0 + \beta_1 a_1 = 0$, $\alpha_0 a_0 + (\gamma_1 - E)a_1 = 0$. Taking also into account (2.8), we obtain:

$$E = \frac{\gamma_1}{2} \pm \sqrt{\frac{\gamma_1^2}{4} + 8(A_1 + 2A_2)^2} \tag{2.9}$$

In a similar way, one gets for the three-dimensional subspace ($L = 2$):

$$E^3 - (\gamma_1 + \gamma_2)E^2 + [\gamma_1\gamma_2 - 16(5A_1^2 + 52A_1A_2 + 140A_2^2)]E + 32\gamma_2(A_1 + 4A_2)^2 = 0 \,. \qquad (2.10)$$

If $A_1 = A_2 = 0$, the low-lying energy levels of an harmonic oscillator are reproduced from (2.9), (2.10). Equation (2.10) can be easily solved in the particular case $A_1 = -4A_2$:

$$E = 0 \,, \quad \frac{\gamma_1 + \gamma_2}{2} \pm \sqrt{\frac{(\gamma_1 - \gamma_2)^2}{4} + 192A_2^2} \,.$$

For odd states in the simplest nontrivial case $M = 1$ we have

$$\begin{aligned}
\tilde{\alpha}_1 &\equiv A_0 + 3A_1 + 9A_2 = 0 \,, \\
\tilde{\alpha}_0 &= A_0 + A_1 + A_2 \,, \ \tilde{\beta}_1 = 6\tilde{\alpha}_0 \,, \\
E &= \frac{\tilde{\gamma}_0 + \tilde{\gamma}_1}{2} \pm \sqrt{\frac{(\tilde{\gamma}_0 + \tilde{\gamma}_1)^2}{4} + \tilde{\alpha}_0\tilde{\beta}_1} \\
&= \frac{\tilde{\gamma}_0 + \tilde{\gamma}_1}{2} \pm \sqrt{\frac{(\tilde{\gamma}_0 + \tilde{\gamma}_1)^2}{4} + 24(A_1 + 4A_2)^2} \,. \qquad (2.11)
\end{aligned}$$

The conditions $\alpha_L = 0$ and $\tilde{\alpha}_M = 0$ are different, so in general the invariant subspace exists only for even or only for odd states. However, it may happen that both conditions are fulfilled. Thus, for $L = 1 = M$ the compatibility of (2.8) and (2.11) demands $A_1 = -5A_2$, $A_0 = 6A_2$. Then we have the simple explicit solutions for 4 levels of Hamiltonian (2.1):

$$E = \frac{\tilde{\gamma}_0 + \tilde{\gamma}_1}{2} \pm \sqrt{\frac{(\tilde{\gamma}_0 + \tilde{\gamma}_1)^2}{4} + 24A_2^2} \,, \quad \frac{\gamma_1}{2} \pm \sqrt{\frac{\gamma_1^2}{4} + 72A_2^2} \,.$$

A similar procedure can be repeated for odd states.

3. Generalization

Consider the operator which is the even polynomial of the forth degree with respect to Bose operators of creation a^+ and annihilation a. It can be written in the form

$$\begin{aligned}
H = {} & a_{++}K_+^2 + a_{--}K_-^2 + a_{00}K_0^2 + a_{0-}K_0K_- + a_{+0}K_+K_0 \\
& + a_0K_0 + a_-K_- + a_+K_+ \,, \qquad (3.1)
\end{aligned}$$

where

$$K_0 = \frac{1}{2}\left(a^+a + \frac{1}{2}\right), K_- = \frac{a^2}{2}, K_+ = \frac{a^{+2}}{2}, \tag{3.2}$$

$$[K_0,\ K_\pm] = \pm K_\pm\,,\ [K_+, K_-] = -2K_0\,. \tag{3.3}$$

The Casimir operator is $C = K_0^2 - \frac{1}{2}(K_+K_- + K_-K_+) \equiv -\frac{3}{16}$.

We will use the coherent state representation in which

$$a \to \frac{\partial}{\partial z}, \quad a^+ \to z \tag{3.4}$$

After substitution into (3.1) Hamiltonian $H(a^+, a)$ becomes a differential operator $H(z, \frac{\partial}{\partial z})$. In our previous article [17] we discussed Bose systems that possess an invariant subspace of the form $F = \text{span}\{z^n\}$ or $\text{span}\{z^{2n}\}$. The first natural step towards generalization consists in considering subspaces (with N fixed)

$$F = \text{span}\{u_n\}\,,\ u_n = z^{2n}u\,,\ n = 0, 1, 2, \ldots, N\,, \tag{3.5}$$

for which the following procedure should be realized. (i) The action of operators of K_i on the functions u_n should lead to the linear combinations of functions from the same set $\{u_n\}$, (ii) by the selection of appropriate coefficients in (3.1), we achieve the subspace F to be closed under the action of Hamiltonian H. We would like to stress that the condition (i) does not forbid u_n with $n > N$ to appear in terms like K_iu_n but the condition (ii) rules out such functions from Hu_n (recall that we consider Hamiltonians which are quadratic-linear combinations of K_i).

It is seen from (3.2), (3.4) that the operators K_i contain z and $\frac{\partial}{\partial z}$. Therefore, it is convenient to assume that differentiation of $u(z)$ gives rise to u up to the factor that contains powers of z. The corresponding choice is not unique. In the present article we restrict ourselves to one of the simplest possibilities that leads to nontrivial solutions. To this end, we choose u that obeys the differential equation

$$u' = A(z)u\,, \qquad A(z) = \left(\frac{\beta}{z} + 2\rho z\right). \tag{3.6}$$

We will show below that the choice (3.6) relates K_iu_n to u_n, $u_{n\pm 1}$ that, in turn, allows us to formulate the conditions of cut off for Hamiltonian in the form of algebraic equations which its coefficients obey. It follows from (3.6) that $u = z^\beta \exp(\rho z^2)$. To ensure asymptotic analytic behavior near $z = 0$, we demand that $\beta = 0, 1, 2, \ldots$. Now let us take into account some basic properties of coherent states (see, e.g. Ch. 7 of Ref. [18]). Our functions $u_n(z)$ must belong to the Bargmann-Fock space. It means that they should obey the conditions of integrability and analyticity. The condition of integrability for any two functions f, g from our space

$$\int dzdz^* f^* g e^{-zz^*} < \infty \,, \tag{3.7}$$

entails, for our choice of u, $|\rho| < 1/2$.

Taking into account eq. (3.6), it is straightforward to show that

$$K_+ u_n = C_+ u_{n+1} \,, \tag{3.8}$$

$$K_- u_n = A_-(n) u_n + B_-(n) u_{n-1} + C_- u_{n+1} \,, \tag{3.9}$$

$$K_0 u_n = A_0(n) u_n + C_0 u_{n+1} \,, \tag{3.10}$$

where

$$\begin{aligned} C_+ &= \frac{1}{2}, \quad C_- = 2\rho^2 \,, \quad C_0 = \rho \,, \\ A_-(n) &= (2\beta + 4n + 1)\rho \,, \quad A_0(n) = \frac{1}{4}(2\beta + 4n + 1) \,, \\ B_-(n) &= \frac{1}{2}(\beta + 2n)(\beta + 2n - 1) \,. \end{aligned}$$

Using eqs.(3.8)–(3.10), one can present the action of the operator (3.1) in the form

$$\begin{aligned} H u_n = D_2 u_{n+2} + D_1(n) u_{n+1} + \tilde{D}_0(n) u_n \\ + \tilde{D}_1(n) u_{n-1} + \tilde{D}_2(n) u_{n-2} \,, \end{aligned} \tag{3.11}$$

where

$$D_2 = \frac{a_{+0}}{2} C_0 + a_{0-} C_- C_0 + a_{00} C_0^2 + a_{--} C_- C_- + \frac{a_{++}}{4} \tag{3.12}$$

$$\begin{aligned} D_1(n) &= a_{--} \left[A_-(n) C_- + C_- A_-(n+1) \right] \\ &\quad + a_{00} \left[A_0(n) C_0 + C_0 A_0(n+1) \right] + \frac{a_{+0}}{2} A_0(n) \\ &\quad + a_{0-} \left[A_-(n) C_0 + C_- A_0(n+1) \right] + a_0 C_0 + a_- C_- + \frac{a_+}{2} \,, \end{aligned} \tag{3.13}$$

$$\begin{aligned} \tilde{D}_0 &= a_{00} A_0^2(n) + a_{0-} \left[A_-(n) A_0(n) + B_-(n) C_0 \right] \\ &\quad + a_0 A_0(n) + a_- A_-(n) \,, \end{aligned} \tag{3.14}$$

$$\begin{aligned} \tilde{D}_1(n) &= a_{--} \left[A_-(n) B_-(n) + B_-(n) A_-(n-1) \right] \\ &\quad + a_{0-} B_-(n) A_0(n-1) + a_- B_-(n) \,, \end{aligned} \tag{3.15}$$

$$\tilde{D}_2 = a_{--} B_-(n) B_-(n-1) \,. \tag{3.16}$$

For the operator (3.1) to be quasi-exactly solvable with the invariant subspace (3.5), it is necessary that the following conditions of cut off be satisfied:

$$D_2 = D_1(N) = \tilde{D}_1(0) = \tilde{D}_2(0) = \tilde{D}_2(1) = 0 \,. \tag{3.17}$$

4. Explicit examples of invariant subspaces

The analysis of the system (3.17) is rather cumbersome. In this section we list shortly the particular examples of results, when the systems turn out to be QES.

1. $u = \cos\omega z$, $v = \sin\omega z$:

$$\begin{aligned} H &= a_{00}K_0^2 + a_{0-}K_0K_- \\ &+ \Big[\frac{a_{0-}}{2}\omega^2 - 2a_{00}(N+1)\Big]K_0 + a_-K_- + \frac{a_{00}}{2}\omega^2 K_+ \end{aligned} \tag{4.1}$$

2. We also found the following QES case: $u = J_n(z)$, $v(z) = J_{n+1}(z)$ (Bessel functions):

$$\begin{aligned} a_- &= \frac{(3+2n)}{4}a_{0-}\,, \quad a_{+0} = 0 = a_{++}\,, \quad a_+ = \frac{a_{00}}{2}\,, \\ a_0 &= \frac{a_{0-}}{2} - \frac{a_{00}}{2}(4N+3)\,. \end{aligned} \tag{4.2}$$

   The effective Hamiltonian is non-Hermitian.

3.

$$\begin{aligned} a_{++} &= 0, \quad a_- = \frac{\alpha\gamma - \rho(4N+3)}{4\rho}a_{0-}, \quad a_{00} = f(\rho)\,a_{0-}, \\ a_0 &= \Big(\rho - \alpha\gamma + \frac{N+1}{\rho}\Big)a_{0-}\,, \quad f(\rho) = -\frac{1+4\rho^2}{2\rho}\,. \end{aligned} \tag{4.3}$$

   Our subspace is

$$\operatorname{span}\{\Psi_n z^{2n}\exp[\Big(\frac{\rho}{2}\Big)z^2],\ \Psi_{n+1}z^{2n+1}\exp[\Big(\frac{\rho}{2}\Big)z^2]\}\,,$$

   where Ψ_n is the wave function of the n-th level of the harmonic oscillator, $\Psi_n = \exp(-\frac{1}{2}\rho z^2)H_n(z\sqrt{\rho})$, H_n is the Hermite polynomial. Here $\beta = 0, 1, \ldots$.

We would like to stress that the corresponding system represents an anharmonic (not harmonic!) Bose oscillator. The functions, which have the same form as those of an harmonic oscillator, appear in this context in the coherent state representation (not in the coordinate one, as would be the case for the usual harmonic oscillator) and represent auxiliary quantities.

5. Concluding remarks

During recent years, the class of QES was extended considerably to include two- and many-dimensional systems, matrix models [23], the QES anharmonic oscillator with complex potentials ([13], p. 192; [26]), etc. Meanwhile, it turned out that, apart from these (sometimes rather sophisticated and exotic) situations, quasi-exact solvability exists in an everyday life around us where anharmonic Bose oscillators can be met at every step. In particular, the results obtained can be exploited in solid state or molecular physics, theory of magnetism, etc. The approach suggested in the present paper, shows the line along which a lot of second-quantized models with algebraization of the part of the spectrum can be constructed. This approach can be also extended to systems with interaction of subsystems of different nature—in particular, between spin and Bose operator, Bose and Fermi oscillators.

References

1. Razavy M 1980 *Am. J. Phys.* **48** 285
2. Razavy M 1980 1981 *Phys. Lett. A* **82** 7
3. Zaslavskii O B, Ulyanov V V and Tsukernik V M 1983 *Fiz. Nizk. Temp.* **9** 511 [*Sov. J. Low Temp. Phys.* 9 259]
4. Zaslavskii O B and V V Ulyanov 1984 *Zh. Eksp. Teor. Phys.* **87** 1724 [*Sov. Phys. JETP* **60** 991]
5. Zaslavskii O B and Ulyanov V V 1987 *Teor. Mat. Fiz.* **71** 260 [*Theor. Math. Phys.* **71** 520]
6. Turbiner A V and Ushveridze A G 1987 *Phys. Lett. A* **126** 181
7. Turbiner A V 1988 *Zh. Eksp. Teor. Fiz.* **94** 33
8. Turbiner A V 1988 *Funct. Anal.* **22** 92
9. Turbiner A V 1988 *Commun. Math. Phys.* **118** 467
10. Zaslavskii O B 1990 *Sov. Phys. J.* **33** 13
11. Turbiner A V 1994 *Contemp. Math.* **160** 263
12. Ushveridze A G 1994 *Quasi-exactly solvable models in quamtum mechanics* (Bristol: Institute of Physics)
13. Ulyanov V V and Zaslavskii O B 1992 *Phys. Reports* **216** 179
14. Zaslavskii O B 1990 *Phys. Lett. A* **149** 365
15. Alvarez G and Alvarez-Estrada R F 1995 *J. Phys. A: Math. Gen.* **28** 5767
16. Debergh N 2000 *J. Phys. A: Math. Gen.* **33** 7109
17. Dolya S N and Zaslavskii O B 2000 *J. of Phys. A: Math. Gen.* **33** L369
18. Klauder J R and Sudarshan E C G 1968 *Fundamentals of quantum optics* (Benjamin W A, Inc. New York Amsterdam)
19. Erdelyi A (ed) 1953 *Higher transcendental functions* (*Bateman Manuscript Project* vol. 1) (New York: Mc-Graw-Hill)
20. Brihaye Y and Kosinski P 1994 *J. Math. Phys.* **35** 3089
21. Zhdanov R 1997 *Phys. Lett. B* **405** 253
22. Zhdanov R 1997 *J. Phys. A: Math. Gen.* **30** 8761
23. Spichak S and Zhdanov R 1999 *J. Phys. A: Math. Gen.* **32** 3815

24. Turbiner A V 1992 *J. Math. Phys.* **33** 3989
25. Post G and Turbiner A V 1995 *Russ. J. Math. Phys.* **3** 113
26. Bender C M and Boettcher S 1998 *J. Phys. A: Math. Gen.* **31** L273

THE RIEMANN AND EINSTEIN–WEYL GEOMETRIES IN THE THEORY OF ORDINARY DIFFERENTIAL EQUATIONS, THEIR APPLICATIONS AND ALL THAT

VALERII DRYUMA (valery@gala.moldova.su)*
Institute of Mathematics and Informatics, AS RM, 5 Academiei Street,
2028 Kishinev, Moldova

Abstract. Some properties of the 4-dimensional Riemannian spaces with metrics

$$ds^2 = 2(za_3 - ta_4)dx^2 + 4(za_2 - ta_3)dxdy + 2(za_1 - ta_2)dy^2 + 2dxdz + 2dydt$$

associated with the second order nonlinear differential equations

$$y'' + a_1(x,y)y'^3 + 3a_2(x,y)y'^2 + 3a_3(x,y)y' + a_4(x,y) = 0$$

with arbitrary coefficients $a_i(x,y)$ are considered. Three-dimensional Einstein–Weyl spaces connected with dual equations

$$b'' = g(a,b,b')$$

where the function $g(a,b,b')$ satisfies the partial differential equation

$$g_{aacc} + 2cg_{abcc} + 2gg_{accc} + c^2 g_{bbcc} + 2cgg_{bccc} + g^2 g_{cccc} + (g_a + cg_b)g_{ccc} - 4g_{abc} - 4cg_{bbc} - cg_c g_{bcc} - 3gg_{bcc} - g_c g_{acc} + 4g_c g_{bc} - 3g_b g_{cc} + 6g_{bb} = 0$$

are also investigated.

The theory of invariants for second order ODE's is applied to the study of nonlinear dynamical systems dependent on a set of parameters.

1. Introduction

Second order ODE's of the type

$$y'' + a_1(x,y)y'^3 + 3a_2(x,y)y'^2 + 3a_3(x,y)y' + a_4(x,y) = 0 \tag{1.1}$$

are connected with nonlinear dynamical systems of the form

$$\frac{dx}{dt} = P(x,y,z,\alpha_i), \quad \frac{dy}{dt} = Q(x,y,z,\alpha_i), \quad \frac{dz}{dt} = R(x,y,z,\alpha_i),$$

* Work supported in part by MURST, Italy

A.B. Shabat et al. (eds.), New Trends in Integrability and Partial Solvability, 115–156.

where α_i are parameters.

For example, the Lorenz system

$$\dot{X} = \sigma(Y - X), \quad \dot{Y} = rX - Y - ZX, \quad \dot{Z} = XY - bZ,$$

which exhibits chaotic behaviour at some values of the parameters, is equivalent to the equation

$$y'' - \frac{3}{y}y'^2 + (\alpha y - \frac{1}{x})y' + \epsilon x y^4 - \beta x^3 y^4 - \beta x^2 y^3 - \gamma y^3 + \delta\frac{y^2}{x} = 0, \quad (1.2)$$

where

$$\alpha = \tfrac{b+\sigma+1}{\sigma}, \quad \beta = \frac{1}{\sigma^2}, \quad \gamma = \frac{b(\sigma+1)}{\sigma^2},$$
$$\delta = \tfrac{(\sigma+1)}{\sigma}, \quad \epsilon = \frac{b(r-1)}{\sigma^2}.$$

The theory of invariants was first used in [1]–[5] for investigating the properties of this system.

Another example is the third-order differential equation

$$\frac{d^3X}{dt^3} + a\frac{d^2X}{dt^2} - \left(\frac{dX}{dt}\right)^2 + X = 0, \quad (1.3)$$

which manifests chaotic properties provided the parameter a satisfies $2.017 < a < 2.082$ [25]. This equation can be rewritten in the form (1.1)

$$y'' + \frac{1}{y}y'^2 + \frac{a}{y}y' + \frac{x}{y^2} - 1 = 0$$

with the help of the standard substitution

$$\frac{dX}{dt} = y(x), \quad \frac{d^2X}{dt^2} = y'y, \quad \frac{d^3X}{dt^3} = y''y^2 + y'^2 y.$$

According to the Liouville theory [6]–[10] all equations of the type (1.1) belong to two different classes

I. $\nu_5 = 0$,

II. $\nu_5 \neq 0$.

Here the value ν_5 is given by

$$\nu_5 = L_2(L_1 L_{2x} - L_2 L_{1x}) + L_1(L_2 L_{1y} - L_1 L_{2y}) \\ - a_1 L_1^3 + 3a_2 L_1^2 L_2 - 3a_3 L_1 L_2^2 + a_4 L_2^3$$

and L_1, L_2 are defined by

$$L_1 = \frac{\partial}{\partial y}(a_{4y} + 3a_2 a_4) - \frac{\partial}{\partial x}(2a_{3y} - a_{2x} + a_1 a_4) - 3a_3(2a_{3y} - a_{2x}) - a_4 a_{1x},$$

$$L_2 = \frac{\partial}{\partial x}(a_{1x} - 3a_1a_3) + \frac{\partial}{\partial y}(a_{3y} - 2a_{2x} + a_1a_4) - 3a_2(a_{3y} - 2a_{2x}) + a_1a_{4y}.$$

For the equations of the class $\nu_5 = 0$, R. Liouville discovered the series of semi-invariants

$$w_{m+2} = L_1\frac{\partial w_m}{\partial y} - L_2\frac{\partial w_m}{\partial x} + mw_m\Big(\frac{\partial L_2}{\partial x} - \frac{\partial L_1}{\partial y}\Big),$$

starting from :

$$w_1 = \frac{1}{L_1^4}\left[L_1^3(\alpha' L_1 - \alpha'' L_2) + R_1(L_1^2)_x - L_1^2 R_{1x} + L_1R_1(a_3L_1 - a_4L_2)\right],$$

$$w_2 = \frac{1}{L_2^4}\left[L_2^3(\alpha' L_2 - \alpha L_1) - R_2(L_2^2)_y + L_2^2 R_{2y} - L_2R_2(a_1L_1 - a_2L_2)\right].$$

Here

$$\begin{aligned}
R_1 &= L_1L_{2x} - L_2L_{1x} + a_2L_1^2 - 2a_3L_1L_2 + a_4L_2^2\\
R_2 &= L_1L_{2y} - L_2L_{1y} + a_1L_1^2 - 2a_2L_1L_2 + a_3L_2^2\\
\alpha &= a_{2y} - a_{1x} + 2(a_1a_3 - a_2^2)\\
\alpha' &= a_{3y} - a_{2x} + a_1a_4 - a_2a_3\\
\alpha'' &= a_{4y} - a_{3x} + 2(a_2a_4 - a_3^2).
\end{aligned}$$

When $w_1 = 0$ there are another series of semi-invariants

$$i_{2m+2} = L_1\frac{\partial i_{2m}}{\partial y} - L_2\frac{\partial i_{2m}}{\partial x} + 2mi_{2m}\Big(\frac{\partial L_2}{\partial x} - \frac{\partial L_1}{\partial y}\Big),$$

where

$$i_2 = \frac{3R_1}{L_1} + \frac{\partial L_2}{\partial x} - \frac{\partial L_1}{\partial y}, \tag{1.4}$$

and a corresponding sequence for absolute invariants

$$j_{2m} = \frac{i_{2m}}{i_2^m}.$$

For $\nu_5 \neq 0$ the semi-invariants have the form

$$\nu_{m+5} = L_1\frac{\partial \nu_m}{\partial y} - L_2\frac{\partial \nu_m}{\partial x} + m\nu_m\Big(\frac{\partial L_2}{\partial x} - \frac{\partial L_1}{\partial y}\Big),$$

and the corresponding series of absolute invariants satisfies

$$[5t_m - (m-2)t_7t_{m-2}]\nu_5^{2/5} = 5\Big(L_1\frac{\partial t_{m-2}}{\partial y} - L_2\frac{\partial t_{m-2}}{\partial x}\Big) \tag{1.5}$$

where

$$t_m = \nu_m \nu_5^{-m/5}.$$

From the equations (1.5) it follows that some relations constraining the invariants are useful for the theory of second order ODE.

In fact, let us consider a relation of the form

$$t_9 = f(t_7).$$

Then we have

$$\begin{aligned}
(5f(t_7) - 7t_7^2)\nu_5^{2/5} &= 5\Big(L_1 \frac{\partial t_7}{\partial y} - L_2 \frac{\partial t_7}{\partial x}\Big), \\
(5(t_{11} - 9t_7 f(t_7))\nu_5^{2/5} &= 5f'(t_7)\Big(L_1 \frac{\partial t_7}{\partial y} - L_2 \frac{\partial t_7}{\partial x}\Big), \\
(5t_{13} - 11t_7 t_{11})\nu_5^{2/5} &= 5\Big(L_1 \frac{\partial t_{11}}{\partial y} - L_2 \frac{\partial t_{11}}{\partial x}\Big),
\end{aligned}$$

and

$$(5f - 7t_7^2) f'_{t7} = 5t_{11} - 9f(t_7)t_7,$$

from which we get $t_{11} = g(t_7)$.

In the simplest case $t_9 = at_7^2$ we have

$$\begin{aligned}
t_{11} &= a(2a-1)t_7^3, \quad t_{13} = a(2a-1)(3a-2)t_7^4, \\
t_{15} &= a(2a-1)(3a-2)(4a-3)t_7^5, \ldots .
\end{aligned}$$

These relations show that some values of the parameter a

$$a = 0, \quad 1/2, \quad 2/3, \quad 3/4, \quad 4/5, \quad \cdots$$

are special for the corresponding second order ODE's.

The first application of the Liouville theory for the study of properties of nonlinear dynamical systems like Lorenz system was presented in the works of the present author [1]–[5].

In particular for the second order differential equation (1.2) (equivalent to the Lorenz system) the ν_5-invariant has the form

$$\nu_5 = Ax^2 + \frac{B}{x^2 y^2} + C$$

where

$$A = \alpha\beta(10\alpha - \alpha^2 - 6\delta), \qquad B = \alpha\Big(\tfrac{4}{9}\alpha^2 + \tfrac{2}{3}\alpha\delta - 2\delta^2\Big),$$
$$C = \alpha\Big(\tfrac{2}{9}\alpha^4 + 6\epsilon\delta - 4\alpha\epsilon - \alpha^2\gamma\Big).$$

In this case the condition

$$\nu_5 = 0$$

corresponds to

$$A = 0, \quad B = 0, \quad C = 0.$$

In particular it leads to the values

$$\sigma = -1/5, \quad b = -16/5, \quad r = -7/5,$$

which have not been considered before in the theory of the Lorenz system.

The consideration of the invariants ν_{m+2} is connected with unwieldy calculations and does not allow us to apply it for the investigation of this system. Here we show that is possible to get a more detailed information for the case of the equation (1.3).

With this aim we transform the equation

$$y'' + \frac{1}{y}y'^2 + \frac{a}{y}y' + \frac{x}{y^2} - 1 = 0$$

to a more convenient form. Firstly, the variable x can be expressed as

$$x = y^2y'' - yy'^2 - ayy' + y^2,$$

and after differentiating this expression we get a third-order ODE which can be reduced to the second order ODE

$$y'' + \frac{1}{y}y'^2 + \Big(\frac{4}{x} + \frac{4}{xy}\Big)y' + \frac{a}{x^2} - \frac{2}{xy} + \frac{1}{x^2y^2} + \frac{y}{x^2} = 0.$$

This equation admits the invariants

$$L_1 = \frac{3y + 2a}{3x^2y^2}, \quad L_2 = \frac{a}{xy^3},$$

and

$$\begin{aligned}
\nu_5 &= -\frac{1}{9}a^3(2a^2y + 18xy - 9)/(x^5y^4),\\
\nu_7 &= \frac{1}{27}a^4(54xy^2 - 27y - 20a^3y - 180axy + 72a)/(x^7y^{15}),\\
\nu_9 &= \frac{2}{81}a^6(702xy^2 - 297y - 140a^3y - 1260axy + 432a)/(x^9y^{19}),\\
\nu_{11} &= \frac{4}{27}a^8(990xy^2 - 369y - 140a^3y + 1260axy + 384a)/(x^{11}y^{23}),\\
\nu_{13} &= \frac{40}{81}a^{10}(2754xy^2 - 927y - 308a^3y - 2772axy + 768a)/(x^{13}y^{27}),\\
\nu_{15} &= \frac{80}{243}a^{12}(42714xy^2 - 13203y - 4004a^3y - 36036axy
\end{aligned}$$

$$+ 9216a)/(x^{15}y^{31}),$$

$$\nu_{25} = \frac{985600}{6561}a^{22}(48428550xy^2 - 11175165y - 2704156a^3y - 24337404axy + 4718592a)/(x^{25}y^{51}).$$

From these expressions we can see that only numerical values of coefficients in the formulas for the invariants are changed under the transition from ν_m to ν_{m+2}.

This fact is useful for studying the relations between the invariants when the parameter a is changed. Notice that the starting equation (1.3) is connected with the Painlevé I equation for the case $a = 0$. The first applications of the Liouville invariants in the theory of Painlevé equations was presented in the works of author [1]–[6]. In particular, for the equations of the Painlevé type it follows that $\nu_5 = 0$. Recently the relations between the invariants for the P-type equations were studied in [30].

2. The Riemann spaces in the theory of ODE's

Here we present the construction of the Riemann spaces connected with the equations of type (1.1).

We start from the equations of geodesical lines of two-dimensional space A_2 equipped with a affine (or Riemann) connection. They have the form

$$\begin{aligned} \ddot{x} + \Gamma^1_{11}\dot{x}^2 + 2\Gamma^1_{12}\dot{x}\dot{y} + \Gamma^1_{22}\dot{y}^2 &= 0, \\ \ddot{y} + \Gamma^2_{11}\dot{x}^2 + 2\Gamma^2_{12}\dot{x}\dot{y} + \Gamma^2_{22}\dot{y}^2 &= 0. \end{aligned}$$

This system of equations is equivalent to the equation

$$y'' - \Gamma^1_{22}y'^3 + (\Gamma^2_{22} - 2\Gamma^1_{12})y'^2 + (2\Gamma^2_{12} - \Gamma^1_{11})y' + \Gamma^2_{11} = 0,$$

which is of type (1.1) for a particular choice of the coefficients $a_i(x,y)$.

The following proposition holds

Proposition 2.1 *The equation (1.1) with the coefficients $a_i(x,y)$ describes the geodesics on a surface with a metric determined by*

$$ds^2 = \frac{1}{\Delta^2}[\psi_1 dx^2 + 2\psi_2 dxdy + \psi_3 dy^2],$$

where $\Delta = \psi_1\psi_3 - \psi_2^2$, provided the relations

$$\begin{aligned} \psi_{1x} + 2a_3\psi_1 - 2a_4\psi_2 &= 0, \\ \psi_{3y} + 2a_1\psi_2 - 2a_2\psi_3 &= 0, \\ \psi_{1y} + 2\psi_{2x} - 2a_3\psi_2 + 4a_2\psi_1 - 2a_4\psi_3 &= 0, \\ \psi_{3x} + 2\psi_{2y} + 2a_2\psi_2 - 4a_3\psi_3 + 2a_1\psi_1 &= 0, \end{aligned}$$

between the coefficients $a_i(x,y)$ and the components of the metrics $\psi_i(x,y)$ are fulfilled.

The equations (1.1) with arbitrary coefficients $a_i(x,y)$ may be considered as the equations describing the geodesics of a 2-dimensional space A_2

$$\begin{aligned} \ddot{x} - a_3\dot{x}^2 - 2a_2\dot{x}\dot{y} - a_1\dot{y}^2 &= 0, \\ \ddot{y} + a_4\dot{x}^2 + 2a_3\dot{x}\dot{y} + a_2\dot{y}^2 &= 0 \end{aligned}$$

equipped with a projective connection with components

$$\mathbf{\Pi}_1 = \begin{pmatrix} -a_3 & -a_2 \\ a_4 & a_3 \end{pmatrix}, \qquad \mathbf{\Pi}_2 = \begin{pmatrix} -a_2 & -a_1 \\ a_3 & a_2 \end{pmatrix}.$$

The curvature tensor of this type of connection is

$$\mathbf{R}_{12} = \frac{\partial \mathbf{\Pi}_2}{\partial x} - \frac{\partial \mathbf{\Pi}_1}{\partial y} + [\mathbf{\Pi}_1, \mathbf{\Pi}_2]$$

and it has the components

$$\begin{aligned} R^1_{112} &= a_{3y} - a_{2x} + a_1a_4 - a_2a_3 = \alpha', \\ R^1_{212} &= a_{2y} - a_{1x} + 2(a_1a_3 - a_2^2) = \alpha, \\ R^2_{112} &= a_{3x} - a_{4y} + 2(a_3^2 - a_2a_4) = -\alpha'', \\ R^2_{212} &= a_{2x} - a_{3y} + a_3a_2 - a_1a_4 = -\alpha'. \end{aligned}$$

In order to construct the Riemann space connected with the equation of type (1.1) we use the Riemannian extension W^4 of the space A_2 with connection Π^k_{ij} [12] . The corresponding metric is

$$ds^2 = -2\Pi^k_{ij}\xi_k dx^i dx^j + 2d\xi_i dx^i.$$

In our case it takes the following form ($\xi_1 = z, \xi_2 = t$)

$$\begin{aligned} ds^2 = 2(za_3 - ta_4)dx^2 &+ 4(za_2 - ta_3)dxdy \\ &+ 2(za_1 - ta_2)dy^2 + 2dxdz + 2dydt. \end{aligned} \tag{2.1}$$

Thus we may state

Proposition 2.2 *For a given equation of type (1.1) there exists a Riemannian space with metric (2.1) whose geodesics contain integral curves of (1).*

The calculation of the geodesics of the space W^4 with the metric (2.1) leads to the system of equations

$$\begin{aligned}
&\frac{d^2x}{ds^2} - a_3\left(\frac{dx}{ds}\right)^2 - 2a_2\frac{dx}{ds}\frac{dy}{ds} - a_1\left(\frac{dy}{ds}\right)^2 = 0,\\
&\frac{d^2y}{ds^2} + a_4\left(\frac{dx}{ds}\right)^2 + 2a_3\frac{dx}{ds}\frac{dy}{ds} + a_2\left(\frac{dy}{ds}\right)^2 = 0,\\
&\frac{d^2z}{ds^2} + [z(a_{4y} - \alpha'') - ta_{4x}]\left(\frac{dx}{ds}\right)^2\\
&\quad +2[za_{3y} - t(a_{3x} + \alpha'')]\frac{dx}{ds}\frac{dy}{ds}\\
&\quad + [z(a_{2y} + \alpha) - t(a_{2x} + 2\alpha')]\left(\frac{dy}{ds}\right)^2 + 2a_3\frac{dx}{ds}\frac{dz}{ds}\\
&\quad - 2a_4\frac{dx}{ds}\frac{dt}{ds} + 2a_2\frac{dy}{ds}\frac{dz}{ds} - 2a_3\frac{dy}{ds}\frac{dt}{ds} = 0,\\
&\frac{d^2t}{ds^2} + [z(a_{3y} - 2\alpha') - t(a_{3x} - \alpha'')]\left(\frac{dx}{ds}\right)^2\\
&\quad +2[z(a_{2y} - \alpha) - ta_{2x}]\frac{dx}{ds}\frac{dy}{ds} + [za_{1y} - t(a_{1x} + \alpha)]\left(\frac{dy}{ds}\right)^2\\
&\quad +2a_2\frac{dx}{ds}\frac{dz}{ds} - 2a_3\frac{dx}{ds}\frac{dt}{ds} + 2a_1\frac{dy}{ds}\frac{dz}{ds} - 2a_2\frac{dy}{ds}\frac{dt}{ds} = 0,
\end{aligned}$$

in which the first two equations for the coordinates $x(s)$, $y(s)$ are equivalent to equation (1.1).

In turn the two last equations for the coordinates $z(s)$ and $t(s)$ have the form of a linear second-order system of differential equations

$$\frac{d^2\Psi}{ds^2} + A(x,y)\frac{d\Psi}{ds} + B(x,y)\Psi = 0 \tag{2.2}$$

where $\Psi(x,y)$ is a two-component vector $\Psi_1 = z(s)$, $\Psi_2 = t(s)$ and $A(x,y)$ and $B(x,y)$ are 2×2 matrix-functions.

Notice that the complete system of equations has the first integral

$$2(za_3 - ta_4)\dot{x}^2 + 4(za_2 - ta_3)\dot{x}\dot{y} + 2(za_1 - ta_2)\dot{y}^2 + 2\dot{x}\dot{z} + 2\dot{y}\dot{t} = 1,$$

which is equivalent to the relation

$$z\dot{x} + t\dot{y} = \frac{s}{2} + \mu,$$

where μ is an integration constant. This allows us to use only one linear second-order differential equation from the matrix system (2.2) in the study of concrete examples.

Thus, we have constructed the four-dimensional Riemannian space with the metric (2.1) and with the connection

$$\mathbf{\Gamma}_1 = \begin{pmatrix} -a_3 & -a_2 & 0 & 0 \\ a_4 & a_3 & 0 & 0 \\ z(a_{4y} - \alpha'') - ta_{4x} & za_{3y} - t(a_{3x} + \alpha'') & a_3 & -a_4 \\ z(a_{3y} - 2\alpha') - t(a_{3x} - \alpha'') & z(a_{2y} - \alpha) - ta_{2x} & a_2 & -a_3 \end{pmatrix},$$

$$\mathbf{\Gamma}_2 = \begin{pmatrix} -a_2 & -a_1 & 0 & 0 \\ a_3 & a_2 & 0 & 0 \\ za_{3y} - t(a_{3x} + \alpha'') & z(a_{2y} + \alpha) - t(a_{2x} + 2\alpha') & a_2 & -a_3 \\ z(a_{2y} - \alpha) - ta_{2x} & za_{1y} - t(a_{1x} + \alpha) & a_1 & -a_2 \end{pmatrix},$$

$$\mathbf{\Gamma}_3 = \begin{pmatrix} 0 & 0 & 0 & 0 \\ 0 & 0 & 0 & 0 \\ a_3 & a_2 & 0 & 0 \\ a_2 & a_1 & 0 & 0 \end{pmatrix}, \qquad \mathbf{\Gamma}_4 = \begin{pmatrix} 0 & 0 & 0 & 0 \\ 0 & 0 & 0 & 0 \\ -a_4 & -a_3 & 0 & 0 \\ -a_3 & -a_2 & 0 & 0 \end{pmatrix}.$$

The curvature tensor of this metric is

$$\begin{aligned}
R^1_{112} &= -R^3_{312} = -R^2_{212} = R^4_{412} = \alpha', \qquad R^1_{212} = -R^4_{312} = \alpha, \\
R^2_{112} &= -R^3_{412} = -\alpha'', \qquad R^1_{312} = R^1_{412} = R^2_{312} = R^2_{412} = 0, \\
R^3_{112} &= 2z(a_2\alpha'' - a_3\alpha') + 2t(a_4\alpha' - a_3\alpha''), \\
R^4_{212} &= 2z(a_3\alpha' - a_2\alpha) + 2t(a_3\alpha - a_2\alpha'), \\
R^3_{212} &= z(\alpha_x - \alpha'_y + a_1\alpha'' - a_3\alpha) + t(\alpha''_y - \alpha'_x + a_4\alpha - a_2\alpha''), \\
R^4_{112} &= z(\alpha'_y - \alpha_x + a_1\alpha'' - a_3\alpha) + t(\alpha'_x - \alpha''_y + a_4\alpha - a_2\alpha'').
\end{aligned}$$

By using the expressions for the components of the projective curvature of the space A_2

$$\begin{aligned}
L_1 &= \alpha''_y - \alpha'_x + a_2\alpha'' + a_4\alpha - 2a_3\alpha', \\
L_2 &= \alpha'_y - \alpha_x + a_1\alpha'' + a_3\alpha - 2a_2\alpha',
\end{aligned}$$

they can be rewritten as

$$\begin{aligned}
R^4_{112} &= z(L_2 + 2a_2\alpha' - 2a_3\alpha) - t(L_1 + 2a_3\alpha' - 2a_4\alpha), \\
R^3_{212} &= z(-L_2 + 2a_1\alpha'' - 2a_2\alpha') + t(L_1 + 2a_3\alpha' - 2a_2\alpha''), \\
R^i_{j34} &= 0,
\end{aligned}$$

$$\mathbf{R}_{13} = \begin{pmatrix} 0 & 0 & 0 & 0 \\ 0 & 0 & 0 & 0 \\ 0 & -\alpha' & 0 & 0 \\ \alpha' & 0 & 0 & 0 \end{pmatrix}, \quad \mathbf{R}_{14} = \begin{pmatrix} 0 & 0 & 0 & 0 \\ 0 & 0 & 0 & 0 \\ 0 & \alpha'' & 0 & 0 \\ -\alpha'' & 0 & 0 & 0 \end{pmatrix},$$

$$\mathbf{R}_{23} = \begin{pmatrix} 0 & 0 & 0 & 0 \\ 0 & 0 & 0 & 0 \\ 0 & -\alpha & 0 & 0 \\ \alpha & 0 & 0 & 0 \end{pmatrix}, \qquad \mathbf{R}_{24} = \begin{pmatrix} 0 & 0 & 0 & 0 \\ 0 & 0 & 0 & 0 \\ 0 & \alpha' & 0 & 0 \\ -\alpha' & 0 & 0 & 0 \end{pmatrix}.$$

The Ricci tensor $R_{ik} = R^{l}_{ilk}$ of our space D^4 has the components

$$R_{11} = 2\alpha'', \quad R_{12} = 2\alpha', \quad R_{22} = 2\alpha,$$

and the scalar curvature $R = g^{in}g^{km}R_{nm}$ of the space D^4 is $R = 0$.

Now we can introduce the tensor

$$L_{ijk} = \nabla_k R_{ij} - \nabla_j R_{ik} = R_{ij;k} - R_{ik;j}.$$

It has the following components

$$L_{112} = -L_{121} = 2L_1, \quad L_{221} = L_{212} = -2L_2$$

which allow us to construct the invariant conditions connected with the equations (1.1) from the covariant derivatives of the curvature tensor and the values of L_1, L_2.

The Weyl tensor of the space D^4 is

$$C_{lijk} = R_{lijk} + \frac{1}{2}(g_{jl}R_{ik} + g_{ik}R_{jl} - g_{jk}R_{il} - g_{il}R_{jk}) + \frac{R}{6}(g_{jk}g_{il} - g_{jl}g_{ik}).$$

It has only one component

$$C_{1212} = tL_1 - zL_2.$$

Observe that the values of L_1 and L_2 in this formulas are the same as the Liouville expressions in theory of invariants of the equations (1.1).

Using the components of the Riemann tensor

$$R_{1412} = \alpha'', \quad R_{2412} = \alpha', \quad R_{2312} = -\alpha, \quad R_{3112} = \alpha',$$

$$R_{1212} = z(\alpha_x - \alpha'_y + a_1\alpha'' - 2a_2\alpha' + a_3\alpha) + t(\alpha''_y - \alpha'_x + a_4\alpha - 2a_3\alpha' - a_2\alpha'')$$

the equation

$$\mid R_{AB} - \lambda g_{AB} \mid = 0$$

for determining the Petrov type of the spaces D^4 has been considered. Here R_{AB} is a symmetric 6×6 matrix constructed from the components of the Riemann tensor R_{ijkl} of the space D^4.

In particular we have checked that all the scalar invariants of the space D^4 of the form

$$R_{ij}R^{ij} = 0, \quad R_{ijkl}R^{ijkl} = 0, \dots$$

constructed from the curvature tensor of the space M^4 and its covariant derivatives vanish.

Remark. The spaces with metrics (2.1) are flat for the equations (1.1) under the conditions

$$\alpha = 0, \quad \alpha' = 0, \quad \alpha'' = 0,$$

for the coefficients $a_i(x, y)$.

Such type of equations imply

$$L_1 = 0, \quad L_2 = 0$$

and they can be reduced to the the form $y'' = 0$ with the help of appropriate point transformations.

On the other hand there are examples of equations (1.1) such that $L_1 = L_2 = 0$ and

$$\alpha \neq 0, \quad \alpha' \neq 0, \quad \alpha'' \neq 0.$$

For such type of equations the curvature of the corresponding Riemann spaces does not vanish.

In fact, the equation

$$y'' + 2e^{\varphi} y'^3 - \varphi_y y'^2 + \varphi_x y' - 2e^{\varphi} = 0 \tag{2.3}$$

verifies $L_1 = L_2 = 0$ and

$$\alpha \neq 0, \quad \alpha' \neq 0, \quad \alpha'' \neq 0.$$

provided the function $\varphi(x, y)$ is a solution of the Wilczynski-Tzitzeika nonlinear equation integrable by the Inverse Transform Method

$$\varphi_{xy} = 4e^{2\varphi} - e^{-\varphi}. \tag{2.4}$$

In particular, there are non-flat Riemannian spaces even for linear second-order differential equations.

Remark. The study of the properties of Riemann spaces with the metric (2.1) associated to equations (1.2) possesing chaotical behaviour at the values of coefficients ($\sigma = 10$, $b = 8/3$, $r > 24$) is an important problem. The spaces with such values of the parameters verify special relations for the components of the curvature tensor.

In order to study this problem we may use the geodesic deviation equation

$$\frac{d^2\eta^i}{ds^2} + 2\Gamma^i_{lm} \frac{dx^m}{ds} \frac{d\eta^l}{ds} + \frac{\partial \Gamma^i_{kl}}{\partial x^j} \frac{dx^k}{ds} \frac{dx^l}{ds} \eta^j = 0$$

where Γ^i_{lm} are the Christoffell coefficients of the metric (2.1) with the coefficients

$$a_1 = 0, \quad a_2 = -\frac{1}{y}, \quad a_3 = \left(\frac{\alpha y}{3} - \frac{1}{3x}\right),$$

$$a_4 = \epsilon x y^4 - \beta x^3 y^4 - \beta x^2 y^3 - \gamma y^3 + \delta \frac{y^2}{x}.$$

For the equations

$$y'' + a_4(x, y) = 0$$

the four-dimensional Riemann spaces with the metric

$$ds^2 = -2ta_4 dx^2 + 2dxdz + 2dydt$$

and geodesic equations

$$\ddot{x} = 0, \quad \ddot{y} + a_4(x, y)(\dot{x})^2 = 0, \quad \ddot{t} + a_{4y}(\dot{x})^2 t = 0$$

$$\ddot{z} - ta_{4x}(\dot{x})^2 - 2ta_{4y}\dot{x}\dot{y} - 2a_4\dot{x}\dot{t} = 0$$

are connected.

It is interesting to note that for the Painleve II equation

$$y'' = 2y^3 + xy + \alpha$$

the system for geodesic deviations of the corresponding Riemann space

$$\begin{aligned}
\frac{d^2\eta^1}{ds^2} &= 0, \qquad \frac{d^2\eta^2}{ds^2} = (6y^2 + s)\eta^2 + 4y^3 + 3sy + \alpha, \\
\frac{d^2\eta^3}{ds^2} &= -(12ty^2 + 2ts)\frac{d\eta}{ds} - (4y^3 + 2sy + 2\alpha)\frac{d\eta^4}{ds} \\
&\quad -(t + 24ty\dot{y} + 2y^2\dot{t} + 2s)\eta^2 - (y + 12y^2\dot{y} + 2s\dot{y})\eta^4 \\
&\quad -2ty - 4ts\dot{y} - 12ts^2\dot{y} - 4y^3\dot{t} - 4sy\dot{t} - 2\alpha\dot{t}, \\
\frac{d^2\eta^4}{ds^2} &= (6y^2 + s)\eta^4 + 12ty^2 + 3ts + 12ty\eta^2
\end{aligned}$$

depends on the parameter α.

For the equations

$$y'' + 3a_3(x, y)y' + a_4(x, y) = 0$$

the corresponding Riemann spaces have the metrics

$$ds^2 = 2(za_3 - ta_4)dx^2 - 4ta_3 dxdy + 2dxdz + 2dydt$$

and the geodesic equations

$$\begin{aligned}
\ddot{x} - a_3\dot{x}^2 &= 0,\\
\ddot{y} + 2a_3\dot{x}\dot{y} + a_4\dot{x}^2 &= 0,\\
\ddot{t} - 2a_3\dot{x}\dot{t} - a_{3y}\dot{x}^2 z + (a_{4y} - 2{a_3}^2 - 2a_{3x})\dot{x}^2 t &= 0,\\
\ddot{z} + 2a_3\dot{x}\dot{z} - 2(a_3\dot{y} + a_4\dot{x})\dot{t} + [(a_{3x} + 2{a_3}^2)\dot{x}^2 + 2a_{3y}\dot{x}\dot{y}]z &\\
-[a_{4x}\dot{x}^2 + 2(a_{4y} - 2a_3^2)\dot{x}\dot{y} + 2a_{3y}\dot{y}^2]t &= 0.
\end{aligned}$$

Let us consider the possibility of submersion of the spaces with metrics (2.1) from the point of view of the theory of the embedding of Riemann spaces into flat spaces

For the Riemann spaces of the first class (which can be embedded into the 5-dimensional Euclidean space) the following conditions are fulfilled

$$R_{ijkl} = b_{ik}b_{jl} - b_{il}b_{jk}, \qquad b_{ij;k} - b_{ik;j} = 0$$

where R_{ijkl} are the components of the curvature tensor associated with the metric $ds^2 = g_{ij}dx^i dx^j$.

The consideration of these relations for the spaces with the metric (2.1) leads to conditions on the values $a_i(x, y)$. Thus one deduces that the embedding in the 5-dimensional flat space with the flat metrics is possible only in the case

$$a_i(x, y) = 0.$$

For the spaces of the second class (which admit the embedding into the 6-dimensional flat space with appropriate signature) the conditions are more complicated. They are

$$\begin{aligned}
R_{abcd} &= e_1(\omega_{ac}\omega_{bd} - \omega_{ad}\omega_{bc}) + e_2(\lambda_{ac}\lambda_{bd} - \lambda_{ad}\lambda_{bc}),\\
\omega_{ab;c} - \omega_{ac;b} &= e_2(\tau_c\lambda_{ab} - \tau_b\lambda_{ac}),\\
\lambda_{ab;c} - \lambda_{ac;b} &= -e_1(\tau_c\omega ab - \tau_b\omega_{ac}),\\
\tau_{a;b} - \tau_{a;c} &= \omega_{ac}\lambda^c_b - \lambda_{ac}\omega^c_b.
\end{aligned}$$

and lead to the relations

$$\begin{aligned}
\epsilon^{abcd}\epsilon^{nmrs}\epsilon^{pqik}R_{abnm}R_{cdpq}R_{rsik} &= 0,\\
\epsilon^{cdmn}R_{abcd}R^{ab}_{mn} &= -8e_1e_2\epsilon^{cdmn}\tau_{c;d}\tau_{m;n}.
\end{aligned}$$

3. On the relationship with the theory of surfaces

The existence of the Riemann metrics for the equations (1.1) may be useful for constructing the corresponding surfaces.

A possible application is the study of two-dimensional subspaces of a given 4-dimensional space which generalize the surfaces of translation. The equations for coordinates $Z^i(u,v)$ of such a type of surfaces are

$$\frac{\partial^2 Z^i}{\partial u \partial v} + \Gamma^i_{jk}\frac{\partial Z^j}{\partial u}\frac{\partial Z^k}{\partial v} = 0. \tag{3.1}$$

where Γ^{jk}_i are the connection components.

Let us consider the system (3.1) in detail. We get the following system of equations for the coordinates $x = x(u,v)$, $y = y(u,v)$, $z = z(u,v)$, $t = t(u,v)$

$$\begin{aligned}
&x_{uv} - a_3 x_u x_v - a_2(x_u y_v + x_v y_u) - a_1 y_u y_v = 0\\
&y_{uv} + a_4 x_u x_v + a_3(x_u y_v + x_v y_u) + a_2 y_u y_v = 0\\
&z_{uv} + z_u(a_2 y_v + a_3 x_v) + z_v(a_2 y_u + a_3 x_u) - t_u(a_3 y_v + a_4 x_v)\\
&- t_v(a_4 x_u + a_3 y_u) + z(2y_u y_v a_1 a_3 - y_v y_u a_{1x} + x_v x_u a_{3x} - 2x_u x_v a_2 a_4\\
&+ y_u x_v a_{3y} - y_v x_u a_{3y} - 2y_v y_u (a_2)^2 + 2y_v y_u a_{2y} + 2x_v y_u (a_3)^2)\\
&+ t(y_v y_u a_{2x} - x_v x_u a_{4_x} - x_v y_u a_{4y} + 2y_u y_v a_2 a_3 - 2y_v x_u a_2 a_4\\
&- 2y_v y_u a_{3y} + 2x_v y_u (a_3)^2 + 2y_v x_u (a_3)^2 - y_v x_u a_{4y} - 2y_u x_v a_2 a_4\\
&- 2y_v y_u a_1 a_4) = 0\\
&t_{uv} + z_u(a_2 x_v + a_1 y_v) + z_v(a_2 x_u + a_1 y_u) - t_u(a_3 x_v + a_2 y_v)\\
&- t_v(a_3 x_u + a_2 y_u) + z(-2x_u y_v a_1 a_3 + x_v y_u a_{1x} + 2x_v x_u a_{2x}\\
&+ 2x_u x_v a_2 a_3 + y_v y_u a_{1y} + 2y_v x_u (a_2)^2 - 2x_v x_u a_4 a_1 - x_v x_u a_{3y}\\
&+ 2x_v y_u (a_2)^2 - 2x_v y_u a_1 a_3 + y_v x_u a_{1_x}) + t(-y_v x_u a_{2x} - x_v y_u a_{2x}\\
&+ x_v x_u a_{4y} + 2x_u x_v a_2 a_4 - 2y_u y_v a_1 a_3 - 2x_v x_u a_{3x} - 2x_v x_u (a_3)^2\\
&+ 2y_v y_u (a_2)^2 - y_v y_u a_{2y}) = 0
\end{aligned}$$

Here the following expressions for the Christoffell coefficients were used

$$\begin{aligned}
&\Gamma^1_{11} = -a_3(x,y), \quad \Gamma^2_{11} = a_4(x,y), \quad \Gamma^1_{12} = -a_2(x,y), \quad \Gamma^2_{12} = a_3(x,y),\\
&\Gamma^3_{13} = a_3(x,y), \Gamma^4_{13} = a_2(x,y), \quad \Gamma^3_{14} = -a_4(x,y), \quad \Gamma^4_{14} = -a_3(x,y),
\end{aligned}$$

$$\begin{aligned}
&\Gamma^1_{22} = -a_1(x,y), \quad \Gamma^2_{22} = -a_2(x,y), \quad \Gamma^3_{23} = a_2(x,y),\\
&\Gamma^4_{23} = a_1(x,y), \quad \Gamma^3_{24} = -a_3(x,y), \quad \Gamma^4_{24} = -a_2(x,y),
\end{aligned}$$

$$\begin{aligned}
\Gamma^3_{11} &= z\frac{\partial a_3(x,y)}{\partial x} - t\frac{\partial a_4(x,y)}{\partial x} + 2za_3(x,y)^2 - 2za_2(x,y)a_4(x,y),\\
\Gamma^4_{11} &= 2z\frac{\partial a_2(x,y)}{\partial x} - 2t\frac{\partial a_3(x,y)}{\partial x} - z\frac{\partial a_3(x,y)}{\partial y} + t\frac{\partial a_4(x,y)}{\partial y}
\end{aligned}$$

$$
\begin{aligned}
&+ 2za_3(x,y)a_2(x,y) - 2za_1(x,y)a_4(x,y)\\
&+ 2ta_2(x,y)a_4(x,y) - 2ta_3(x,y)^2,\\
\Gamma^3_{12} &= z\frac{\partial a_3(x,y)}{\partial y} - t\frac{\partial a_4(x,y)}{\partial y} + 2ta_3(x,y)^2 - 2ta_2(x,y)a_4(x,y),\\
\Gamma^4_{12} &= z\frac{\partial a_1(x,y)}{\partial x} - t\frac{\partial a_2(x,y)}{\partial x} + 2za_2(x,y)^2 - 2za_1(x,y)a_3(x,y),\\
\Gamma^4_{22} &= z\frac{\partial a_1(x,y)}{\partial y} - t\frac{\partial a_2(x,y)}{\partial y} + 2ta_3(x,y)^2 - 2ta_1(x,y)a_3(x,y),\\
\Gamma^3_{22} &= 2z\frac{\partial a_2(x,y)}{\partial y} - 2t\frac{\partial a_3(x,y)}{\partial y} - z\frac{\partial a_1(x,y)}{\partial x} + t\frac{\partial a_2(x,y)}{\partial x}\\
&+ 2za_3(x,y)a_1(x,y) - 2ta_1(x,y)a_4(x,y)\\
&+ 2ta_2(x,y)a_3(x,y) - 2za_2(x,y)^2.
\end{aligned}
$$

We notice that the two last equations are linear and reduce to the 2×2 matrix Laplace equations

$$\frac{\partial^2 \Psi}{\partial u \partial v} + A\frac{\partial \Psi}{\partial u} + B\frac{\partial \Psi}{\partial v} + C\Psi = 0, \tag{3.2}$$

which can be integrated by means of a generalization of the Laplace transformation [26].

For this aim we use the transformations

$$\Psi_1 = (\partial_v + A)\Psi, \quad (\partial_u + B)\Psi_1 = h\Psi,$$

where the Laplace invariants are

$$H = A_u + BA - C, \quad K = B_v + AB - C$$

and then construct a new equation of type (2.3) for the function Ψ_1 with new invariants given by

$$
\begin{aligned}
A_1 &= HAH^{-1} - H_v H^{-1}, \qquad B_1 = B,\\
C_1 &= B_v - H + (HAH^{-1} - H_v H^{-1})B,
\end{aligned}
$$

Let us consider some examples. For the first example with set

$$
\begin{gathered}
x = x, \qquad y = y, \qquad u = x, \qquad v = y,\\
z = z(u,v) = z(x,y), \qquad t = t(u,v) = t(x,y).
\end{gathered}
$$

From the first equations of the full system we get

$$a_2(x,y) = 0, \quad a_3(x,y) = 0,$$

and from the next two we have the system of equations

$$\frac{\partial^2 z}{\partial x \partial y} - \frac{\partial a_4(x,y)}{\partial y} t - a_4(x,y) \frac{\partial t}{\partial y} = 0,$$

$$\frac{\partial^2 t}{\partial x \partial y} + \frac{\partial a_1(x,y)}{\partial x} z + a_1(x,y) \frac{\partial z}{\partial x} = 0,$$

They are equivalent to the independent relations

$$\frac{\partial z}{\partial x} - t a_4(x,y) = 0,$$

$$\frac{\partial t}{\partial y} + z a_1(x,y) = 0,$$

or

$$\begin{aligned} \frac{\partial^2 z}{\partial x \partial y} - \frac{1}{a_4} \frac{\partial a_4(x,y)}{\partial y} \frac{\partial z}{\partial x} + a_1(x,y) a_4(x,y) z &= 0, \\ \frac{\partial^2 t}{\partial x \partial y} - \frac{1}{a_1} \frac{\partial a_1(x,y)}{\partial x} \frac{\partial t}{\partial y} + a_1(x,y) a_4(x,y) t &= 0. \end{aligned}$$

Any solution of this system of equations supplies examples of surfaces corresponding to ODE's of the form

$$\frac{d^2 y}{dx^2} + a_1(x,y) \left(\frac{dy}{dx} \right)^3 + a_4(x,y) = 0$$

The next example assumes the conditions:

$$x = u + v, \quad y = uv.$$

From the first relation we get

$$a_3 + x a_2 = -y a_1, \quad y a_2 + x a_3 = -1 - a_4$$

from which it follows

$$a_2 = \frac{1 + a_4(x,y) - x y a_1(x,y)}{x^2 - y}, \quad a_3 = \frac{y^2 a_1(x,y) - x - x a_4(x,y)}{x^2 - y}.$$

As a result we get the equations

$$\frac{d^2 y}{dx^2} + a_1 \left(\frac{dy}{dx} \right)^3 + 3 \frac{(1 + a_4 - x y a_1)}{x^2 - y} \left(\frac{dy}{dx} \right)^2 + 3 \frac{(y^2 a_1 - x - x a_4)}{x^2 - y} \frac{dy}{dx} + a_4 = 0$$

In the particular case

$$a_1(x,y) = 0, \quad a_4(x,y) = \frac{-x^2}{y}$$

we get the equation

$$\frac{d^2y}{dx^2} - \frac{3}{y}\left(\frac{dy}{dx}\right)^2 + \frac{3x}{y}\frac{dy}{dx} - \frac{x^2}{y} = 0 \tag{3.3}$$

and the following equations for the coordinates of the corresponding surfaces

$$\begin{aligned} t_{uv} - \frac{t_u}{u} - \frac{t_v}{v} - \frac{1}{uv}(z_u + z_v) + \frac{u+v}{u^2v^2}z + \frac{uv+u^2+v^2}{u^2v^2}t &= 0, \\ z_{uv} + \frac{z_u}{u} + \frac{z_v}{v} + (u+v)\left(\frac{t_v}{v} + \frac{t_u}{u}\right) - \frac{uv+u^2+v^2}{u^2v^2}z & \\ - \frac{u^2v+u^3+v^3+v^2u}{u^2v^2}t &= 0. \end{aligned}$$

Observe that the equation (3.3) can be rewritten as

$$\frac{d^2z}{d\rho^2} - \frac{3}{z}\left(\frac{dz}{d\rho}\right)^2 + \left(\frac{3}{z} - 9\right)\frac{dz}{d\rho} - 10z + 6 - \frac{1}{z} = 0 \tag{3.4}$$

with the help of the substitution

$$y(x) = x^2 z(\log(x)).$$

Another possible application for the study of two-dimensional surfaces in the space with metrics (2.1) is connected with the choice

$$x = x, \quad y = y, \quad z = z(x,y), \quad t = t(x,y).$$

Using the expressions

$$dz = z_x dx + z_y dy, \quad dt = t_x dx + t_y dy$$

we get the metric

$$\begin{aligned} ds^2 = 2(z_x + za_3 - ta_4)dx^2 + 2(t_x + z_y + 2za_2 - 2ta_3)dxdy \\ +2(t_y + za_1 - ta_2)dy^2. \end{aligned}$$

We can use this representation for the analysis of particular cases of equations (1.1).

The constraints

$$\begin{aligned} z_x + za_3 - ta_4 &= 0, \\ t_x + z_y + 2za_2 - 2ta_3 &= 0 \\ t_y + za_1 - ta_2 &= 0 \end{aligned}$$

are connected with flat surfaces and reduce under the substitution

$$z = \Phi_x, \qquad t = \Phi_y$$

to the system

$$\begin{aligned} \Phi_{xx} &= a_4\Phi_y - a_3\Phi_x, \\ \Phi_{xy} &= a_3\Phi_y - a_2\Phi_x, \\ \Phi_{yy} &= a_2\Phi_y - a_1\Phi_x, \end{aligned}$$

which is compatible with the conditions

$$\alpha = 0, \qquad \alpha' = 0, \qquad \alpha'' = 0.$$

Remark. The choice of the functions $z = \Phi_x, t = \Phi_y$ satisfying to the system of equations

$$\begin{aligned} \Phi_{xx} &= a_4\Phi_y - a_3\Phi_x, \\ \Phi_{yy} &= a_2\Phi_y - a_1\Phi_x \end{aligned}$$

corresponds to coefficients $a_i(x, y)$ which are given by

$$a_4 = R_{xxx}, \quad a_3 = -R_{xyy}, \quad a_2 = R_{xyy}, \quad a_1 = R_{yyy}$$

where the function $R(x, y)$ is the solution of WDVV-equation

$$R_{xxx}R_{yyy} - R_{xxy}R_{xyy} = 1,$$

and are associated to the equation (1.1)

$$y'' - R_{yyy}y'^3 + 3R_{xyy}y'^2 - 3R_{xxy}y' + R_{xxx} = 0.$$

The following choice of the coefficients a_i

$$a_4 = -2\omega, \quad a_1 = 2\omega, \quad a_3 = \frac{\omega_x}{\omega}, \quad a_2 = -\frac{\omega_y}{\omega}$$

leads to the system

$$\Phi_{xx} + \frac{\omega_x}{\omega}\Phi_x + 2\omega\Phi_y = 0,$$

$$\Phi_{yy} + 2\omega\Phi_x + \frac{\omega_y}{\omega}\Phi_y = 0$$

the compatibility condition of which

$$\frac{\partial^2 \log\omega}{\partial x \partial y} = 4\omega^2 + \frac{\kappa}{\omega}$$

is the Wilczynski-Tzitzeika-equation.

Remark. The linear system of equations for the WDVV-equation determines some surfaces in three-dimensional projective space. In canonical form it becomes [13]

$$\begin{aligned}\Phi_{xx} - R_{xxx}\Phi_y + \Big(\frac{R_{xxxy}}{2} - \frac{R_{xxy}^2}{4} - \frac{R_{xxx}R_{xxy}}{2}\Big)\Phi &= 0,\\ \Phi_{yy} - R_{yyy}\Phi_x + \Big(\frac{R_{yyyx}}{2} - \frac{R_{xyy}^2}{4} - \frac{R_{yyy}R_{xxy}}{2}\Big)\Phi &= 0.\end{aligned}$$

The relations between Wilczynsky invariants for the linear system corresponds to several types of surfaces. Some of them are connected with solutions of the WDVV equation.

Remark. From an elementary point of view the surfaces connected with a system of equations like the Lorenz system can be constructed from the above scheme.

Under the assumption

$$z = z(x, y)$$

we get

$$\sigma(y - x)z_x + (rx - y - zx)z_y = xy - bz.$$

The solutions of this equation give us examples of surfaces $z = z(x, y)$.

The Riemann metric of the space associated with the equation 2 has the form

$$\begin{aligned}\frac{1}{2}ds^2 &= \Big(\frac{\alpha}{3}\,zy - \frac{z}{3x} - \delta y^2\,\frac{t}{x} - \epsilon txy^4 + \beta tx^3y^4 + \beta tx^2y^3 + \gamma ty^3\Big)dx^2\\ &\quad + 2\Big(-\frac{z}{y} - \frac{\alpha}{3}ty + \frac{t}{3x}\Big)dxdy + \frac{t}{y}\,dy^2 + dxdz + dydt.\end{aligned}$$

The properties of a space with such a metric depend on the parameters α, β, γ δ, ϵ and may be very specifical when the Lorenz dynamical system has a strange attractor.

4. Symmetry, the Laplace-Beltrami equation, tetradic presentation

Let us consider the system of equations

$$\xi_{i,j} + \xi_{j,i} = 2\Gamma^k_{ij}\xi_k$$

for the Killing vectors of the metric (2.1). It has the form

$$\begin{aligned}
\xi_{1x} &= -a_3\xi_1 + a_4\xi_2 + (zA - ta_{4x})\xi_3 + (zE + tF)\xi_4,\\
\xi_{2y} &= -a_1\xi_1 + a_2\xi_2 + (zC + tD)\xi_3 + (za_{1y} - tH)\xi_4,\\
\xi_{1y} + \xi_{2x} &= 2[-a_2\xi_1 + a_3\xi_2 + (za_{3y} - tB)\xi_3 + (zG - ta_{2x})\xi_4,\\
\xi_{1z} + \xi_{3x} &= 2[a_3\xi_3 + a_2\xi_4], \qquad \xi_{1t} + \xi_{4x} = 2[-a_4\xi_3 - a_3\xi_4],\\
\xi_{2z} + \xi_{3y} &= 2[a_2\xi_3 + a_1\xi_4], \qquad \xi_{2t} + \xi_{4y} = -2[a_3\xi_3 - a_2\xi_4],\\
\xi_{3z} &= \xi_{4t} = 0.
\end{aligned}$$

In the particular case

$$\xi_3 = \xi_4 = 0, \quad \xi_i = \xi_i(x, y)$$

we get the system of equations

$$\xi_{1x} = -a_3\xi_1 + a_4\xi_2, \quad \xi_{2y} = -a_1\xi_1 + a_2\xi_2,$$

$$\xi_{1y} + \xi_{2x} = 2[-a_2\xi_1 + a_3\xi_2]$$

which is equivalent to the system for the $z = z(x, y)$ and $t = t(x, y)$

Remark. The Laplace-Beltrami operator

$$\Delta = g^{ij}\Big(\frac{\partial^2}{\partial x^i \partial x^j} - \Gamma^k_{ij}\frac{\partial}{\partial x^k}\Big)$$

can be used for investigating the properties of the metric (2.1). For example the equation

$$\Delta\Psi = 0$$

has the form

$$(ta_4 - za_3)\Psi_{zz} + 2(ta_3 - za_2)\Psi_{zt} + (ta_2 - za_1)\Psi_{tt} + \Psi_{xz} + \Psi_{yt} = 0.$$

Some solutions of this equation are connected with the geometry of the metric (2.1).

By substituting

$$\Psi = \exp(zA + tB)$$

into the equation

$$\Delta\Psi = 0,$$

we get the conditions

$$A = \Phi_y, \quad B = -\Phi_x,$$

and

$$\begin{aligned} a_4\Phi_y^2 - 2a_3\Phi_x\Phi_y + a_2\Phi_x^2 - \Phi_y\Phi_{xx} + \Phi_x\Phi_{xy} &= 0, \\ a_3\Phi_y^2 - 2a_2\Phi_x\Phi_y + a_1\Phi_x^2 - \Phi_y\Phi_{xy} + \Phi_x\Phi_{yy} &= 0. \end{aligned}$$

Another possible application for the study of the properties of a given Riemann space is connected with the computation of the heat invariants of the Laplace-Beltrami operator. For that aim the fundamental solution $K(\tau, x, y)$ of the heat equation

$$\frac{\partial \Psi}{\partial \tau} = g^{ij}\left(\frac{\partial^2 \Psi}{\partial x^i \partial x^j} - \Gamma^k_{ij}\frac{\partial \Psi}{\partial x^k}\right)$$

is considered. The diagonal of the function $K(\tau, x, y)$ has the following asymptotic expansion as $t \to 0_+$

$$K(\tau, x, x) \sim \sum_{n=0}^{\infty} a_n(x)\tau^{n-2}$$

where the coefficients $a_n(x)$ are local invariants (heat invariants) of the Riemann space D^4 with the metric (2.1).

In turn the eikonal equation

$$g^{ij}\frac{\partial F}{\partial x^i}\frac{\partial F}{\partial x^j} = 0$$

or

$$F_xF_z + F_yF_t - (ta_4 - za_3)F_zF_z - 2(ta_3 - za_2)F_zF_t - (ta_2 - za_1)F_tF_t = 0,$$

can also be used for investigating the properties of isotropic surfaces in the space with metric (2.1).

In particular, the solutions of the eikonal equation of the form

$$F = A(x,y)z^2 + B(x,y)zt + C(x,y)t^2 + D(x,y)z + E(x,y)t$$

lead to the following constraints for the coefficients

$$2AA_x + BA_y - a_1B^2 - 4a_2AB - 4a_3A^2 = 0,$$

$$2AB_x + BA_x + 2CA_y + BB_y - 4a_1BC - a_2(B^2 + 8AC) + 4a_4A^2 = 0,$$

$$2CB_y + BC_y + 2AC_x + BB_x - 4a_1C^2 + a_3(B^2 + 8AC) + 4a_4AB = 0,$$

$$2CC_y + BC_x + 4a_2C^2 + 4a_3BC + a_4B^2 = 0,$$

$$2AD_x + DA_x + EA_y + BD_y - 2a_1BE - 2a_2(BD + 2AE) - 4a_3AD = 0,$$

$$2CD_y + (BD)_x + 2AE_x + (BE)_y - 4a_1EC - 4a_2CD + 4a_3AE + 4a_4AD = 0,$$

$$2CE_y + CE_y + DC_x + BE_x - 4a_2CE + 2a_3(BE + 2CD) + 2a_4BD = 0,$$

$$DD_x + ED_y - a_1E^2 - 2a_2DE - a_3D^2 = 0,$$

$$EE_y + DE_x + a_2E^2 + 2a_3DE + a_4D^2 = 0$$

which may be useful for the theory of the equations (1.1).

Remark. The metric (2.1) has a tetradic presentation

$$g_{ij} = \omega_i^a \omega_j^b \eta_{ab}$$

where

$$\eta_{ab} = \begin{pmatrix} 0 & 0 & 1 & 0 \\ 0 & 0 & 0 & 1 \\ 1 & 0 & 0 & 0 \\ 0 & 1 & 0 & 0 \end{pmatrix}.$$

For example we get

$$ds^2 = 2\omega^1\omega^3 + 2\omega^2\omega^4$$

where

$$\omega^1 = dx + dy, \quad \omega^2 = dx + dy + \frac{1}{t(a_2 - a_4)}(dz - dt),$$

$$\omega^4 = -t(a_4dx + a_2dy), \quad \omega^3 = z(a_3dx + a_1dy) + \frac{1}{(a_2 - a_4)}(a_2dz - a_4dt).$$

and

$$a_1 + a_3 = 2a_2, \quad a_2 + a_4 = 2a_3.$$

Remark. Some of the equations on the curvature tensors in the space M^4 are connected with ODE's. For example, the equation

$$R_{ij;k} + R_{jk;i} + R_{ki;j} = 0$$

leads to the following conditions on the coefficients $a_i(x, y)$

$$\begin{aligned} \alpha''_x + 2a_3\alpha'' - 2a_4\alpha' &= 0, \\ \alpha_y + 2a_1\alpha' - 2a_2\alpha &= 0, \\ \alpha''_y + 2\alpha'_x + 4a_2\alpha'' - 2a_4\alpha - 2a_3\alpha' &= 0, \\ \alpha_x + 2\alpha'_y - 4a_3\alpha + 2a_2\alpha' + 2a_1\alpha'' &= 0. \end{aligned}$$

The solutions of this system give us the second-order equations associated to the space M^4 with a given condition on the Ricci tensor. The simplest examples are

$$y'' - \frac{3}{2y}y'^2 + y^3 = 0,\ y'' - \frac{3}{y}y'^2 + y^4 = 0,\ y'' + 3(2+y)y' + y^3 + 6y^2 - 16 = 0.$$

It is interesting to note that the above system is the Liouville system for geodesics in the Proposition 1.

The study of invariant conditions like

$$R_{ij;k} - R_{jk;i} = R^n_{ijk;n}, \quad \Box R_{ijkl} = 0, \quad \Box R_{ijkl;m} = 0$$

is also of interest for the theory of the equations (1.1).

Remark. The construction of the Riemannian extension of two-dimensional spaces connected with ODE's of type (1.1) can be generalized to the three-dimensional case by equations of the form

$$\begin{aligned}
\ddot{x} + A_1(\dot{x})^2 + 2A_2\dot{x}\dot{y} + 2A_3\dot{x}\dot{z} + A_4(\dot{y})^2 + 2A_5\dot{y}\dot{z} + A_6(\dot{z})^2 &= 0,\\
\ddot{y} + B_1(\dot{x})^2 + 2B_2\dot{x}\dot{y} + 2B_3\dot{x}\dot{z} + B_4(\dot{y})^2 + 2B_5\dot{y}\dot{z} + B_6(\dot{z})^2 &= 0,\\
\ddot{z} + C_1(\dot{x})^2 + 2C_2\dot{x}\dot{y} + 2C_3\dot{x}\dot{z} + C_4(\dot{y})^2 + 2C_5\dot{y}\dot{z} + C_6(\dot{z})^2 &= 0,
\end{aligned}$$

or

$$\begin{aligned}
& x'' + a_0 + a_1x' + a_2y' + a_3{x'}^2 + a_4x'y' + a_5{y'}^2\\
& + x'(b_0 + b_1x' + b_2y' + b_3{x'}^2 + b_4x'y' + b_5{y'}^2) = 0,\\
& y'' + c_0 + c_1x' + c_2y' + c_3{x'}^2 + c_4x'y' + c_5{y'}^2\\
& + y'(b_0 + b_1x' + b_2y' + b_3{x'}^2 + b_4x'y' + b_5{y'}^2) = 0,
\end{aligned}$$

where a_i, b_i, c_i are functions of variables x, y, z.

The corresponding expression for the 6-dimensional metric is:

$$\begin{aligned}
\frac{1}{2}ds^2 &= -(A_1u + B_1v + C_1w)dx^2 - 2(A_2u + B_2v + C_2w)dxdy\\
&\quad - 2(A_3u + B_3v + C_3w)dxdz - (A_4u + B_4v + C_4w)dy^2\\
&\quad - 2(A_5u + B_5v + C_5w)dydz - (A_6u + B_6v + C_6w)dz^2\\
&\quad + dxdu + dydv + dzdw
\end{aligned}$$

This gives us the possibility of studying the properties of such type of equations from a geometrical point of view.

Let us consider some examples.

5. The Riemann metrics of zero curvature and the KdV equation

Let us consider the system of matrix equations

$$\begin{aligned}
\frac{\partial \mathbf{\Gamma}_2}{\partial x} - \frac{\partial \mathbf{\Gamma}_1}{\partial y} + [\mathbf{\Gamma}_1, \mathbf{\Gamma}_2] &= 0, \\
\frac{\partial \mathbf{\Gamma}_3}{\partial x} - \frac{\partial \mathbf{\Gamma}_1}{\partial z} + [\mathbf{\Gamma}_1, \mathbf{\Gamma}_3] &= 0, \\
\frac{\partial \mathbf{\Gamma}_3}{\partial y} - \frac{\partial \mathbf{\Gamma}_2}{\partial z} + [\mathbf{\Gamma}_2, \mathbf{\Gamma}_3] &= 0,
\end{aligned} \tag{5.1}$$

where $\mathbf{\Gamma}_i(x, y, z)$ is the 3×3 matrix function with matrix elements $(\mathbf{\Gamma}_i)^k_j = \Gamma^k_{ij} = \Gamma^k_{ji}$.

This system can be regarded as the zero-curvature condition of some three-dimensional spaces with the affine connection given by the coefficients $\mathbf{\Gamma}(x, y, z)$.

If $\Gamma^k_{ij}(x, y, z)$ are of the form

$$\begin{aligned}
2\mathbf{\Gamma}_1 &= y^2 B_1(x, z) + y A_1(x, z) + C_1(x, z) + \frac{1}{y} D_1(x, z), \\
\mathbf{\Gamma}_3 &= y^2 B_3(x, z) + y A_3(x, z) + C_3(x, z) + \frac{1}{y} D_3(x, z) + \frac{1}{y^2} E_3(x, z), \\
\mathbf{\Gamma}_2 &= C_2(x, z) + \frac{1}{y} D_2(x, z),
\end{aligned}$$

then after substitution of these expressions in formulas (5.1) we get the system of nonlinear equations for the components of the affine connection. Some of these equations may be useful for applications.

Let us consider the space with metric

$$g_{ik} = \begin{pmatrix} y^2 & 0 & y^2 l(x, z) + m(x, z) \\ 0 & 0 & 1 \\ y^2 l + m & 1 & y^2 l^2 - 2y l_x + 2l\, m + 2n \end{pmatrix},$$

where l, m, n are functions of (x, z). By using the corresponding connection

$$\Gamma^k_{ij} = \frac{1}{2} g^{kl} (\partial_i g_{jl} + \partial_j g_{il} - \partial_l g_{ij})$$

we get the following matrices $\mathbf{\Gamma}_i$

$$\mathbf{\Gamma}_1 = \begin{pmatrix} yl + \frac{m}{y} & \frac{1}{y} & yl^2 + \frac{lm}{y} \\ -\gamma_1 & -\frac{m}{y} & -\gamma_2 \\ -y & 0 & -ly \end{pmatrix},$$

$$\mathbf{\Gamma}_2 = \begin{pmatrix} \frac{1}{y} & 0 & \frac{l}{y} \\ -\frac{m}{y} & 0 & -\frac{lm}{y} - l_x \\ 0 & 0 & 0 \end{pmatrix},$$

$$\mathbf{\Gamma}_3 = \begin{pmatrix} l^2 + \frac{m}{y} & \frac{l}{y} & \gamma_3 \\ \gamma_2 & -l_x - \frac{lm}{y} & -\gamma_4 \\ -yl & 0 & -yl^2 + l_x \end{pmatrix},$$

where

$$\begin{aligned}
\gamma_1 &= y^2 l_x - 2yn + m^2/y - m_x \\
\gamma_2 &= y^2 l l_x - 2yln + lm^2/y + yl_{xx} - ml_x - lm_x - n_x \\
\gamma_3 &= l_z + m_z/y^2 - 2ll_x + l_{xx}/y - 2ml_x/y^2 - lm_x/y^2 - n_x/y^2 + yl^3 \\
&\quad + l^2 m/y \\
\gamma_4 &= y^2 l^2 l_x + yll_{xx} - 2lml_x - l^2 m_x - ln_x + mm_z/y^2 + ml_{xx}/y \\
&\quad - 2m^2 l_x/y^2 - lmm_x/y^2 - mn_x/y^2 - 2yl_x^2 - 2ynl^2 + 2nl_x \\
&\quad + m^2 l^2/y + yl_{zx} - n_z.
\end{aligned}$$

If $l(x,z) = n(x,z)$ we get

$$\begin{aligned}
R_{1313} &= \Big(\frac{\partial^3 l}{\partial x^3} - 3l\frac{\partial l}{\partial x} + \frac{\partial l}{\partial z}\Big)y^2 + \Big(\frac{\partial^2 m}{\partial x \partial z} - 2m\frac{\partial^2 l}{\partial x^2} - l\frac{\partial^2 m}{\partial x^2} \\
&\quad - 3\frac{\partial m}{\partial x}\frac{\partial m}{\partial x} - \frac{\partial^2 l}{\partial x^2}\Big)y - m\frac{\partial m}{\partial z} + 2m^2\frac{\partial l}{\partial x} + m\frac{\partial l}{\partial x} \\
&\quad + ml\frac{\partial m}{\partial x} - m\frac{\partial m}{\partial z} + 2m^2\frac{\partial l}{\partial x} + m\frac{\partial l}{\partial x} + ml\frac{\partial m}{\partial x},
\end{aligned}$$

and

$$R_{1323} = \frac{1}{y}\Big(-\frac{\partial m}{\partial z} + 2m\frac{\partial l}{\partial x} + l\frac{\partial m}{\partial x} + \frac{\partial l}{\partial x}\Big)$$

From the condition $R_{ijkl} = 0$ it follows that the function $l(x,z)$ solves the KdV-equation

$$\frac{\partial^3 l}{\partial x^3} - 3l\frac{\partial l}{\partial x} + \frac{\partial l}{\partial z} = 0,$$

so that all flat metrics of such a type are determined from solutions of this equation.

Notice that starting from the Riemannian extensions of a space with a given metric the corresponding metric of the six-dimensional space can be written. The equations of the geodesics of such 6-dimensional space contains the linear second order ODE (Schrödinger operator) which applies for integrating the KdV equation.

6. Applications in the theory of general Relativity

The Riemann extensions of a given metric can be used for the study of Riemannian spaces satisfying the Einstein conditions

$$R_{ij} = g^{kl} R_{ijkl} = 0$$

on the curvature tensor R_{ijkl}.

Let us consider some examples. Let

$$ds^2 = -t^{2p_1} dx^2 - t^{2p_2} dy^2 - t^{2p_3} dz^2 + dt^2 \tag{6.1}$$

be the metric of the Kasner type in the classical theory of gravitation.

The Ricci tensor $\mathbf{R} \equiv (R_{ij})$ of this metric is

$$\mathbf{R} = \begin{pmatrix} \frac{p_1+p_2+p_3-1}{t^{2p_1-2}} & 0 & 0 & 0 \\ 0 & \frac{p_1+p_2+p_3-1}{t^{2p_2-2}} & 0 & 0 \\ 0 & 0 & \frac{p_3(p_1+p_2+p_3-1)}{t^{2p_3-2}} & 0 \\ 0 & 0 & 0 & \frac{p_1+p_2+p_3-p_1^2-p_2^2-p_3^2}{t^2} \end{pmatrix},$$

and in the particular case $\mathbf{R} = 0$ we get the well-known Kasner solution of the vacuum Einsten equations.

Now we shall consider the construction of Riemann extension for the metric (6.1). One gets the eight-dimensional space with local coordinates (x, y, z, t, P, Q, R, S) and the metric

$$ds^2 = -2\Gamma^k_{ij}\xi_k dx^i dx^j + 2dxdP + 2dydQ + 2dzdR + 2dtdS \tag{6.2}$$

were Γ^k_{ij} are the Christoffel coefficients of the metric (6.1) and $\xi_k = (P, Q, R, S)$.

They are given by:

$$\Gamma^4_{11} = p_1 t^{2p_1-1}, \quad \Gamma^4_{22} = p_2 t^{2p_2-1}, \quad \Gamma^4_{33} = p_3 t^{2p_3-1},$$
$$\Gamma^1_{14} = p_1/t, \quad \Gamma^2_{24} = p_2/t, \quad \Gamma^3_{34} = p_3/t.$$

As a result we find the metric of the space K^8 in the form

$$\begin{aligned} ds^2 &= -2p_1 t^{2p_1-1} S dx^2 - 2p_2 t^{2p_2-1} S dy^2 - 2p_3 t^{2p_3-1} S dz^2 \\ &\quad -4p_1/tPdxdt - 4p_2/tQdydt - 4p_3/tRdzdt + 2dxdP \\ &\quad +2dydQ + 2dzdR + 2dtdS. \end{aligned}$$

The nonzero components of the Ricci tensor ${}^8R_{ij}$ are

$$R_{11} = 2p_1 t^{2p_1-2}(p_1 + p_2 + p_3 - 1), \quad R_{22} = 2p_2 t^{2p_2-2}(p_2 + p_2 + p_3 - 1),$$

$$R_{33} = 2p_3 t^{2p_3-2}(p_2 + p_2 + p_3 - 1), \quad R_{44} = 2(p_2 + p_2 + p_3 - p_1^2 - p_2^2 - p_3^2)/t^2$$

which coincide with the components of the Ricci tensor ${}^4R_{ij}$ of the space K^4.

Therefore the geometry of the Riemann space before and after extension is the same.

In turn the equations of the geodesics of the extended space

$$\frac{d^2x}{ds^2}+2\frac{p_1}{t}\frac{dx}{ds}\frac{dt}{ds}=0,\quad \frac{d^2y}{ds^2}+2\frac{p_2}{t}\frac{dy}{ds}\frac{dt}{ds}=0,\quad \frac{d^2z}{ds^2}+2\frac{p_3}{t}\frac{dz}{ds}\frac{dt}{ds}=0$$

$$\frac{d^2t}{ds^2}+p_1t^{2p_1-1}\left(\frac{dx}{ds}\right)^2+p_2t^{2p_2-1}\left(\frac{dy}{ds}\right)^2+p_3t^{2p_3-1}\left(\frac{dz}{ds}\right)^2=0,$$

$$\frac{d^2R}{ds^2}-2\frac{p_3}{t}\frac{dt}{ds}\frac{dR}{ds}-2p_3t^{2p_3-1}\frac{dz}{ds}\frac{dS}{ds}+\left[2\frac{p_1p_3t^{2p_1-1}}{t}\left(\frac{dx}{ds}\right)^2\right.$$
$$\left.+2\frac{p_2p_3t^{2p_2-1}}{t}\left(\frac{dy}{ds}\right)^2+2\frac{p_3^2t^{2p_3-1}}{t}\left(\frac{dz}{ds}\right)^2+2\frac{p_3}{t^2}\left(\frac{dt}{ds}\right)^2\right]R$$
$$+2\frac{p_3t^{2p_3-1}}{t}\frac{dz}{ds}\frac{dt}{ds}S=0,$$

$$\frac{d^2Q}{ds^2}-2\frac{p_2}{t}\frac{dt}{ds}\frac{dQ}{ds}-2p_2t^{2p_2-1}\frac{dy}{ds}\frac{dS}{ds}+\left[2\frac{p_1p_2t^{2p_1-1}}{t}\left(\frac{dx}{ds}\right)^2\right.$$
$$\left.+2\frac{p_2p_3t^{2p_3-1}}{t}\left(\frac{dz}{ds}\right)^2+2\frac{p_2^2t^{2p_2-1}}{t}\left(\frac{dy}{ds}\right)^2+2\frac{p_2}{t^2}\left(\frac{dt}{ds}\right)^2\right]Q$$
$$+2\frac{p_2t^{2p_2-1}}{t}\frac{dy}{ds}\frac{dt}{ds}S=0,$$

$$\frac{d^2P}{ds^2}-2\frac{p_1}{t}\frac{dt}{ds}\frac{dP}{ds}-2p_1t^{2p_1-1}\frac{dx}{ds}\frac{dS}{ds}+\left[2\frac{p_1p_3t^{2p_1-1}}{t}\left(\frac{dz}{ds}\right)^2\right.$$
$$\left.+2\frac{p_2p_1t^{2p_2-1}}{t}\left(\frac{dy}{ds}\right)^2+2\frac{p_1^2t^{2p_1-1}}{t}\left(\frac{dx}{ds}\right)^2+2\frac{p_1}{t^2}\left(\frac{dt}{ds}\right)^2\right]P$$
$$+2\frac{p_1t^{2p_1-1}}{t}\frac{dx}{ds}\frac{dt}{ds}S=0,$$

$$\frac{d^2S}{ds^2}-2\frac{p_3}{t}\frac{dz}{ds}\frac{dR}{ds}-2\frac{p_2}{t}\frac{dy}{ds}\frac{dQ}{ds}-2\frac{p_1}{t}\frac{dx}{ds}\frac{dP}{ds}+\frac{4p_2^2}{t^2}\frac{dy}{ds}\frac{dt}{ds}Q$$
$$+\frac{4p_3^2}{t}\frac{dz}{ds}\frac{dt}{ds}R+\left[\frac{p_1(2p_1-1)t^{2p_1-1}}{t}\left(\frac{dx}{ds}\right)^2\right.$$
$$\left.+\frac{p_2(2p_2-1)t^{2p_2-1}}{t}\left(\frac{dy}{ds}\right)^2+\frac{p_3(2p_3-1)t^{2p_3-1}}{t}\left(\frac{dz}{ds}\right)^2\right]S=0,$$

contain the linear 4×4 matrix system of second-order ODE's for the additional coordinates (P,Q,R,S)

$$\frac{d^2\Psi}{ds^2}=A(x,y,z,t)\frac{d\Psi}{ds}+B(x,y,z,t)\Psi.$$

Here A, B are 4×4 matrix functions depending on the coordinates (x, y, z, t). This fact allows us to use the methods of soliton theory for integrating the full system of geodesics and its corresponding Einstein equations.

Observe that the signature of the space ${}^8D$ is 0, i.e. it has the form $(++++----)$. From this it follows that starting from the Riemann space with the Lorentz signature $(---+)$ and after the extension, we get the additional subspace with local coordinates P, Q, R, S having the signature $(-+++)$.

Remark. For the Schwarzschild metrics

$$g_{ij} = \begin{pmatrix} -\frac{1}{1-m/x} & 0 & 0 & 0 \\ 0 & -x^2 & 0 & 0 \\ 0 & 0 & -x^2\sin^2 y & 0 \\ 0 & 0 & 0 & 1-\frac{m}{x} \end{pmatrix}$$

the Christoffell coefficients are

$$\Gamma^1_{11} = \tfrac{m}{2x(x+m)}, \quad \Gamma^1_{22} = -(x+m), \quad \Gamma^1_{33} = -(x+m)\sin^2 y,$$
$$\Gamma^1_{44} = -\tfrac{(x+m)m}{2x^3}, \quad \Gamma^2_{12} = \tfrac{1}{x}, \quad \Gamma^2_{33} = -\sin y\cos y,$$
$$\Gamma^3_{13} = \tfrac{1}{x}, \quad \Gamma^3_{23} = \tfrac{\cos y}{\sin y}, \quad \Gamma^4_{14} = -\tfrac{m}{2x(x+m)}.$$

The corresponding system (3.1) for the surfaces of translations $x(u, v)$, $y(u, v)$, $z(u, v)$, $t(u, v)$ is nonlinear. After the extension and with the help of the new coordinates (P, Q, R, S) we get the S^8 space with metric ds^2 given by

$$\begin{aligned} -\frac{1}{2}\,ds^2 &= \Gamma^1_{11}P dx^2 + \Gamma^1_{22}P dy^2 + \Gamma^1_{33}P dz^2 + \Gamma^1_{44}P dt^2 \\ &\quad + \Gamma^2_{33}Q dz^2 + 2\Gamma^2_{12}Q dx\,dy + 2\Gamma^3_{13}R\,dx\,dz \\ &\quad + 2\Gamma^3_{23}R\,dx\,dy\,dz + 2\Gamma^4_{14}S\,dx\,dt. \end{aligned}$$

7. Anti-Self-Dual Kahler metrics and second order ODE's

Here we discuss the relations of the equations (1.1) with the theory of ASD-Kahler spaces [27].

It is known that all ASD null Kahler metrics are locally given by

$$ds^2 = -\Theta_{tt}dx^2 + 2\Theta_{zt}dxdy - \Theta_{zz}dy^2 + dxdz + dydt$$

where the function $\Theta(x, y, z, t)$ is the solution of the equation

$$\Theta_{xz} + \Theta_{yt} + \Theta_{zz}\Theta_{tt} - \Theta^2_{zt} = \Lambda(x, y, z, t),$$

$$\Lambda_{xz} + \Lambda_{yt} + \Theta_{tt}\Lambda_{zz} + \Theta_{zz}\Lambda_{tt} - 2\Theta_{zt}\Lambda_{tz} = 0.$$

This system of equations has the solution

$$\Theta = -\frac{1}{6}a_1(x,y)z^3 + \frac{1}{2}a_1(x,y)z^2t - \frac{1}{2}a_3(x,y)zt^2 + \frac{1}{6}a_4(x,y)t^3$$

which leads to the metric

$$ds^2 = 2(za_3 - ta_4)dx^2 + 4(za_2 - ta_3)dxdy + 2(za_1 - ta_2)dy^2 + 2dxdz + 2dydt$$

with geodesics determined by the equation

$$y'' + a_1(x,y)y'^3 + 3a_2(x,y)y'^2 + 3a_3(x,y)y' + a_4(x,y) = 0 \tag{7.1}$$

In this case the coefficients $a_i(x,y)$ are not arbitrary but satisfy the conditions

$$L_1 = \frac{\partial}{\partial y}(a_{4y} + 3a_2a_4) - \frac{\partial}{\partial x}(2a_{3y} - a_{2x} + a_1a_4) - 3a_3(2a_{3y} - a_{2x}) - a_4a_{1x} = 0,$$

$$L_2 = \frac{\partial}{\partial x}(a_{1x} - 3a_1a_3) + \frac{\partial}{\partial y}(a_{3y} - 2a_{2x} + a_1a_4) - 3a_2(a_{3y} - 2a_{2x}) + a_1a_{4y} = 0.$$

According to the Liouville theory this means that these equations can be transformed into the equation

$$y'' = 0$$

by means of point transformations.

Notice that the conditions $L_1 = 0$, $L_2 = 0$ are connected with an integrable nonlinear p.d.e. (as the equation (2.4) for example) and from this we can get a lot of examples of ASD–spaces.

8. Dual equations and the Einstein–Weyl geometry in the theory of second order ODE's

In the theory of second-order ODE's

$$y'' = f(x,y,y')$$

we have the following fundamental diagram:

$$\begin{array}{ccccc} & & F(x,y,a,b) = 0 & & \\ & \swarrow\nearrow & & \searrow\nwarrow & \\ y'' = f(x,y,y') & & & & b'' = g(a,b,b') \\ \Updownarrow & & & & \Updownarrow \\ M^3(x,y,y') & & \Longleftrightarrow & & N^3(a,b,b') \end{array}$$

which shows the relations between a given second-order ODE $y'' = f(x, y, y')$, its general integral $F(x, y, a, b) = 0$ and the so-called dual equation $b'' = g(a, b, b')$ which can be obtained from the general integral when the variables x and y are considered as parameters.

In particular for the equations of type (1.1) the dual equation

$$b'' = g(a, b, b') \tag{8.1}$$

it follows that the function $g(a, b, b')$ satisfies the partial differential equation

$$\begin{aligned} g_{aacc} + 2cg_{abcc} + 2gg_{accc} + c^2 g_{bbcc} + 2cgg_{bccc} + g^2 g_{cccc} \\ + (g_a + cg_b)g_{ccc} - 4g_{abc} - 4cg_{bbc} - cg_c g_{bcc} - 3gg_{bcc} \\ - g_c g_{acc} + 4g_c g_{bc} - 3g_b g_{cc} + 6g_{bb} = 0. \end{aligned} \tag{8.2}$$

According to E. Cartan the expressions for the curvature of the space of linear elements $(x,\ y,\ y')$ connected with equation (1.1) is

$$\Omega^1_2 = a[\omega^2 \wedge \omega^2_1], \quad \Omega^0_1 = b[\omega^1 \wedge \omega^2], \quad \Omega^0_2 = h[\omega^1 \wedge \omega^2] + k[\omega^2 \wedge \omega^2_1]$$

where:

$$a = -\frac{1}{6}\frac{\partial^4 f}{\partial y'^4}, \quad h = \frac{\partial b}{\partial y'}, \quad k = -\frac{\partial \mu}{\partial y'} - \frac{1}{6}\frac{\partial^2 f}{\partial^2 y'}\frac{\partial^3 f}{\partial^3 y'},$$

and

$$\begin{aligned} 6b = {} & f_{xxy'y'} + 2y' f_{xyy'y'} + 2f f_{xy'y'y'} + y'^2 f_{yyy'y'} + 2y' f f_{yy'y'y'} \\ & + f^2 f_{y'y'y'y'} + (f_x + y' f_y) f_{y'y'y'} - 4f_{xyy'} - 4y' f_{yyy'} - y' f_{y'} f_{yy'y'} \\ & - 3f f_{yy'y'} - f_{y'} f_{xy'y'} + 4f_{y'} f_{yy'} - 3f_y f_{y'y'} + 6f_{yy}\,. \end{aligned}$$

Two types of equations arise in a natural way: the first type derives from the condition $a = 0$ and the second type from the condition $b = 0$.

The first condition $a = 0$ leads to the equation in form (1.1) and the second condition leads to the equations (8.1), where the function $g(a, b, b')$ satisfies the above p.d.e..

E. Cartan also proved that the Einstein–Weyl 3-manifolds parametrize the families of curves of equation (8.1) which are dual to the equation (1.1).

Some examples of solutions of equation (8.1) were first obtained in [2]. For instance if we take

$$g = a^{-\gamma} A(ca^{\gamma-1}),$$

we get the equation

$$\begin{aligned} [A + (\gamma - 1)\xi]^2 A^{IV} + 3(\gamma - 2)[A + (\gamma - 1)\xi]A^{III} \\ + (2 - \gamma)A^I A^{II} + (\gamma^2 - 5\gamma + 6)A^{II} = 0. \end{aligned}$$

One solution of this equation is

$$A = (2-\gamma)[\xi(1+\xi^2) + (1+\xi^2)^{3/2}] + (1-\gamma)\xi,$$

which corresponds to the equation

$$b'' = \frac{1}{a}[b'(1+b'^2) + (1+b'^2)^{3/2}]$$

with general integral

$$F(x,y,a,b) = (y+b)^2 + a^2 - 2ax = 0$$

In this case the dual equation has the form

$$y'' = -\frac{1}{2x}(y'^3 + y')$$

Remark. For more general classes of form-invariant equations the dual equation is introduced in an analogous way.

For example, for the form-invariant equation

$$P_n(b')b'' - P_{n+3}(b') = 0,$$

where $P_n(b')$ is a polynomial of degree n in b' with coefficients depending on the variables a, b, the dual equation

$$b'' = g(a,b,b')$$

implies

$$\begin{vmatrix} \psi_{n+4} & \psi_{n+3} & \cdots & \psi_4 \\ \psi_{n+5} & \psi_{n+4} & \cdots & \psi_5 \\ \vdots & \vdots & & \vdots \\ \psi_{2n+4} & \psi_{2n+3} & \cdots & \psi_{n+4} \end{vmatrix} = 0$$

where the functions ψ_i are determined with the help of the relations

$$\begin{aligned} 4!\,\psi_4 &= -\frac{d^2}{da^2}g_{cc} + 4\frac{d}{da}g_{bc} - g_c\Big(4g_{bc} - \frac{d}{da}g_{cc}\Big) + 3g_b g_{cc} - 6g_{bb}, \\ i\,\psi_i &= \frac{d}{da}\psi_{i-1} - (i-3)g_c\psi_{i-1} + (i-5)g_b\psi_{i-2}, \quad i > 4 \end{aligned}$$

For example, the equation

$$2yy'' - y'^4 - y'^2 = 0$$

with solution

$$x = a(t + \sin t) + b, \quad y = a(1 - \cos t)$$

has a dual equation

$$b'' = -\frac{1}{a}\tan(b'/2).$$

According to the above formulas, at $n = 1$ we get

$$\begin{aligned}
4!\,\psi_4 &= \frac{3}{2a^3}\tan\frac{c}{2}\Big(1+\tan^2\frac{c}{2}\Big)^3,\\
5!\,\psi_5 &= -\frac{15}{4a^4}\tan\frac{c}{2}\Big(1+\tan^2\frac{c}{2}\Big)^4,\\
6!\,\psi_6 &= \frac{90}{8a^5}\tan\frac{c}{2}\Big(1+\tan^2\frac{c}{2}\Big)^5,
\end{aligned}$$

and the relation

$$\begin{vmatrix} \psi_5 & \psi_4 \\ \psi_6 & \psi_5 \end{vmatrix} = 0$$

is satisfied.

We consider next some properties of the Einstein–Weyl spaces [15].

A Weyl space is a smooth manifold equipped with a conformal metric $g_{ij}(x)$, and a symmetric connection

$$G^k_{ij} = \Gamma^k_{ij} - \frac{1}{2}(\omega_i\delta^k_j + \omega_j\delta^k_i - \omega_l g^{kl} g_{ij})$$

satisfying a condition of the form

$$D_i g_{kj} = \omega_i g_{kj}$$

where $\omega_i(x)$ are the components of a given vector field.

The Weyl connection G^k_{ij} has a curvature tensor W^i_{jkl} and a Ricci tensor W^i_{jil} which is not symmetrical $W^i_{jil} \neq W^i_{lij}$ in general.

A Weyl space satisfying the Einstein condition

$$\frac{1}{2}(W_{jl} + W_{lj}) = \lambda(x) g_{jl}(x),$$

for some function $\lambda(x)$, is called an Einstein–Weyl space.

Let us consider some examples.

1. The components of the Weyl connection of the three-dimensional flat space:

$$ds^2 = dx^2 + dy^2 + dz^2$$

are

$$2G_1 = \begin{pmatrix} -\omega_1 & -\omega_2 & -\omega_3 \\ \omega_2 & -\omega_1 & 0 \\ \omega_3 & 0 & -\omega_1 \end{pmatrix}, \quad 2G_2 = \begin{pmatrix} -\omega_2 & \omega_1 & 0 \\ -\omega_1 & -\omega_2 & -\omega_3 \\ 0 & \omega_3 & -\omega_2 \end{pmatrix},$$

$$2G_3 = \begin{pmatrix} -\omega_3 & 0 & \omega_1 \\ 0 & -\omega_3 & \omega_2 \\ -\omega_1 & -\omega_2 & -\omega_3 \end{pmatrix}.$$

From the equations of the Einstein–Weyl spaces

$$W_{[ij]} = \frac{W_{ij} + W_{ji}}{2} = \lambda g_{ij}$$

we get the system of equations

$$\omega_{3x} + \omega_{1z} + \omega_1\omega_3 = 0, \quad \omega_{3y} + \omega_{2z} + \omega_2\omega_3 = 0, \quad \omega_{2x} + \omega_{1y} + \omega_1\omega_2 = 0,$$

$$2\omega_{1x} + \omega_{2y} + \omega_{3z} - \frac{\omega_2^2 + \omega_3^2}{2} = 2\lambda, \quad 2\omega_{2y} + \omega_{1x} + \omega_{3z} - \frac{\omega_1^2 + \omega_3^2}{2} = 2\lambda,$$

$$2\omega_{3z} + \omega_{2y} + \omega_{1x} - \frac{\omega_1^2 + \omega_2^2}{2} = 2\lambda.$$

Notice that the first three equations lead to the Chazy equation [16]

$$R''' + 2RR'' - 3R'^2 = 0$$

for the function

$$R = R(x + y + z) = \omega_1 + \omega_2 + \omega_3$$

where $\omega_i = \omega_i(x + y + z)$.

The Einstein–Weyl geometry of the metric $g_{ij} = \operatorname{diag}(1, -e^U, -e^U)$ and vector $\omega_i = (2U_z, 0, 0)$ is determined by the solutions of equation [17]

$$U_{xx} + U_{yy} = (e^U)_{zz}.$$

After substituting $U = U(x + y = \tau, z)$ this equation becomes

$$U_\tau = (e^{U/2})_z.$$

The consideration of the E–W structure for the metrics

$$ds^2 = dy^2 - 4dxdt - 4U(x, y, t)dt^2$$

leads to the dispersionless KP equation [18]

$$(U_t - UU_x)_x = U_{yy}.$$

2. For the four-dimensional Minkovskii space

$$ds^2 = dx^2 + dy^2 + dz^2 - dt^2,$$

the components of the Weyl connection are

$$2G_1 = \begin{pmatrix} -\omega_1 & -\omega_2 & -\omega_3 & -\omega_4 \\ \omega_2 & -\omega_1 & 0 & 0 \\ \omega_3 & 0 & -\omega_1 & 0 \\ -\omega_4 & 0 & 0 & -\omega_1 \end{pmatrix}, \quad 2G_2 = \begin{pmatrix} -\omega_2 & \omega_1 & 0 & 0 \\ -\omega_1 & -\omega_2 & -\omega_3 & -\omega_4 \\ 0 & \omega_3 & -\omega_2 & 0 \\ 0 & -\omega_4 & 0 & -\omega_2 \end{pmatrix},$$

$$2G_3 = \begin{pmatrix} -\omega_3 & 0 & \omega_1 & 0 \\ 0 & -\omega_3 & \omega_2 & 0 \\ -\omega_1 & -\omega_2 & -\omega_3 & -\omega_4 \\ 0 & 0 & -\omega_4 & -\omega_3 \end{pmatrix}, \quad 2G_4 = \begin{pmatrix} -\omega_4 & 0 & 0 & \omega_1 \\ 0 & -\omega_4 & 0 & -\omega_2 \\ 0 & 0 & -\omega_4 & -\omega_3 \\ -\omega_1 & -\omega_2 & -\omega_3 & -\omega_4 \end{pmatrix}.$$

The Einstein–Weyl condition

$$W_{[ij]} = \frac{W_{ij} + W_{ji}}{2} = \lambda g_{ij}$$

where

$$W_{ij} = W^l_{ilj}$$

and

$$W^k_{ilj} = \frac{\partial G^k_{ij}}{\partial x^l} - \frac{\partial G^k_{il}}{\partial x^j} + G^k_{in} G^n_{lj} - G^k_{jn} G^n_{il}$$

leads to the system of equations

$$\begin{aligned} \omega_{3x} + \omega_{1z} + \omega_1\omega_3 = 0, \quad \omega_{3y} + \omega_{2z} + \omega_2\omega_3 &= 0, \\ \omega_{2x} + \omega_{1y} + \omega_1\omega_2 = 0, \quad \omega_{4x} + \omega_{1t} + \omega_1\omega_4 &= 0, \\ \omega_{4y} + \omega_{2t} + \omega_2\omega_4 = 0, \quad \omega_{4z} + \omega_{3t} + \omega_3\omega_4 &= 0, \\ 3\omega_{1x} + \omega_{2y} + \omega_{3z} - \omega_{4t} + \omega_4^2 - \omega_2^2 - \omega_3^2 &= 2\lambda, \\ 3\omega_{2y} + \omega_{1x} + \omega_{3z} - \omega_{4t} + \omega_4^2 - \omega_1^2 - \omega_3^2 &= 2\lambda, \\ 3\omega_{3z} + \omega_{2y} + \omega_{1x} - \omega_{4t} + \omega_4^3 - \omega_1^2 - \omega_2^2 &= 2\lambda, \\ 3\omega_{4t} - \omega_{2y} - \omega_{1x} - \omega_{3z} + \omega_3^2 + \omega_1^2 + \omega_2^2 &= 2\lambda. \end{aligned}$$

9. On the solutions of dual equations

The equation (3.2) can be rewritten in the compact form

$$\frac{d^2 g_{cc}}{da^2} - g_c \frac{dg_{cc}}{da} - 4\frac{dg_{bc}}{da} + 4g_c g_{bc} - 3g_b g_{cc} + 6g_{bb} = 0 \tag{9.1}$$

where

$$\frac{d}{da} = \frac{\partial}{\partial a} + c\frac{\partial}{\partial b} + g\frac{\partial}{\partial c}.$$

It admits many types of the reductions, the simplest of them being

$$\begin{gathered} g = c^\alpha \omega[ac^{\alpha-1}], \quad g = c^\alpha \omega[bc^{\alpha-2}], \quad g = c^\alpha \omega[ac^{\alpha-1}, bc^{\alpha-2}], \\ g = a^{-\alpha}\omega[ca^{\alpha-1}], \quad g = b^{1-2\alpha}\omega[cb^{\alpha-1}], \quad g = a^{-1}\omega(c - b/a), \\ g = a^{-3}\omega[b/a, b - ac], \quad g = a^{\beta/\alpha-2}\omega[b^\alpha/a^\beta, c^\alpha/a^{\beta-\alpha}]. \end{gathered}$$

The corresponding equations can be integrated in some particular cases. For instance, let us assume that

$$g = g(a, c)\,.$$

From the condition (9.1) we get

$$\frac{d^2 g_{cc}}{da^2} - g_c \frac{dg_{cc}}{da} = 0 \tag{9.2}$$

where

$$\frac{d}{da} = \frac{\partial}{\partial a} + g\frac{\partial}{\partial c}.$$

By inserting the relation

$$g_{ac} = -g g_{cc} + \chi(g_c)$$

in (9.2), we get the following equation for $\chi(\xi)$, $\xi = g_c$

$$\chi(\chi'' - 1) + (\chi' - \xi)^2 = 0.$$

It has the solutions

$$\chi = \frac{1}{2}\xi^2, \quad \chi = \frac{1}{3}\xi^2$$

So we get two reductions of the equation (9.1) given by

$$g_{ac} + g g_{cc} - \frac{g_c^2}{2} = 0$$

and

$$g_{ac} + g g_{cc} - \frac{g_c^2}{3} = 0.$$

Remark. The first reduction of equation (9.1) follows from its representation in the form

$$\begin{aligned} g_{ac} + g g_{cc} - \frac{1}{2}{g_c}^2 + c g_{bc} - 2g_b &= h, \\ h_{ac} + g h_{cc} - g_c h_c + c h_{bc} - 3h_b &= 0 \end{aligned}$$

and was already considered in [3].

In the particular case $h = 0$ we get

$$g_{ac} + g g_{cc} - \frac{1}{2} g_c^2 + c g_{bc} - 2g_b = 0$$

which is the equation (9.1) for the function $g = g(a, c)$. It can be integrated by means of a Legendre transformation (see [3]).

The solutions of the equation

$$u_{xy} = uu_{xx} + \epsilon u_x^2$$

were constructed in [19]. In [20] it was showed that they can be written as

$$u = B'(y) + \int [A(z) - \epsilon y]^{(1-\epsilon)/\epsilon} dz,$$

$$x = -B(y) + \int [A(z) - \epsilon y]^{1/\epsilon} dz.$$

To integrate the above equations we apply the parametric representation

$$g = A(a) + U(a, \tau), \quad c = B(a) + V(a, \tau).$$

By using the formulas

$$g_c = \frac{g_\tau}{c_\tau}, \quad g_a = g_a + g_\tau \tau_a$$

and after substitution in (9.1), we get the conditions

$$A(a) = \frac{dB}{da}$$

and

$$U_{a\tau} - \left(\frac{V_a U_\tau}{V_\tau}\right)_\tau + U\left(\frac{U_\tau}{V_\tau}\right)_\tau - \frac{1}{2}\frac{U_\tau^2}{V_\tau} = 0.$$

So we obtain one equation for two functions $U(a, \tau)$ and $V(a, \tau)$. Any solution of this equation determines a solution of equation (9.1).

Let us consider the example

$$A = B = 0, \quad U = 2\tau - \frac{a\tau^2}{2}, \quad V = a\tau - 2\log(\tau).$$

By using the representation

$$U = \tau\omega_\tau - \omega, \quad V = \omega_\tau$$

it is possible to obtain other solutions of this equation.

The problem of integration of the dual equation with a right-hand side $g = g(a, b')$ depending on the two variables a and b' was solved in work [28]. Here we present the construction of such type of solutions.

Proposition 9.1 *If $h \neq 0$ and $g = g(a, c)$, the equation (9.1) is equivalent to the equation*

$$\Theta_a \left(\frac{\Theta_a}{\Theta_c}\right)_{ccc} - \Theta_c \left(\frac{\Theta_a}{\Theta_c}\right)_{acc} = 1, \tag{9.3}$$

where

$$g = -\frac{\Theta_a}{\Theta_c}, \quad h_c = \frac{1}{\Theta_c}.$$

To integrate this equation we use the representation

$$c = \Omega(\Theta, a).$$

From the relations

$$1 = \Omega_\Theta \Theta_c, \quad 0 = \Omega_\Theta \Theta_a + \Omega_c$$

we obtain

$$\Theta_c = \frac{1}{\Omega_\Theta}, \quad \Theta_a = -\frac{\Omega_a}{\Omega_\Theta}$$

and

$$\frac{\Omega_a}{\Omega_\Theta}(\Omega_a)_{ccc} + \frac{1}{\Omega_\Theta}(\Omega_a)_{cca} = 1$$

Now we get

$$\Omega_{ac} = \frac{\Omega_{a\Theta}}{\Omega_\Theta} = (\log \Omega_\Theta)_a = K, \quad \Omega_{acc} = \frac{K_\Theta}{\Omega_\Theta}$$

$$\Omega_{accc} = \left(\frac{K_\Theta}{\Omega_\Theta}\right)_\Theta \frac{1}{\Omega_\Theta}, \quad (\Omega_{acc})_a = \left(\frac{K_\Theta}{\Omega_\Theta}\right)_a - \frac{\Omega_a}{\Omega_\Theta}\left(\frac{K_\Theta}{\Omega_\Theta}\right)_\Theta.$$

As a result the equation (9.3) takes the form

$$\left[\frac{(\log \Omega_\Theta)_{a\Theta}}{\Omega_\Theta}\right]_a = \Omega_\Theta$$

and can be integrated with the help of the substitution

$$\Omega(\Theta, a) = \Lambda_a.$$

So we get the following Abel type equation for the function Λ_Θ

$$\Lambda_{\Theta\Theta} = \frac{1}{6}\Lambda_\Theta^3 + \alpha(\Theta)\Lambda_\Theta^2 + \beta(\Theta)\Lambda(\Theta) + \gamma(\Theta) \tag{9.4}$$

with arbitrary coefficients α, β, γ.

As an example, let us discuss the case.

$$\alpha = \beta = \gamma = 0.$$

The solution of equation (9.4) is

$$\Lambda = A(a) - 6\sqrt{B(a) - \frac{1}{3}\Theta}$$

and we get

$$c = A' - \frac{3B'}{\sqrt{B - \frac{1}{3}\Theta}}$$

or

$$\Theta = 3B - 27\frac{B'^2}{(c - A')^2}$$

This solution corresponds to the equation

$$b'' = -\frac{\Theta_a}{\Theta_c} = -\frac{1}{18B'}b'^3 + \frac{A'}{6B'}b'^2 + \left(\frac{B''}{B'} - \frac{A'^2}{6B'}\right)b' + A'' + \frac{A'^3}{18B'} - \frac{A'B''}{B'}$$

with a cubic dependence on the first derivative b' and arbitrary coefficients $A(a)$, $B(a)$. This equation is equivalent to the equation

$$b'' = 0$$

under application of appropriate point transformations.

In fact from the formulas

$$L_1 = \frac{\partial}{\partial y}(a_{4y} + 3a_2a_4) - \frac{\partial}{\partial x}(2a_{3y} - a_{2x} + a_1a_4) - 3a_3(2a_{3y} - a_{2x}) - a_4a_{1x},$$

$$L_2 = \frac{\partial}{\partial x}(a_{1x} - 3a_1a_3) + \frac{\partial}{\partial y}(a_{3y} - 2a_{2x} + a_1a_4) - 3a_2(a_{3y} - 2a_{2x}) + a_1a_{4y},$$

derived from the components of a projective curvature of the space of linear elements for

$$y'' + a_1(x,y)y'^3 + 3a_2(x,y)y'^2 + 3a_3(x,y)y' + a_4(x,y) = 0,$$

we obtain

$$a_1(x,y) = \frac{1}{18B'}, \quad a_2(x,y) = -\frac{A'}{18B'},$$
$$a_3(x,y) = \frac{A'^2}{18B'} - \frac{B''}{3B'}, \quad a_4(x,y) = \frac{A'B''}{B'} - \frac{A'^3}{18B'} - A'',$$

as well as the conditions

$$L_1 = 0, \quad L_2 = 0.$$

This means that our equation is determined by a projective flat structure in the space of elements (x, y, y').

Remark. The conditions

$$L_1 = 0, \quad L_2 = 0$$

correspond to the solutions of the equation (1.3) of the form

$$g(a,b,b') = A(a,b)b'^3 + 3B(a,b)b'^2 + 3C(a,b)b' + D(a,b).$$

10. Third-order ODE's and Weyl geometry

The geometry of the equation

$$b''' = g(a, b, b', b'')$$

with a general integral of the form

$$F(a, b, X, Y, Z) = 0,$$

was studied by E. Cartan.

It was showed that there are a lot of types of geometrical structures connected with this type of equations.

More recently [21, 22] the geometry of third-order ODE's was considered in the context of the null-surface formalism and it was discovered that the function $g(a, b, b', b'')$ satisfies the conditions:

$$\frac{d^2 g_r}{da^2} - 2g_r \frac{dg_r}{da} - 3\frac{dg_c}{da} + \frac{4}{9}g_r^3 + 2g_c g_r + 6g_b = 0, \quad (10.1)$$

$$\frac{d^2 g_{rr}}{da^2} - \frac{dg_{cr}}{da} + g_{br} = 0, \quad (10.2)$$

where

$$\frac{d}{da} = \frac{\partial}{\partial a} + c\frac{\partial}{\partial b} + r\frac{\partial}{\partial c} + g\frac{\partial}{\partial r}.$$

We present here some solutions of the equations (21,22) which are connected with the theory of second order ODE's.

Following the notations of E. Cartan we study the third-order differential equations

$$y''' = F(x, y, y', y'')$$

where the function F satisfies the system of equations

$$\frac{d^2 F_2}{dx^2} - 2F_2\frac{dF_2}{dx} - 3\frac{dF_1}{dx} + \frac{4}{9}F_2^3 + 2F_1F_2 + 6F_0 = 0 \quad (10.3)$$

$$\frac{d^2 F_{22}}{dx^2} - \frac{dF_{12}}{dx} + F_{02} = 0 \quad (10.4)$$

where $F_k \equiv \partial F / \partial y^{(k)}$ and

$$\frac{d}{dx} = \frac{\partial}{\partial x} + y'\frac{\partial}{\partial y} + y''\frac{\partial}{\partial y'} + F\frac{\partial}{\partial y''}.$$

In particular the third order equation

$$y''' = \frac{3y'y''^2}{1 + y'^2}$$

of all cycles on the plane is a good example which turns out to be connected with the Einstein–Weyl geometry.

We consider the case

$$y''' = F(x, y', y'').$$

Here $F_0 = 0$ and from the second equation we have

$$H_{x2} + y''H_{12} + FH_{22} = 0,$$

where

$$F_{x2} + FF_{22} - \frac{F_2^2}{2} + y''F_{12} - 2F_1 = H.$$

Taking into account this relation, the first equation gives us the condition

$$H_x + y''(H_1 - F_{11}) - FF_{12} - \frac{1}{18}F_2^3 - F_{x1} = 0.$$

If

$$H = H(F_2), \quad \text{with} \quad F \equiv F(x, y'')$$

then we get

$$F_{x2} + FF_{22} - \frac{F_2^2}{3} = 0.$$

The corresponding third-order equation is

$$y''' = F(x, y'')$$

and it is connected with the second-order equation

$$z'' = g(x, z').$$

Another example is the solution of the system for a function $F = F(x, y', y'')$ obeying the equation

$$F_{x2} + FF_{22} - \frac{F_2^2}{2} + y''F_{12} - 2F_1 = 0.$$

In this case $H = 0$ and we get the system of equations

$$\begin{aligned} F_{x2} + FF_{22} - \frac{F_2^2}{2} + y''F_{12} - 2F_1 &= 0, \\ y''F_{11} + FF_{12} + \frac{1}{18}F_2^3 + F_{x1} &= 0, \end{aligned}$$

with the condition of compatibility

$$\left(\frac{F_2^2}{6} - F_1\right)F_{22} + 2F_2F_{12} + 3F_{11} = 0.$$

Acknowledgements

The author would like to thank the Cariplo Foundation (Center Landau–Volta, Como, Italy) as well as the INTAS-99-01782 Programme and the Royal Swedish Academy of Sciences for financial support.

References

1. V. Dryuma, *Application of the E. Cartan method for studying of nonlinear dynamical systems*, Matematicheskie issledovaniya, Kishinev, Stiintsa, 1987, vol. 92, 49–68.
2. V. Dryuma, *Projective duality in theory of the second order differential equations*, Mathematical Researches, Kishinev, Stiintsa, 1990, vol. 112, 93–103.
3. V.S. Dryuma, *On Initial values problem in theory of the second order ODE's*, Proceedings of the "Workshop on Nonlinearity, Integrability and all that: Twenty years after NEEDS'79", Gallipoli(Lecce), Italy, July 1–July 10, 1999, ed. M. Boiti, L. Martina, F. Pempinelli, B. Prinari and G. Soliani. World Scientific, Singapore, 2000, 109–116.
4. V. Dryuma, *On geometry of the second order differential equations*, Proceedings of Conference Nonlinear Phenomena, ed. K.V. Frolov, Moscow, Nauka, 1991, 41–48.
5. V. Dryuma, *Geometrical properties of multidimensional differential equations and the Finsler metrics of dynamical systems*, Theoretical and Mathematical Physics, **99**, 1994, 241–249.
6. V.S. Dryuma, *Geometrical properties of nonlinear dynamical systems*, "Proceedings of the First Workshop on Nonlinear Physics", Le Sirenuse, Gallipoli (Lecce), Italy, June 29–July 7, 1995, ed. E. Alfinito, M. Boiti, L. Martina and F. Pempinelli. World Scientific, Singapore, 1996, 83–93.
7. R. Liouville, *Sur les invariants de certaines équations différentielles et sur leurs applications*, J. de l'École Polytechnique **59**, 7–76, 1889.
8. A. Tresse, *Détermination des Invariants ponctuels de l'équation differentielle ordinaire de second ordre:* $y'' = w(x,y,y')$. Preisschriften der fürstlichen Jablonowski'schen Gesellschaft XXXII, Leipzig, S. Hirzel, 1896.
9. A. Tresse, *Sur les invariants différentiels des groupes continus de transformations*, Acta Math. **18**, 1–88, 1894.
10. E. Cartan, *Sur les variétés a connexion projective*, Bulletin de la Société Mathémat. de France **52**, 205–241, 1924.
11. G. Thomsen, *Über die topologischen Invarianten der Differentialgleichung* $y'' = f(x,y)y'^3 + g(x,y)y'^2 + h(x,y)y' + k(x,y)$, Abhandlungen aus dem mathematischen Seminar der Hamburgischen Universität, **7**, 301–328, 1930.
12. E.M. Paterson and A.G. Walker, *Riemann extensions*, Quart. J. Math. Oxford **3**, 19–28, 1952.
13. E. Wilczynski, *Projective Differential Geometry of Curved Surfaces*, Transactions of the American Mathematical Society, **9**, 103–128, 1908.
14. E. Cartan, *Sur une classe d'espaces de Weyl*, Ann. Ec. Norm. Sup. **14**, 1–16, 1943.
15. H. Pedersen and K.P. Tod, *Three-dimensional Einsten–Weyl Geometry* Advances in Mathematics **97**, 71–109, 1993.
16. J. Chazy, *Sur les equations differentielle dont l'integrale possede une coupure essentielle mobile*, C.R. Acad. Sc. Paris **150**, 456–458, 1910.

17. R. Ward, *Einstein–Weyl spaces and Toda fields*, Classical and Quantum Gravitation **7**, L45–L48, 1980.
18. M. Dunajski, L. Mason and K.P. Tod, *Einstein–Weyl geometry, the dKP equation and twistor theory*, arXiv:math. DG/0004031, 6 Apr. 2000.
19. F. Calogero, *A solvable nonlinear wave equation*, Studies in Applied Mathematics **LXX**, N3, 189–199, 1984.
20. M. Pavlov, *The Calogero equation and Liouville type equations*, arXiv:nlin. SI/0101034, 19 Jan. 2001.
21. M. Tanimoto, *On the null surface formalism*, arXiv:gr-qc/9703003, 1997.
22. D.M. Forni, M. Iriondo and C.N. Kozameh, *Null surface formalism in 3D*, arXiv:gr-qc/0005120, 26 May 2000.
23. K.P. Tod, *Einstein–Weyl spaces and third-order differential equations*, J. Math. Phys., N9, 2000.
24. S. Frittelli, C.N. Kozameh and E.T. Newman, *Differential geometry from differential equations*, arXiv:gr-qc/0012058, 15 Dec. 2000.
25. J.C. Sprot, *Symplest dissipative chaotic flow*, Physics Letters A **228**, 271–274, 1997.
26. B. Konopelchenko, *The non-abelian (1+1)-dimensional Toda lattice*, Physics Letters A **156**, 221–222, 1991.
27. M. Dunajski, *Anti-self-dual four-manifolds with a parallel spinor*, arXiv:math. DG/0102225, 18 Oct. 2001.
28. V. Dryuma and M. Pavlov. *On initial value problem in theory of the second order ODE*, Buletinul AS RM, Matematica **242**, 51–58, 2003.
29. V. Dryuma, *Applications of the Riemann and the Einstein–Weyl geometry in theory of the second order ODE*, Theoretical and Mathematical Physics, **128**, N1, 15–26, 2001.
30. J. Hietarinta and V. Dryuma, *Is my ODE a Painlevé equation in disguise?*, Journal of Nonlinear Mathematical Physics, **9**, Suppl. 1, 67–74, 2002.

DUNKL OPERATORS AND CALOGERO–SUTHERLAND MODELS

F. FINKEL, D. GÓMEZ-ULLATE, A. GONZÁLEZ-LÓPEZ, M.A. RODRÍGUEZ and R. ZHDANOV
Dpto. de Física Teórica II, Universidad Complutense, 28040 Madrid, Spain

Abstract. We describe a general method for constructing (scalar or spin) Calogero–Sutherland models of A_N or BC_N type, which are either exactly or quasi-exactly solvable. Our approach is based on the simultaneous use of three different families of Dunkl operators of each type, one of which was recently introduced by the authors. We perform a complete classification of the models which can be constructed by our method. We obtain in this way several new families of (quasi-)exactly solvable Calogero–Sutherland models, some of them with elliptic interactions.

1. Introduction

The quantum Calogero [6] and Sutherland [61, 62] models describe a system of N particles on the line and the circle with pairwise interactions falling off as the inverse square of the distance between them. Both models are characterized by two key mathematical properties, namely *complete integrability* and *exact solvability*, which have probably been one of the main motivations for the vast attention devoted to these models since their introduction in the early seventies. The physical and mathematical interest of these models is reflected by their appearance in such diverse areas as fractional statistics and anyons [51, 5, 30, 31, 7, 41], quantum Hall liquids [1], Yang–Mills theories [29, 12], random matrix theory [64, 37], integrable PDEs with soliton solutions [43, 55], and orthogonal polynomials [44, 2, 13, 16].

A landmark in the development of this field was in fact Olshanetsky and Perelomov's proof of the complete integrability of the Calogero and Sutherland models [48]. In their construction, the integrability of the models follows by expressing the Hamiltonian as one of the radial parts of the Laplace–Beltrami operator in a symmetric space associated with the A_N root system. Furthermore, these authors also constructed new families of

A.B. Shabat et al. (eds.), New Trends in Integrability and Partial Solvability, 157–189.

integrable many-body Hamiltonians associated with other root systems, and showed that the most general interaction potential for these models is proportional to the Weierstrass $\wp$ function. We shall use in what follows the customary term *Calogero–Sutherland* (CS) to collectively refer to these models.

As previously mentioned, the original models introduced by Calogero and Sutherland are *exactly solvable* (ES) in the sense that the Hamiltonian preserves a *known* infinite increasing sequence (or flag) of finite-dimensional subspaces of smooth functions [57, 66]. This property is shared by some but not all the CS models introduced in Olshanetsky and Perelomov's paper. Some of these models —such as the elliptic model of BC_N type [38, 23]— are in fact *quasi-exactly solvable* (QES), which means that the Hamiltonian leaves invariant some known finite-dimensional functional space, so that a subset of the spectrum can be computed by diagonalizing the finite-dimensional matrix of the restriction of the Hamiltonian to this space, see [65, 59, 67, 27, 49] and references therein. In contrast, the possible (quasi-)exact solvability of the elliptic model of A_N type remains an open question.

Quasi-exactly solvable models are typically constructed by a two-step procedure: i) take a quadratic combination of the generators of an algebra of first-order differential operators with a finite-dimensional module of smooth functions; ii) solve the *equivalence problem*, that is, determine whether this second-order operator can be reduced to the Schrödinger form by a transformation preserving the QES character. This procedure —sometimes called the *hidden symmetry algebra* approach— is most easily implemented in the case of a spinless particle in one spatial dimension. In this case the only relevant algebra is the realization of the Lie algebra $\mathfrak{sl}(2)$ spanned by $\{\partial_z, z\partial_z - \frac{m}{2}, z^2\partial_z - mz\}$, which for integer m preserve the space of polynomials in z of degree at most m [42]. Moreover, every second degree polynomial in these generators is equivalent to a Schrödinger operator by a suitable change of the independent variable and a gauge transformation. This basic property enabled González-López, Kamran and Olver [25, 27] to perform a complete classification of the resulting QES potentials.

The extension of this construction to several variables (with or without internal degrees of freedom) is considerably more involved. In the first place, there are many algebras of first-order differential operators admitting an invariant finite-dimensional functional space. For instance, in two variables there are exactly 17 inequivalent families of finite-dimensional Lie algebras of first-order differential operators (plus 5 additional real forms) with an invariant module, [24, 28]. (No such classification in more than two variables has been completed so far.) Secondly, the corresponding equivalence problem cannot be explicitly solved for $N > 1$ variables [11, 17]. This explains why even in two variables there is no exhaustive classification of QES models, but rather an extensive collection of examples [60, 26, 70]. In N variables, some QES deformations of the CS models in Olshanetsky and Perelomov's paper have been recently constructed using a realization of $\mathfrak{sl}(N+1)$ [47, 36, 22] as the hidden symmetry algebra.

A good deal of effort has been devoted over the past decade to the study Calogero–Sutherland models with spin, due in part to their intimate connection with integrable spin chains of Haldane–Shastry type [33, 58]. Spin CS models have been explored by several techniques, such as the exchange operator formalism [52, 46], the Dunkl operators approach [4, 10, 16], reduction by discrete symmetries [56], and construction of Lax pairs [34, 35, 40]. Historically, the first CS spin models discussed in the literature were the trigonometric [32] and rational [46] spin versions of the original Sutherland and Calogero models of A_N type. The exact solvability and integrability of both models can be established by relating the Hamiltonian to a quadratic combination of *either* the Dunkl [15] *or* the Dunkl–Cherednik [9, 10] operators of A_N type, respectively. To the best of our knowledge, the only additional spin CS model of A_N type discussed in the literature prior to 2001 was the hyperbolic Sutherland model with an external Morse potential [39]. In the BC_N case, only the rational and trigonometric spin models had been constructed [68] prior to that date. In the latter paper, the spectrum of the rational BC_N spin model was explicitly determined, and its integrability was shown by means of the Lax pair approach. An alternative proof of the integrability of this model using Dunkl operators of B_N type was later presented in [69]. The same operators were later employed by Dunkl to construct a complete basis of eigenfunctions [16]. In contrast, the solvability and/or integrability of the trigonometric/hyperbolic BC_N model was still unproved by 2001.

In this paper, we review the systematic method for constructing Calogero–Sutherland models with spin using Dunkl operators introduced in Refs. [19, 20]. Our construction is reminiscent of the hidden symmetry algebra approach to (Q)ES scalar models, in the sense that it also involves general quadratic combinations of *different* (families of) Dunkl operators, which leave invariant a common polynomial space of finite dimension. Our contributions can be summarized as follows:

1. We have introduced a third new family of Dunkl operators (in their A_N and B_N versions), which, together with the two previously known families leave invariant a finite-dimensional polynomial subspace.
2. We have shown that certain quadratic combinations involving the three families of Dunkl operators of either A_N or B_N type are equivalent to a Schrödinger operator.
3. We have performed a complete classification of the (Q)ES spin CS mod-

els of A_N and BC_N types which can be constructed by this method. All the previously known families of spin CS models appear in the classification. Several families of quantum spin CS models with elliptic interactions are obtained for the first time.

4. In the A_N case, we have shown that each spin CS model yields a corresponding scalar one, which is (Q)ES by construction. Moreover, we have shown that each one-body (Q)ES potential in the González-López, Kamran and Olver's classification [25, 26] is associated with a (Q)ES Calogero–Sutherland model (with or without spin).
5. In the BC_N case, we have proved that all the potentials in the classification are expressible in a unified way in terms of the Weierstrass $\wp$ function with suitable (sometimes infinite) half-periods. This provides a natural spin counterpart of Olshanetsky and Perelomov's formula for a general scalar potential related to the BC_N root system.

The rest of the paper is organized as follows. In Section 2 we present the results for the A_N root system. To this end, we first introduce the relevant families of Dunkl operators and discuss their properties. We then explain the details of the construction of the many-body Hamiltonians. Finally, we describe the classification of the models. Section 3 is devoted to the parallel construction of the models of BC_N type. We point out the main differences with respect to the A_N case and derive the general formula for the potentials in terms of the Weierstrass $\wp$ function. In Section 4, we present some concluding remarks and related open questions. We refer the reader to Refs. [19, 20] for a complete account of the intermediate results and proofs not included in this review.

2. Calogero–Sutherland spin models of A_N type

2.1. A NEW FAMILY OF DUNKL OPERATORS OF A_N TYPE

The Dunkl operators

$$J_i^- = \frac{\partial}{\partial z_i} + a \sum_{j \neq i} \frac{1}{z_i - z_j} (1 - K_{ij}), \qquad i = 1, \ldots, N, \tag{2.1}$$

were introduced in Ref. [15] in connection with the theory of orthogonal polynomials in several variables. In the latter expression, a is a real parameter and the sum runs over $1, \ldots, i-1, i+1, \ldots, N$. In general, all sums and products in this paper will range from 1 to N unless otherwise restricted. The operators $K_{ij} = K_{ji}$ are coordinate permutation operators, whose action on a function $f(\mathbf{z})$, $\mathbf{z} = (z_1, \ldots, z_N) \in \mathbb{R}^N$, is given by

$$(K_{ij} f)(z_1, \ldots, z_i, \ldots, z_j, \ldots, z_N) = f(z_1, \ldots, z_j, \ldots, z_i, \ldots, z_N). \tag{2.2}$$

The Dunkl operators (2.1) form a commuting family, and together with the permutation operators K_{ij} span a degenerate affine Hecke algebra [10]. These properties play an essential role in the proof of the integrability and the computation of the spectrum of the rational spin model of A_N type [4, 3]. The very same properties are also satisfied by the operators

$$\widetilde{J}_i^0 = z_i \frac{\partial}{\partial z_i} + a \sum_{j<i} \frac{z_i}{z_i - z_j} (1 - K_{ij}) + a \sum_{j>i} \frac{z_j}{z_i - z_j} (1 - K_{ij}) + 1 - i, \quad (2.3)$$

$i = 1, \dots, N$, introduced by Cherednik in connection with the trigonometric model of A_N type. We shall employ the slightly more symmetric (but non-commuting) form of the Cherednik operators given by[1]

$$J_i^0 = z_i \frac{\partial}{\partial z_i} - \frac{m}{2} + \frac{a}{2} \sum_{j \neq i} \frac{z_i + z_j}{z_i - z_j} (1 - K_{ij}), \qquad i = 1, \dots, N, \quad (2.4)$$

where m is a nonnegative integer whose significance will become clear shortly.

The third new family of Dunkl operators of A_N type introduced by the authors in [19] reads

$$J_i^+ = z_i^2 \frac{\partial}{\partial z_i} - m z_i + a \sum_{j \neq i} \frac{z_i z_j}{z_i - z_j} (1 - K_{ij}), \qquad i = 1, \dots, N. \quad (2.5)$$

These operators also form a commuting family and span a degenerate affine Hecke algebra. As far as the solvability of the resulting models is concerned, the important commutation relations are

$$[K_{ij}, J_k^\epsilon] = 0, \qquad K_{ij} J_i^\epsilon = J_j^\epsilon K_{ij}, \quad (2.6)$$

where $\epsilon = \pm, 0$, and the indices i, j, k are all different.

The key property in our construction of (quasi-)exactly solvable spin CS models is the fact that the operators J_i^ϵ preserve a common polynomial subspace of finite-dimension. Indeed, consider the polynomial spaces

$$\mathcal{R}_n(\mathbf{z}) = \langle z_1^{l_1} \cdots z_N^{l_N} \mid l_1 \le n, \dots, l_N \le n \rangle, \quad (2.7)$$

$$\mathcal{T}_n(\mathbf{z}) = \langle z_1^{l_1} \cdots z_N^{l_N} \mid l_1 + \cdots + l_N \le n \rangle, \quad (2.8)$$

which shall be referred to as the *rectangular* and *triangular* modules, respectively, by analogy with the two variable case [18].

---

[1] Note that the operators $\widetilde{J}_i^0$ and J_i^0 differ by a linear combination with constant coefficients of the permutation operators K_{ij}. Therefore, the operators J_i^0 and K_{ij} also span a degenerate affine Hecke algebra.

Theorem 2.1 *The operators J_i^- and J_i^0 preserve the modules $\mathcal{T}_n(\mathbf{z})$ and $\mathcal{R}_n(\mathbf{z})$ for any nonnegative integer n. The operators J_i^+ preserve the module $\mathcal{R}_m(\mathbf{z})$, but do not preserve the modules $\mathcal{T}_n(\mathbf{z})$ and $\mathcal{R}_n(\mathbf{z})$ for $n \neq m$.*

Corollary 2.2 *Any polynomial in the operators J_i^ϵ preserves the rectangular module $\mathcal{R}_m(\mathbf{z})$. In addition, if the polynomial does not depend on J_i^+, it preserves the modules $\mathcal{R}_n(\mathbf{z})$ and $\mathcal{T}_n(\mathbf{z})$ for all n.*

2.2. CONSTRUCTION OF SPIN CS MODELS OF A_N TYPE

Let $\mathfrak{S}$ be the Hilbert space of the particles' internal degrees of freedom or "spin", given explicitly by

$$\mathfrak{S} = \operatorname{span}\{\, |s_1, \dots, s_N\rangle\}_{-M \le s_i \le M}\,, \qquad M \in \frac{1}{2}\mathbb{N}\,.$$

Consider the spin permutation operators S_{ij}, $i, j = 1, \dots, N$, whose action on a spin state $|s_1, \dots, s_N\rangle$ is given by

$$S_{ij}|s_1, \dots, s_i, \dots, s_j, \dots, s_N\rangle = |s_1, \dots, s_j, \dots, s_i, \dots, s_N\rangle\,. \tag{2.9}$$

The operators S_{ij} are represented in $\mathfrak{S}$ by $(2M+1)^N$-dimensional symmetric matrices. The permutation operators S_{ij} can be easily expressed in terms of one-particle spin operators. For instance, if $M = 1/2$ we have $S_{ij} = 2\,\sigma_i\!\cdot\!\sigma_j + \frac{1}{2}$, where $\sigma_i \equiv (\sigma_i^1, \sigma_i^2, \sigma_i^3)$ are the usual SU(2) spin operators associated with the i-th particle.

The starting point in our construction is the following quadratic combination involving *all* three families of Dunkl operators of A_N type:

$$\begin{aligned} -\overline{H} = \sum_i \Big(& c_{++}(J_i^+)^2 + c_{00}(J_i^0)^2 + c_{--}(J_i^-)^2 + \frac{c_{0+}}{2}\{J_i^0, J_i^+\} \\ & + \frac{c_{0-}}{2}\{J_i^0, J_i^-\} + c_+ J_i^+ + c_0 J_i^0 + c_- J_i^- \Big)\,, \end{aligned} \tag{2.10}$$

where $c_{\epsilon\epsilon'}, c_\epsilon\,,\ \epsilon, \epsilon' = \pm, 0$, are arbitrary real constants. The operator $\overline{H}$ possesses the following obvious properties:

i) It preserves $\mathcal{R}_m$ (and also $\mathcal{R}_n$ and $\mathcal{T}_n$ for all n if $c_{++} = c_{0+} = c_+ = 0$).
ii) It commutes with K_{ij} for all values of i, j.
iii) It is completely determined by the pair of polynomials (P, Q) given by

$$\begin{aligned} P(z) &= c_{++}z^4 + c_{0+}z^3 + c_{00}z^2 + c_{0-}z + c_{--}\,, \\ Q(z) &= c_+z^2 + c_0 z + c_-\,. \end{aligned} \tag{2.11}$$

Let Λ be the projector on states antisymmetric under the simultaneous interchange of any two particles' coordinates and spins. In terms of the total permutation operators $\Pi_{ij} \equiv K_{ij}S_{ij}$, the projector Λ is defined by the relations $\Lambda^2 = \Lambda$ and $\Pi_{ij}\Lambda = -\Lambda$, $j > i = 1, \dots, N$. Since $K_{ij}^2 = 1$, the latter relations are equivalent to

$$K_{ij}\Lambda = -S_{ij}\Lambda, \qquad j > i = 1, \dots, N. \tag{2.12}$$

The antisymmetrizer Λ is given by a polynomial in the total permutation operators Π_{ij}. For instance, for $N = 1, 2$ we have:

$$\begin{aligned} N = 2: &\qquad \Lambda = \frac{1}{2}\,(1 - \Pi_{12})\,, \\ N = 3: &\qquad \Lambda = \frac{1}{6}\,(1 - \Pi_{12} - \Pi_{13} - \Pi_{23} + \Pi_{12}\Pi_{13} + \Pi_{12}\Pi_{23})\,. \end{aligned}$$

Since $\overline{H}$ commutes with K_{ij} and acts trivially on $\mathfrak{S}$, it follows that $\overline{H}$ commutes with Λ and therefore preserves $\Lambda(\mathcal{R}_m \otimes \mathfrak{S})$.

Given a scalar differential operator D, define the linear mapping * by the requirement that

$$(DK_{i_1j_1} \cdots K_{i_rj_r})^* = (-1)^r\, DS_{i_rj_r} \cdots S_{i_1j_1}\,. \tag{2.13}$$

It follows from this definition and Eq. (2.12) that

$$\overline{H}\Lambda = \overline{H}^*\Lambda\,,$$

and thus $\overline{H}^*$ also preserves $\Lambda(\mathcal{R}_m \otimes \mathfrak{S})$.

Inserting the expressions (2.1), (2.4) and (2.5) into Eq. (2.10), after some algebra we obtain

$$\begin{aligned} -\overline{H}^* = &\sum_i \Big(P(z_i)\partial_{z_i}^2 + \tilde{Q}(z_i)\partial_{z_i} + R(z_i)\Big) + ac_{++}(1-m)\sum_{i\neq j} z_i z_j \\ &+ 2a\sum_{i\neq j}\frac{P(z_i)}{z_{ij}^-}\,\partial_{z_i} + \frac{a}{2}\sum_{i\neq j}\Big(c_{++}(z_{ij}^+)^2 + c_{0+}z_{ij}^+ + c_{00}\Big)\,S_{ij} \\ &- a\sum_{i\neq j}\frac{P(z_i)}{(z_{ij}^-)^2}\,(1 + S_{ij}) + \frac{a^2}{12}\,c_{00}\sum_{i,j,k}{}' \,(1 - S_{ij}S_{ik}), \end{aligned}$$

where P and Q are given in Eq. (2.11),

$$\begin{aligned} \tilde{Q}(z) &= Q(z) + \Big(1 - \tfrac{\ell}{2}\Big)\,P'(z)\,, \qquad \ell = 1 + m + a(N-1)\,, \\ R(z) &= c_{++}(\ell + m(m-2) - 1)z^2 + \Big[c_{0+}\Big[m\Big(m - 1 - \tfrac{\ell}{2}\Big) \\ &\quad + \ell - 1\Big] - mc_+\Big]\,z + \tfrac{c_{00}}{4}\,(2(\ell-1) + m(m-2)) - \tfrac{m}{2}\,c_0\,, \end{aligned} \tag{2.14}$$

$z_{ij}^{\pm} \equiv z_i \pm z_j$ and $\sum_{i,j,k}{}'$ denotes summation in i, j, k with $i \neq j \neq k \neq i$. Note that the terms in first derivatives do not depend on the spin permutation operators S_{ij}. This fact is essential for the existence of a gauge transformation and a change of the independent variables mapping $\overline{H}^*$ into a Schrödinger operator, i.e.,

$$\mu(\mathbf{z}) \cdot \overline{H}^*(\mathbf{z}) \cdot \mu^{-1}(\mathbf{z})\Big|_{\mathbf{z}=\zeta(\mathbf{x})} \equiv H = -\sum_i \partial_{x_i}^2 + V(\mathbf{x}), \tag{2.15}$$

where $V(\mathbf{x})$ is a Hermitian matrix-valued function.

Theorem 2.3 *The operator* $\overline{H}^*$ *can be reduced to a Schrödinger operator* H *by the change of variables*

$$x_i = \zeta^{-1}(z_i) = \int^{z_i} \frac{\mathrm{d}y}{\sqrt{P(y)}}, \qquad i = 1, \dots, N, \tag{2.16}$$

and the gauge transformation with gauge factor

$$\mu(\mathbf{z}) = \prod_{i<j} |z_{ij}^-|^a \prod_i P(z_i)^{-\frac{1}{4}} \exp \int^{z_i} \frac{\tilde{Q}(y)}{2P(y)} \,\mathrm{d}y. \tag{2.17}$$

The potential is given by

$$\begin{aligned} V(\mathbf{x}) = {} & a^2 \sum_{i,j,k}{}' \frac{P(z_i)}{z_{ij}^- \, z_{ik}^-} + a \sum_{i \neq j} \frac{\tilde{Q}(z_i)}{z_{ij}^-} + a \sum_{i \neq j} \frac{P(z_i)}{(z_{ij}^-)^2} (a + S_{ij}) \\ & - \frac{a}{2} \sum_{i \neq j} \Big(c_{++} (z_{ij}^+)^2 + c_{0+} z_{ij}^+ + c_{00} \Big) S_{ij} - \frac{a^2}{12} c_{00} \sum_{i,j,k}{}' (1 - S_{ij} S_{ik}) \\ & - a c_{++} (1 - m) \sum_{i \neq j} z_i z_j + \frac{1}{4} \sum_i W(z_i) \Big|_{\mathbf{z}=\zeta(\mathbf{x})}, \end{aligned} \tag{2.18}$$

where

$$W(z) = \frac{1}{P(z)} \Big(\frac{3}{4} P'(z)^2 + \tilde{Q}(z)^2 - 2\tilde{Q}(z) P'(z) \Big) - P''(z) + 2\tilde{Q}'(z) - 4R(z).$$

Corollary 2.4 *The Schrödinger operator* (2.15) *with potential* (2.18) *leaves invariant the finite-dimensional space*

$$\mathfrak{R}_m = \boldsymbol{\mu}(\boldsymbol{\zeta}(\mathbf{x})) \boldsymbol{\Lambda} \Big(\mathcal{R}_m(\boldsymbol{\zeta}(\mathbf{x})) \otimes \mathfrak{S} \Big). \tag{2.19}$$

In addition, if $c_{++} = c_{0+} = c_+ = 0$, *it preserves the space* $\mathfrak{R}_n$ *and*

$$\mathfrak{T}_n = \boldsymbol{\mu}(\boldsymbol{\zeta}(\mathbf{x})) \boldsymbol{\Lambda} \Big(\mathcal{T}_n(\boldsymbol{\zeta}(\mathbf{x})) \otimes \mathfrak{S} \Big), \tag{2.20}$$

for any nonnegative integer n.

Remark 2.5. The change of variables (2.16) is defined up to an arbitrary translation for each coordinate x_i, $i = 1, \ldots, N$. This arbitrariness can be removed in some cases by requiring that the resulting potential be invariant under sign reversals of any coordinate x_i.

2.3. CLASSIFICATION OF THE MODELS OF A_N TYPE

We have seen that any quadratic combination (2.10) yields a (Q)ES Schrödinger operator or *Hamiltonian* of the form (2.15), which is totally determined (up to coordinate translations) by the pair (P, Q). However, different pairs (P, Q) may give rise to the same potential. This is a consequence of the invariance of the vector spaces $\langle J_i^-, J_i^0, J_i^+\rangle$, $i = 1, \ldots, N$, under the projective action of $\mathrm{GL}(2, I\!\!R)$. The image of the operator $\overline{H}$ under this action is still of the form (2.10), albeit with different coefficients $\hat{c}_{\epsilon\epsilon'}$ and $\hat{c}_\epsilon$. We shall make use of this fact to completely classify the resulting spin CS models.

Consider the mapping

$$J_i^\epsilon(\mathbf{w}) \mapsto \widehat{J_i^\epsilon}(\mathbf{z}) = \mu_m(\mathbf{z}) \cdot J_i^\epsilon(\mathbf{w}(\mathbf{z})) \cdot \mu_m^{-1}(\mathbf{z}), \tag{2.21}$$

where $\mathbf{w} = (w_1, \ldots, w_N)$ is given by the projective (Möbius) transformation

$$w_i = \frac{\alpha z_i + \beta}{\gamma z_i + \delta}, \qquad i = 1, \ldots, N, \quad \Delta = \alpha\delta - \beta\gamma \neq 0, \tag{2.22}$$

and the gauge factor $\mu_m(\mathbf{z})$ is defined by

$$\mu_m(\mathbf{z}) = \prod_{i=1}^N (\gamma\, z_i + \delta)^m. \tag{2.23}$$

Lemma 2.6 *The mapping* (2.21)–(2.23) *acts linearly on the vector spaces* $\langle J_i^-, J_i^0, J_i^+\rangle$, $i = 1, \ldots, N$, *as*

$$\begin{pmatrix} \widehat{J_i^+}(\mathbf{z}) \\ \widehat{J_i^0}(\mathbf{z}) \\ \widehat{J_i^-}(\mathbf{z}) \end{pmatrix} = \frac{1}{\Delta} \begin{pmatrix} \alpha^2 & 2\alpha\beta & \beta^2 \\ \alpha\gamma & \alpha\delta + \beta\gamma & \beta\delta \\ \gamma^2 & 2\gamma\delta & \delta^2 \end{pmatrix} \begin{pmatrix} J_i^+(\mathbf{z}) \\ J_i^0(\mathbf{z}) \\ J_i^-(\mathbf{z}) \end{pmatrix}. \tag{2.24}$$

Remark 2.7. The irreducible multiplier representation $\rho_{n,i}$ of $\mathrm{GL}(2, I\!\!R)$ on the space of univariate polynomials of degree at most n is defined [50] by the linear transformations

$$p(z) \mapsto \hat{p}(z) = \Delta^i (\gamma z + \delta)^n\, p\left(\frac{\alpha z + \beta}{\gamma z + \delta}\right).$$

Thus $\langle J_i^-, J_i^0, J_i^+\rangle$, $i = 1, \ldots, N$, is a carrier space of the representation $\rho_{2,-1}$.

It follows from the previous lemma that the operator $\widehat{H}$ defined by

$$\widehat{H}(\mathbf{z}) = \mu_m(\mathbf{z}) \cdot \overline{H}(\mathbf{w}(\mathbf{z})) \cdot \mu_m^{-1}(\mathbf{z})$$

is also a second degree polynomial in the Dunkl operators $J_i^\epsilon(\mathbf{z})$. In fact, it may be easily shown that the operator $\widehat{H}$ is determined by the transform of the pair (P, Q) under the representation $\rho_{4,-2} \oplus \rho_{2,-1}$.

Using the Möbius transformations (2.21)–(2.23), together with the *complex* projective (linear) transformation $w_j = \mathrm{i}\, z_j$, $j = 1, \dots, N$, the polynomial P can be conveniently chosen as one of the following canonical forms (see [19] for more details):

$$\begin{array}{llll} 1) & 1\,, & 6) & \nu(1+z^2)^2\,, \\ 2) & z\,, & 7) & \nu(1-z^2)(1-k^2 z^2)\,, \\ 3) & \nu(z^2-1)\,, & 8) & \nu(z^2-1)(1-k'^2 z^2)\,, \\ 4) & \nu(1-z^2)\,, & 9) & \nu(1-z^2)(k'^2+k^2 z^2)\,, \\ 5) & \nu z^2\,, & & \end{array}$$

where $\nu > 0$, $0 < k < 1$ and $k'^2 = 1 - k^2$. Each canonical form yields a corresponding family of (Q)ES spin potentials. Taking advantage of the arbitrariness mentioned in Remark 2.5 and dropping some inessential constant terms, the resulting potentials can be written in all cases as

$$V(\mathbf{x}) = \sum_i U(x_i) + V_{\mathrm{int}}(\mathbf{x})\,, \tag{2.25}$$

where U represents a scalar external field, and the interaction potential V_{int} is of the form

$$V_{\mathrm{int}}(\mathbf{x}) = \sum_{i \neq j} \Big(V^-(x_{ij}^-) + V^+(x_{ij}^+) \Big)\, a(a + S_{ij}), \qquad x_{ij}^\pm \equiv x_i \pm x_j, \tag{2.26}$$

with either $V^+ = 0$ or $V^+ = V^-$.

We shall now present the list of potentials for a suitably chosen value of the parameter ν. This is justified, since the scaling $(c_{\epsilon\epsilon'}, c_\epsilon) \mapsto (\lambda c_{\epsilon\epsilon'}, \lambda c_\epsilon)$ induces the mapping

$$V(\mathbf{x}\,; c_{\epsilon\epsilon'}, c_\epsilon) \;\mapsto\; V(\mathbf{x}\,; \lambda\, c_{\epsilon\epsilon'}, \lambda\, c_\epsilon) = \lambda\, V(\sqrt{\lambda}\,\mathbf{x}\,; c_{\epsilon\epsilon'}, c_\epsilon) \tag{2.27}$$

of the corresponding potentials. The explicit expressions of the coefficients A, B, C and D are given in Table I.

Case 1. $P(z) = \dfrac{1}{4}$.
External potential:

$$U(x) = A\,x^4 + B\,x^3 + C\,x^2 + D\,x\,.$$

Interaction potential:

$$V_{\text{int}}(\mathbf{x}) = \sum_{i \neq j} (x_{ij}^-)^{-2}\, a(a + S_{ij})\,.$$

Case 2. $P(z) = 4z$.
External potential:

$$U(x) = A\, x^6 + B\, x^4 + C\, x^2 + \frac{D}{x^2}\,.$$

Interaction potential:

$$V_{\text{int}}(\mathbf{x}) = \sum_{i \neq j} \Big((x_{ij}^-)^{-2} + (x_{ij}^+)^{-2} \Big)\, a(a + S_{ij})\,.$$

Case 3. $P(z) = 4(z^2 - 1)$.
External potential:

$$U(x) = A\, \cosh^2 2x + B\, \cosh 2x + C\, \cosh 2x\, \sinh^{-2} 2x + D\, \sinh^{-2} 2x\,.$$

Interaction potential:

$$V_{\text{int}}(\mathbf{x}) = \sum_{i \neq j} \Big(\sinh^{-2} x_{ij}^- + \sinh^{-2} x_{ij}^+ \Big)\, a(a + S_{ij})\,.$$

Case 4. $P(z) = 4(1 - z^2)$.
External potential:

$$U(x) = A\, \cos^2 2x + B\, \cos 2x + C\, \cos 2x\, \sin^{-2} 2x + D\, \sin^{-2} 2x\,.$$

Interaction potential:

$$V_{\text{int}}(\mathbf{x}) = \sum_{i \neq j} \Big(\sin^{-2} x_{ij}^- + \sin^{-2} x_{ij}^+ \Big)\, a(a + S_{ij})\,.$$

Case 5. $P(z) = 4z^2$.
External potential:

$$U(x) = A\, \mathrm{e}^{4x} + B\, \mathrm{e}^{2x} + C\, \mathrm{e}^{-2x} + D\, \mathrm{e}^{-4x}\,.$$

Interaction potential:

$$V_{\text{int}}(\mathbf{x}) = \sum_{i \neq j} \sinh^{-2} x_{ij}^-\, a(a + S_{ij})\,.$$

Case 6. $P(z) = (1 + z^2)^2$.
External potential:

$$U(x) = A \cos 4x + B \cos 2x + C \sin 4x + D \sin 2x\,.$$

Interaction potential:

$$V_{\text{int}}(\mathbf{x}) = \sum_{i \neq j} \sin^{-2} x_{ij}^- \, a(a + S_{ij})\,.$$

Case 7. $P(z) = 4(1 - z^2)(1 - k^2 z^2)$.
Here (and also in Cases 8 and 9) the functions $\operatorname{sn} x \equiv \operatorname{sn}(x|k)$, $\operatorname{cn} x \equiv \operatorname{cn}(x|k)$, and $\operatorname{dn} x \equiv \operatorname{dn}(x|k)$ are the usual Jacobian elliptic functions of *modulus* k, and $k' = \sqrt{1 - k^2}$ is the *complementary modulus.*

External potential:

$$U(x) = A \operatorname{sn}^2 2x + B \operatorname{cn} 2x \operatorname{dn} 2x + \operatorname{sn}^{-2} 2x \,(C + D \operatorname{cn} 2x \operatorname{dn} 2x)\,.$$

Interaction potential:

$$V_{\text{int}}(\mathbf{x}) = \sum_{i \neq j} \left(\frac{\operatorname{cn}^2 x_{ij}^- \operatorname{dn}^2 x_{ij}^-}{\operatorname{sn}^2 x_{ij}^-} + \frac{\operatorname{cn}^2 x_{ij}^+ \operatorname{dn}^2 x_{ij}^+}{\operatorname{sn}^2 x_{ij}^+} \right) a(a + S_{ij})\,.$$

Case 8. $P(z) = 4(z^2 - 1)(1 - k'^2 z^2)$.
External potential:

$$U(x) = \operatorname{sn}^{-2} 2x \,(A + B \operatorname{dn} 2x) + \operatorname{cn}^{-2} 2x \,(C + D \operatorname{dn} 2x)\,.$$

Interaction potential:

$$V_{\text{int}}(\mathbf{x}) = \sum_{i \neq j} \left(\frac{\operatorname{dn}^2 x_{ij}^-}{\operatorname{sn}^2 x_{ij}^- \operatorname{cn}^2 x_{ij}^-} + \frac{\operatorname{dn}^2 x_{ij}^+}{\operatorname{sn}^2 x_{ij}^+ \operatorname{cn}^2 x_{ij}^+} \right) a(a + S_{ij})\,.$$

Case 9. $P(z) = 4(1 - z^2)(k'^2 + k^2 z^2)$.
External potential:

$$U(x) = \operatorname{dn}^{-2} 2x \,(A + B \operatorname{cn} 2x) + \operatorname{sn}^{-2} 2x \,(C + D \operatorname{cn} 2x)\,.$$

Interaction potential:

$$V_{\text{int}}(\mathbf{x}) = \sum_{i \neq j} \left(\frac{\operatorname{cn}^2 x_{ij}^-}{\operatorname{sn}^2 x_{ij}^- \operatorname{dn}^2 x_{ij}^-} + \frac{\operatorname{cn}^2 x_{ij}^+}{\operatorname{sn}^2 x_{ij}^+ \operatorname{dn}^2 x_{ij}^+} \right) a(a + S_{ij})\,.$$

Remark 2.8. Some of the potentials that appear in this classification have been previously studied by other authors. Case 1 with $A = B = 0$ yields the rational Calogero A_N spin model. Case 5 with $A = B = 0$ or $C = D = 0$ is the model studied by Inozemtsev [39], while for $A = B = C = D = 0$ the hyperbolic Sutherland A_N spin model is obtained. Case 6 with $A = B = C = D = 0$ is the trigonometric Sutherland A_N spin model. The remaining potentials were first introduced in Ref. [19].

Remark 2.9. In Cases 1–5, the potential is ES if $A = B = 0$. The only ES potential in Case 6 is the trigonometric Sutherland potential ($A = B = C = D = 0$). The remaining potentials, including all the elliptic potentials in Cases 7–9, are QES.

Remark 2.10. We emphasize that the spin potentials in the above list reduce to (Q)ES scalar many-body potentials by the substitution $S_{ij} \to -1$. This is a straightforward consequence of the fact that the operator $\overline{H}$ preserves the space $\mathcal{S}_m(\mathbf{z})$ of symmetric polynomials in $\mathbf{z}$ of degree at most m, where the permutation operators K_{ij} act as the identity. This implies that the *scalar* Hamiltonian

$$H|_{S_{ij}\to -1} = \mu(\mathbf{z}) \cdot \overline{H}(\mathbf{z})|_{K_{ij}\to 1} \cdot \mu^{-1}(\mathbf{z})\Big|_{\mathbf{z}=\zeta(\mathbf{x})}$$

preserves the space $\boldsymbol{\mu}(\boldsymbol{\zeta}(\mathbf{x}))\boldsymbol{\mathcal{S}_m}(\boldsymbol{\zeta}(\mathbf{x}))$. The list of potentials thus obtained includes all the previously known examples of scalar (Q)ES CS models.

Remark 2.11. All the potentials in the classification diverge on the hyperplanes $x_i = x_j$, $1 \le i < j \le N$, as $(x_{ij}^-)^{-2}$. In some cases there are additional inverse-square singularities on other hyperplanes. In order to qualify as physical wavefunctions, the eigenfunctions of H in $\mathfrak{R}_m$ must vanish at all these hyperplanes faster than the square root of the distance to the hyperplane. In addition, in Cases 1–3 and 5 the eigenfunctions are required to be square-integrable over a suitable domain of $\mathbb{R}^N$. Both requirements impose certain constraints on the parameters of the potential (see Ref. [25] for a complete treatment of this problem in the one-particle case). For example, in Case 5 the change of variable and gauge factor are respectively given by

$$z = \mathrm{e}^{2x}, \quad \mu(\mathbf{x}) = \prod_{i<j} |\sinh x_{ij}^-|^a \prod_i \mathrm{e}^{(\frac{c_0}{4}-m)x_i} \exp\left[\frac{1}{8}\Big(c_+\mathrm{e}^{2x_i} - c_-\mathrm{e}^{-2x_i}\Big)\right].$$

It follows that the spin functions $\psi \in \mathfrak{R}_m$ vanish at the singularities $x_{ij}^- = 0$ of the potential faster than $|x_{ij}^-|^{1/2}$ and are square-integrable over the domain $\{\mathbf{x} \in \mathbb{R}^N : x_1 > \cdots > x_N\}$ *provided* $a > 1/2$, $\alpha_+ < 0$ and $\alpha_- > 0$.

Remark 2.12. The Hamiltonian H describes a system of identical particles, whose corresponding states must thus be either completely symmetric or antisymmetric under particle permutations. Assume for definiteness that we are dealing with a system of fermions[2]. By construction, the elements of $\mathfrak{R}_m$ are completely antisymmetric and hence qualify as physical states provided the parameters of the potential satisfy certain constraints (see Remark 2.11). We can also construct completely antisymmetric states of the Hamiltonian H in the following alternative way. Consider the symmetrizer $\widetilde{\Lambda}$ defined by the relations $\widetilde{\Lambda}^2 = \widetilde{\Lambda}$ and $\Pi_{ij}\widetilde{\Lambda} = \widetilde{\Lambda}$. (Re)define the star mapping substituting Eq. (2.13) with the requirement that

$$(DK_{i_1 j_1} \cdots K_{i_r j_r})^* = DS_{i_r j_r} \cdots S_{i_1 j_1}\,, \tag{2.28}$$

where D is a scalar differential operator. Consider the gauge factor $\tilde{\mu}$ defined as the antisymmetric extension to $\mathbb{R}^N$ of the gauge factor μ restricted to the domain $\Omega \equiv \{\mathbf{z} \in \mathbb{R}^N \,:\, z_1 > \cdots > z_N\}$. In other words,

$$\tilde{\mu}(\mathbf{z}; a) = (-1)^{\varepsilon(i_1,\ldots,i_N)} \mu(\mathbf{z}; a)\,,$$

where $\mu(\mathbf{z}; a)$ is given in Eq. (2.23), $(i_1, \ldots, i_N) \in S_N$ is the permutation such that $(z_{i_1}, \ldots, z_{i_N}) \in \Omega$, and ε is the parity. We observe that the hamiltonian H is the image of the operator $\overline{H}(-a)$ (with $\overline{H}(a)$ given by Eq. (2.10)) under the new star mapping (2.28), the gauge transformation with gauge factor $\tilde{\mu}(\mathbf{z}; -a)$, and the change of variables (2.16), namely,

$$H = \tilde{\mu}(\mathbf{z}; -a) \cdot \overline{H}^*(\mathbf{z}; -a) \cdot \tilde{\mu}^{-1}(\mathbf{z}; -a)\Big|_{\mathbf{z}=\zeta(\mathbf{x})}\,.$$

It follows that H preserves the finite-dimensional space of completely antisymmetric spin functions

$$\widetilde{\mathfrak{R}}_m = \tilde{\mu}(\zeta(\mathbf{x}); -a)\, \widetilde{\Lambda}\Big(\mathcal{R}_m(\zeta(\mathbf{x})) \otimes \mathfrak{S}\Big)\,. \tag{2.29}$$

Note that the Hamiltonian H may not possess physical wavefunctions in both spaces $\mathfrak{R}_m$ and $\widetilde{\mathfrak{R}}_m$ at the same time. For instance, in Case 5, the functions in $\widetilde{\mathfrak{R}}_m$ do not vanish in general faster than $|x_{ij}^-|^{1/2}$ at the singularities $x_{ij}^- = 0$ of the potential if $a > 1/2$ (see Remark 2.11).

[2] This assumption covers the physically most interesting case for which the internal degrees of freedom are naturally interpreted as the particles' spin.

3. Calogero–Sutherland spin models of BC_N type

3.1. DUNKL OPERATORS OF B_N TYPE

In the BC_N case, our construction makes use of the following three families of Dunkl operators[3]:

$$\begin{aligned} J_i^- &= \partial_{z_i} + a\Bigg(\sum_{j\neq i}\frac{1}{z_{ij}^-}\,(1-K_{ij}) + \sum_{j\neq i}\frac{1}{z_{ij}^+}\,(1-\tilde K_{ij})\Bigg) + \frac{b}{z_i}\,(1-K_i)\,,\\ J_i^0 &= z_i\partial_{z_i} - \frac{m}{2} + \frac{a}{2}\Bigg(\sum_{j\neq i}\frac{z_{ij}^+}{z_{ij}^-}\,(1-K_{ij}) + \sum_{j\neq i}\frac{z_{ij}^-}{z_{ij}^+}\,(1-\tilde K_{ij})\Bigg), \qquad (3.1)\\ J_i^+ &= z_i^2\partial_{z_i} - m\,z_i + a\Bigg(\sum_{j\neq i}\frac{z_i z_j}{z_{ij}^-}\,(1-K_{ij}) - \sum_{j\neq i}\frac{z_i z_j}{z_{ij}^+}\,(1-\tilde K_{ij})\Bigg)\\ &\quad - b' z_i(1-(-1)^m K_i), \end{aligned}$$

where $i = 1,\dots,N$, a, b, b' are real parameters, $\tilde K_{ij} \equiv K_i K_j K_{ij}$, and K_i are sign-reversing operators. The action of the latter operators on a function $f(\mathbf{z})$ is given by

$$(K_i f)(z_1,\dots,z_i,\dots,z_N) = f(z_1,\dots,-z_i,\dots,z_N)\,.$$

We note that the operators K_{ij} and K_i span the Weyl group of type B_N. The first two families in Eq (3.1) were introduced by Dunkl in Ref. [16], while the third one was defined by the authors in Ref. [19]. Each family, together with the operators K_{ij} and K_i span a degenerate affine Hecke algebra. They also satisfy the commutation relations (2.6), along with

$$[K_i, J_j^\epsilon] = 0\,, \qquad K_i J_i^\epsilon = (-1)^\epsilon J_i^\epsilon K_i\,, \qquad (3.2)$$

where $i \neq j$. Just as in the A_N case, the key property regarding the (quasi)-exact solvability of the resulting models is the invariance of the module $\mathcal{R}_m$ under these operators. In fact, Theorem 2.1 and its Corollary 2.2 are both satisfied by the B_N-type Dunkl operators (3.1).

3.2. CONSTRUCTION OF SPIN CS MODELS OF BC_N TYPE

In the B_N case, we shall use the spin permutation operators S_{ij} and the spin reversing operators S_i, whose action on a spin state $|s_1,\dots,s_N\rangle$ is given by

$$S_i|s_1,\dots,s_i,\dots,s_N\rangle = |s_1,\dots,-s_i,\dots,s_N\rangle\,.$$

[3] No confusion should arise by the fact *both* the A_N and B_N-type Dunkl operators are denoted as J_i^ϵ, $\epsilon = \pm, 0$.

We shall also employ the customary notation $\tilde{S}_{ij} = S_i S_j S_{ij}$.

In the BC_N case, the outset of our construction is the following quadratic combination of the B_N-type Dunkl operators (3.1):

$$-\overline{H} = \sum_i \Big(c_{++}(J_i^+)^2 + c_{00}(J_i^0)^2 + c_{--}(J_i^-)^2 + c_0\, J_i^0 \Big) . \tag{3.3}$$

The terms $\sum_i \{J_i^\pm, J_i^0\}$ and $\sum_i J_i^\pm$ which appear in the A_N case (see Eq. (2.10)) have now been discarded in order to ensure that $\overline{H}$ commutes with both the permutation and sign-reversing operators K_i. Note also that $\overline{H}$ preserves the space $\mathcal{R}_m(\mathbf{z})$ (or the spaces $\mathcal{R}_n(\mathbf{z})$ and $\mathcal{T}_n(\mathbf{z})$ for arbitrary n, if $c_{++} = 0$).

Let Λ' be the total antisymmetrizer with respect to permutations and sign-reversals of the particles' coordinates and spins, which is determined by the relations

$$\Lambda'^2 = \Lambda', \qquad K_{ij}\Lambda' = -S_{ij}\Lambda', \qquad K_i\Lambda' = -S_i\Lambda', \tag{3.4}$$

for $j > i = 1, \dots, N$. It may be easily shown that

$$\Lambda' = \frac{\Lambda}{2^N} \prod_i (1 - \Pi_i) ,$$

where $\Pi_i \equiv K_i S_i$. Since $\overline{H}$ commutes with K_{ij} and K_i and acts trivially in $\mathfrak{S}$, it commutes with Λ' and thus preserves $\Lambda'(\mathcal{R}_m \otimes \mathfrak{S})$. We now define the linear mapping * by the condition

$$(DK_{\alpha_1} \cdots K_{\alpha_r})^* = (-1)^r\, DS_{\alpha_r} \cdots S_{\alpha_1} , \tag{3.5}$$

where D is a scalar differential operator and the subindex α stands either for ij or i. Just as before, this definition and Eq. (3.4) imply that $\overline{H}\Lambda' = \overline{H}^*\Lambda'$. It follows that $\overline{H}^*$ also preserves $\Lambda'(\mathcal{R}_m \otimes \mathfrak{S})$ (or $\Lambda'(\mathcal{R}_n \otimes \mathfrak{S})$ and $\Lambda'(\mathcal{T}_n \otimes \mathfrak{S})$ for arbitrary n, if $c_{++} = 0$).

Inserting the expressions of the Dunkl operators (3.1) into (3.3), one obtains

$$\begin{aligned}
-\overline{H}^* = {} & \sum_i \Big(P(z_i)\partial_{z_i}^2 + \tilde{Q}(z_i)\partial_{z_i} + R(z_i) \Big) + 4a \sum_{i \neq j} \frac{z_i P(z_i)}{z_i^2 - z_j^2}\, \partial_{z_i} \\
& - \sum_i \Big(\frac{b\, c_{--}}{z_i^2} (1 + S_i) + b' c_{++} z_i^2 (1 + (-1)^m S_i) \Big) \\
& - a \sum_{i \neq j} P(z_i) \Big(\frac{1 + S_{ij}}{(z_{ij}^-)^2} + \frac{1 + \tilde{S}_{ij}}{(z_{ij}^+)^2} \Big) \\
& + \frac{a\, c_{++}}{2} \sum_{i \neq j} \Big((z_{ij}^+)^2 (1 + S_{ij}) + (z_{ij}^-)^2 (1 + \tilde{S}_{ij}) \Big) + \overline{C} ,
\end{aligned} \tag{3.6}$$

where

$$
\begin{aligned}
P(z) &= c_{++}z^4 + c_{00}z^2 + c_{--}\,,\\
\tilde{Q}(z) &= 2c_{++}(1-m-b'+2a(1-N))z^3\\
&\qquad + \Big(c_0 + c_{00}(1-m+2a(1-N))\Big)z + \frac{2bc_{--}}{z}\,,\\
R(z) &= c_{++}m(m-1+2b')z^2\,,\\
\overline{C} &= c_{00}\Bigg[\frac{Nm^2}{4} + \frac{a^2}{12}\bigg(\sum_{i,j,k}{}'[4-(S_{ij}+\tilde{S}_{ij})(S_{ik}+\tilde{S}_{ik})]\\
&\qquad + 6\sum_{i\neq j}(1-S_iS_j)\bigg) + \frac{a}{2}\sum_{i\neq j}(2+S_{ij}+\tilde{S}_{ij})\Bigg] - \frac{Nmc_0}{2}\,.
\end{aligned}
$$

We observe that the first-derivatives terms of $\overline{H}^*$ do not involve the spin operators, which otherwise makes it very difficult (if not impossible) to find a gauge transformation reducing $\overline{H}^*$ to a Schrödinger operator[4]. The existence of such gauge factor is one of the main ingredients in our construction.

Theorem 3.1 *The operator $\overline{H}^*$ in Eq. (3.6) can be reduced to a Schrödinger operator H by the change of variables*

$$
x_i = \zeta^{-1}(z_i) = \int^{z_i} \frac{\mathrm{d}y}{\sqrt{P(y)}}\,, \qquad i = 1,\dots,N, \tag{3.7}
$$

and the gauge transformation generated by

$$
\mu(\mathbf{z}) = \prod_{i<j} |z_i^2 - z_j^2|^a \prod_i P(z_i)^{-\frac14} \exp\int^{z_i} \frac{\tilde{Q}(y_i)}{2P(y_i)}\,\mathrm{d}y_i\,. \tag{3.8}
$$

The corresponding potential is

$$
\begin{aligned}
V(\mathbf{x}) = a\sum_{i\neq j} P(z_i)\bigg(&\frac{a+S_{ij}}{(z_{ij}^-)^2} + \frac{a+\tilde{S}_{ij}}{(z_{ij}^+)^2}\bigg)\\
&- \frac{c_{++}}{2}\Big((z_{ij}^+)^2(1+S_{ij}) + (z_{ij}^-)^2(1+\tilde{S}_{ij})\Big)\\
&+ \sum_i \bigg(b'c_{++}\,z_i^2(1+(-1)^m S_i) + \frac{b\,c_{--}}{z_i^2}\,(1+S_i) + W(z_i)\bigg)\\
&+ aN(N-1)\Big[c_0 - \frac{c_{00}}{3}\,(a(2N-1)+3(m-1))\Big] - \overline{C}\,\bigg|_{\mathbf{z}=\zeta(\mathbf{x})}\,,
\end{aligned} \tag{3.9}
$$

[4] For this reason, we have omitted the term $\sum_i [J_i^+, J_i^-]$ in (3.3).

where

$$W(z) = \frac{1}{2}\Big(\tilde{Q}' - \frac{P''}{2}\Big) + \frac{1}{4P}\Big(\tilde{Q} - \frac{P'}{2}\Big)\Big(\tilde{Q} - \frac{3P'}{2}\Big) + c\,z^2\,,$$

and

$$\begin{aligned} c = & -c_{++}\Big(2a^2(N-1)(2N-1) + 4a(N-1)(b'+m-1) \\ & + m(2b'+m-1)\Big)\,. \end{aligned}$$

Corollary 3.2 *The Schrödinger operator* (2.15) *with potential* (3.9) *leaves invariant the finite-dimensional space*

$$\mathfrak{R}'_m = \mu(\zeta(\mathbf{x}))\boldsymbol{\Lambda}'\Big(\mathcal{R}_m(\zeta(\mathbf{x})) \otimes \mathfrak{S}\Big) \tag{3.10}$$

In addition, if $c_{++} = 0$*, it preserves the space* $\mathfrak{R}'_n$ *and*

$$\mathfrak{T}'_n = \mu(\zeta(\mathbf{x}))\boldsymbol{\Lambda}'\Big(\mathcal{T}_n(\zeta(\mathbf{x})) \otimes \mathfrak{S}\Big), \tag{3.11}$$

for any nonnegative integer n.

Remark 3.3. As in the A_N case, each physical coordinate x_i is defined up to an arbitrary translation, which can be fixed in most cases by requiring that the resulting potentials be invariant under the realization of the Weyl group of B_N type spanned by Π_{ij} and Π_i.

3.3. CLASSIFICATION OF THE MODELS OF BC_N TYPE

We shall now perform a complete classification of the (Q)ES spin Calogero–Sutherland models determined by quadratic combinations of the form (3.3). In contrast to the A_N case, the vector spaces $\langle J_i^-, J_i^0, J_i^+\rangle$, $i = 1, \dots, N$, are no longer invariant under the full projective action (2.21)–(2.23) of $\mathrm{GL}(2, \mathbb{R})$. However, these vector spaces are still *form*-invariant under inversions and scalings. Indeed, consider the mappings

$$J_i^\epsilon(\mathbf{w}) \;\mapsto\; \widehat{J_i^\epsilon}(\mathbf{z}) = \Big(\prod_k z_k^m\Big) J_i^\epsilon(\mathbf{w}(\mathbf{z}))\Big(\prod_k z_k^{-m}\Big), \quad z_j = \frac{1}{w_j}\,; \tag{3.12a}$$

$$J_i^\epsilon(\mathbf{w}) \;\mapsto\; \widehat{J_i^\epsilon}(\mathbf{z}) = J_i^\epsilon(\mathbf{w}(\mathbf{z})), \quad z_j = \lambda w_j\,, \tag{3.12b}$$

where $i, j = 1, \dots, N$ and $\lambda \neq 0$ is real or purely imaginary.

Lemma 3.4 *The transformed Dunkl operators* $\widehat{J_i^\epsilon}$ *under the mappings* (3.12) *are respectively given by*

$$\widehat{J_i^-} = -J_i^+\Big|_{b'\to b}\,, \qquad \widehat{J_i^0} = -J_i^0\,, \qquad \widehat{J_i^+} = -J_i^-\Big|_{b\to b'}\,; \tag{3.13a}$$

$$\widehat{J_i^-} = \lambda\, J_i^-\,, \qquad \widehat{J_i^0} = J_i^0\,, \qquad \widehat{J_i^+} = \lambda^{-1} J_i^+\,. \tag{3.13b}$$

It follows from the previous lemma that the transformed of the operator $\overline{H}$ under the mappings (3.12) is still of the form (3.3), with (in general) different coefficients $\hat{c}_{++}$, $\hat{c}_{00}$, $\hat{c}_{--}$ and $\hat{c}_0$. Using these transformations, the polynomial P in (3.6) can be reduced to one of the following canonical forms:

1) 1,
2) $\pm\,\nu z^2$,
3) $\pm\,\nu(1+z^2)$,
4) $\pm\,\nu(1-z^2)^2$,
5) $\nu(e^{2i\theta}-z^2)(e^{-2i\theta}-z^2)$,
6) $\pm\,\nu(1-z^2)(1-k^2z^2)$,
7) $\nu(1-z^2)(1-k^2+k^2z^2)$,

where $\nu>0$, $0<k<1$, and $0<\theta\leq\pi/4$. It turns out that for each of these canonical forms except the second ones, the resulting potentials are invariant under the Weyl group of B_N type provided the arbitrary constants which appear in the change of variables (3.7) are suitably chosen. The potential in each case can be expressed as

$$V(\mathbf{x}) = \sum_i U(x_i) + V_{\text{spin}}(\mathbf{x})\,,$$

where the first term does not contain the spin operators S_{ij}, S_i and thus can be viewed as the contribution of a scalar external field. Without lost of generality, cf. Eq. (2.27), we shall present the list of potentials for a particular value of the parameter ν. We shall express the external potentials in terms of three parameters, α, β, β', where

$$\alpha = a(N-1) + \frac{1}{2}\,(b+b'+m)\,, \tag{3.14}$$

and β, β' are given in Table II. We shall also drop some inessential constant terms which commute Λ' and therefore preserve $\mathfrak{R}'_n$ and $\mathfrak{T}'_n$ for all n. We refer the reader to Ref. [20] for the explicit expressions of the change of variable and gauge factor in each case.

Case 1. $P(z)=1\,.$
Scalar external potential:

$$U(x) = \beta x^2\,.$$

Spin potential:

$$\begin{aligned} V_{\text{spin}}(\mathbf{x}) &= a \sum_{i \neq j} \Big[(x_{ij}^-)^{-2} \, (a + S_{ij}) + (x_{ij}^+)^{-2} \, (a + \tilde{S}_{ij}) \Big] \\ &\quad + b \sum_i x_i^{-2} \, (b + S_i) \, . \end{aligned}$$

Case 2a. $P(z) = 4\, z^2$.
Scalar external potential:

$$U(x) = 0 \, .$$

Spin potential:

$$V_{\text{spin}}(\mathbf{x}) = a \sum_{i \neq j} \Big[\sinh^{-2} x_{ij}^- \, (a + S_{ij}) - \cosh^{-2} x_{ij}^- \, (a + \tilde{S}_{ij}) \Big] \, .$$

Case 2b. $P(z) = -4\, z^2$.
Scalar external potential:

$$U(x) = 0 \, .$$

Spin potential:

$$V_{\text{spin}}(\mathbf{x}) = a \sum_{i \neq j} \Big[\sin^{-2} x_{ij}^- \, (a + S_{ij}) + \cos^{-2} x_{ij}^- \, (a + \tilde{S}_{ij}) \Big] \, .$$

Case 3a. $P(z) = 4\,(1 + z^2)$.
Scalar external potential:

$$U(x) = -4\beta(\beta - 1) \cosh^{-2} 2x \, .$$

Spin potential:

$$\begin{aligned} V_{\text{spin}}(\mathbf{x}) &= a \sum_{i \neq j} \Big[(\, \sinh^{-2} x_{ij}^- - \cosh^{-2} x_{ij}^+) \, (a + S_{ij}) \\ &\qquad + (\, \sinh^{-2} x_{ij}^+ - \cosh^{-2} x_{ij}^-) \, (a + \tilde{S}_{ij}) \Big] \\ &\quad + 4b \sum_i \sinh^{-2} 2x_i \, (b + S_i) \, . \end{aligned}$$

Case 3b. $P(z) = -4\,(1 + z^2)$.
Scalar external potential:

$$U(x) = 4\beta(\beta - 1) \cos^{-2} 2x \, .$$

Spin potential:

$$\begin{aligned} V_{\text{spin}}(\mathbf{x}) &= a\sum_{i\neq j}\Big[(\sin^{-2}x_{ij}^- + \cos^{-2}x_{ij}^+)\,(a+S_{ij}) \\ &\qquad\qquad + (\sin^{-2}x_{ij}^+ + \cos^{-2}x_{ij}^-)\,(a+\tilde S_{ij})\Big] \\ &\quad + 4b\sum_i \sin^{-2}2x_i\,(b+S_i)\,. \end{aligned}$$

Case 4a. $P(z) = (1-z^2)^2$.
Scalar external potential:

$$U(x) = 2\beta^2\cosh 4x + 4\beta(1+2\alpha)\cosh 2x\,.$$

Spin potential:

$$\begin{aligned} V_{\text{spin}}(\mathbf{x}) &= a\sum_{i\neq j}\Big[\sinh^{-2}x_{ij}^-\,(a+S_{ij}) + \sinh^{-2}x_{ij}^+\,(a+\tilde S_{ij})\Big] \\ &\quad + b\sum_i \sinh^{-2}x_i\,(b+S_i) - b'\sum_i \cosh^{-2}x_i\,(b'+(-1)^m S_i). \end{aligned}$$

Case 4b. $P(z) = -(1-z^2)^2$.
Scalar external potential:

$$U(x) = -2\beta^2\cos 4x - 4\beta(1+2\alpha)\cos 2x\,.$$

Spin potential:

$$\begin{aligned} V_{\text{spin}}(\mathbf{x}) &= a\sum_{i\neq j}\Big[\sin^{-2}x_{ij}^-\,(a+S_{ij}) + \sin^{-2}x_{ij}^+\,(a+\tilde S_{ij})\Big] \\ &\quad + b\sum_i \sin^{-2}x_i\,(b+S_i) + b'\sum_i \cos^{-2}x_i\,(b'+(-1)^m S_i)\,. \end{aligned}$$

Case 5. $P(z) = (\mathrm{e}^{2\mathrm{i}\theta} - z^2)(\mathrm{e}^{-2\mathrm{i}\theta} - z^2)$.

The modulus of the elliptic functions in this case is $k = \cos\theta$.

Scalar external potential:

$$U(x) = 4k'^2\,\mathrm{dn}^{-2}2x\left[\beta^2 - \alpha(\alpha+1) - \frac{k\beta}{k'}(1+2\alpha)\,\mathrm{cn}\,2x\right].$$

Spin potential:

$$\begin{aligned} V_{\mathrm{spin}}(\mathbf{x}) &= a\sum_{i\neq j}\bigg[\bigg(\frac{\mathrm{dn}^2x_{ij}^-}{\mathrm{sn}^2x_{ij}^-} - k^2k'^2\,\frac{\mathrm{sn}^2x_{ij}^+}{\mathrm{dn}^2x_{ij}^+}\bigg)\,(a+S_{ij}) \\ &\qquad + \bigg(\frac{\mathrm{dn}^2x_{ij}^+}{\mathrm{sn}^2x_{ij}^+} - k^2k'^2\,\frac{\mathrm{sn}^2x_{ij}^-}{\mathrm{dn}^2x_{ij}^-}\bigg)\,(a+\tilde S_{ij})\bigg] \\ &\quad + b\sum_i\Big(\frac{\mathrm{cn}\,x_i}{\mathrm{sn}\,x_i\,\mathrm{dn}\,x_i}\Big)^2(b+S_i) \\ &\quad + b'\sum_i\Big(\frac{\mathrm{sn}\,x_i\,\mathrm{dn}\,x_i}{\mathrm{cn}\,x_i}\Big)^2(b'+(-1)^mS_i)\,. \end{aligned}$$

Case 6a. $P(z) = 4(1-z^2)(1-k^2\,z^2)\,.$

Here and in the remaining cases the elliptic functions have modulus k.

Scalar external potential:

$$U(x) = 4k'^2\Big[\beta(\beta-1)\,\mathrm{cn}^{-2}2x - \beta'(\beta'-1)\,\mathrm{dn}^{-2}2x\Big]\,.$$

Spin potential:

$$\begin{aligned} V_{\mathrm{spin}}(\mathbf{x}) &= a\sum_{i\neq j}\bigg[\bigg(\frac{\mathrm{cn}^2x_{ij}^-\mathrm{dn}^2x_{ij}^-}{\mathrm{sn}^2x_{ij}^-} + k'^4\frac{\mathrm{sn}^2x_{ij}^+}{\mathrm{cn}^2x_{ij}^+\mathrm{dn}^2x_{ij}^+}\bigg)\,(a+S_{ij}) \\ &\qquad + \bigg(\frac{\mathrm{cn}^2x_{ij}^+\mathrm{dn}^2x_{ij}^+}{\mathrm{sn}^2x_{ij}^+} + k'^4\frac{\mathrm{sn}^2x_{ij}^-}{\mathrm{cn}^2x_{ij}^-\mathrm{dn}^2x_{ij}^-}\bigg)\,(a+\tilde S_{ij})\bigg] \\ &\quad + 4b\sum_i\mathrm{sn}^{-2}2x_i\,(b+S_i) + 4k^2b'\sum_i\mathrm{sn}^22x_i\,(b'+(-1)^mS_i)\,. \end{aligned}$$

Case 6b. $P(z) = -4(1-z^2)(1-k'^2\,z^2)\,.$

Scalar external potential:

$$U(x) = 4\Big[\beta(\beta-1)\,k^2\mathrm{sn}^22x - \beta'(\beta'-1)\,k'^2\mathrm{dn}^{-2}2x\Big]\,.$$

Spin potential:

$$\begin{aligned} V_{\mathrm{spin}}(\mathbf{x}) &= a\sum_{i\neq j}\bigg[\bigg(\frac{\mathrm{dn}^2x_{ij}^-}{\mathrm{sn}^2x_{ij}^-\mathrm{cn}^2x_{ij}^-} + k^4\,\frac{\mathrm{sn}^2x_{ij}^+\mathrm{cn}^2x_{ij}^+}{\mathrm{dn}^2x_{ij}^+}\bigg)\,(a+S_{ij}) \\ &\qquad + \bigg(\frac{\mathrm{dn}^2x_{ij}^+}{\mathrm{sn}^2x_{ij}^+\mathrm{cn}^2x_{ij}^+} + k^4\,\frac{\mathrm{sn}^2x_{ij}^-\mathrm{cn}^2x_{ij}^-}{\mathrm{dn}^2x_{ij}^-}\bigg)\,(a+\tilde S_{ij})\bigg] \\ &\quad + 4b\sum_i\mathrm{sn}^{-2}2x_i\,(b+S_i) + 4k'^2b'\sum_i\mathrm{cn}^{-2}2x_i\,(b'+(-1)^mS_i)\,. \end{aligned}$$

Case 7. $P(z) = 4(1-z^2)(k'^2 + k^2\, z^2)\,.$
Scalar external potential:

$$U(x) = 4\Big[\beta'(\beta'-1)\,k^2\mathrm{sn}^2 2x + \beta(\beta-1)\,k'^2\mathrm{cn}^{-2}2x\Big]\,.$$

Spin potential:

$$\begin{aligned} V_{\mathrm{spin}}(\mathbf{x}) &= a\sum_{i\neq j}\Bigg[\bigg(\frac{\mathrm{cn}^2x_{ij}^-}{\mathrm{sn}^2x_{ij}^-\mathrm{dn}^2x_{ij}^-}+\frac{\mathrm{sn}^2x_{ij}^+\mathrm{dn}^2x_{ij}^+}{\mathrm{cn}^2x_{ij}^+}\bigg)(a+S_{ij}) \\ &\qquad + \bigg(\frac{\mathrm{cn}^2x_{ij}^+}{\mathrm{sn}^2x_{ij}^+\mathrm{dn}^2x_{ij}^+}+\frac{\mathrm{sn}^2x_{ij}^-\mathrm{dn}^2x_{ij}^-}{\mathrm{cn}^2x_{ij}^-}\bigg)(a+\tilde{S}_{ij})\Bigg] \\ &\quad + 4b\sum_i \mathrm{sn}^{-2}2x_i\,(b+S_i) - 4k'^2b'\sum_i \mathrm{dn}^{-2}2x_i\,(b'+(-1)^mS_i)\,. \end{aligned}$$

Remark 3.5. The potentials in Cases 1, 2, and 3 are ES for all values of the parameters. The potentials of type 4 are also ES for $\beta = 0$. The elliptic potentials in Cases 5–7 are all QES. The potentials in Case 2 are *not* invariant under the B_N Weyl group, due to the fact that a sign reversal in the variable z_i yields a translation in the corresponding physical variable x_i.

Remark 3.6. The potential in Case 1 is the rational B_N-type model introduced by Yamamoto [68] and later studied by Dunkl [16]. Case 4b for $\beta = 0$ is Yamamoto's B_N-type trigonometric potential with $\lambda_1 = -b$ (in the notation of Ref. [68]), and either $\lambda_1' = -b'$ for m even or $\lambda_1' = b'$ for m odd. The remaining potentials were introduced in [20]. The spectrum and integrability of the hyperbolic Yamamoto model (Case 4a) have been recently studied in detail in Ref. [21].

Remark 3.7. The potentials in the classification diverge as $(x_{ij}^-)^{-2}$ at the hyperplanes $x_i = x_j$, $1 \le i < j \le N$. Depending on the case, the potential may possess inverse-square singularities at some other hyperplanes. The physical eigenfunctions are required to vanish faster than the square root distance to any of these hyperplanes, which yield some constraints on the parameters defining the potential. In addition, when the singular hyperplanes do not divide $\mathbb{R}^N$ into bounded disjoint sets, one has to impose further constraints in order to ensure the square-integrability of the eigenfunctions (see Ref. [20] for the complete list of constraints in each case). For example, in Case 3a the change of variables and gauge factor can be respectively taken as

$$z = \sinh 2x\,, \tag{3.15}$$

$$\mu(\mathbf{x}) = \prod_{i<j} \left| \sinh 2x_{ij}^- \sinh 2x_{ij}^+ \right|^a \prod_i |\sinh 2x_i|^b \cosh^\beta 2x_i\,, \tag{3.16}$$

where β is given in Table II. It follows that the spin functions $\psi \in \mathfrak{R}'_m$ vanish at the singularities $x_{ij}^\pm = 0$ and $x_i = 0$ of the potential faster than the square root distance to these hyperplanes and are square-integrable over the domain $\{\mathbf{x} \in \mathbb{R}^N \,:\, x_1 > \cdots > x_N > 0\}$ if and only if $a > 1/2$, $b > 1/2$ and $\beta < -(2a(N-1)+b+m)$.

Remark 3.8. In all cases except the second one, we can obtain physical wavefunctions of a given parity by a suitable definition of the gauge factor in each of the domains determined by the singular hyperplanes. For instance, in Case 3a, the wavefunctions in $\mathfrak{R}'_m$ (with the gauge factor (3.16)) are anti-symmetric under simultaneous sign-reversal of any particle's coordinate and spin. Note that we can also construct odd-parity fermionic wavefunctions by replacing Λ' by any of the projectors

$$\frac{\Lambda}{2^N}\prod_i (1+\Pi_i)\,, \qquad \frac{\widetilde{\Lambda}}{2^N}\prod_i (1\pm\Pi_i)\,,$$

where $\widetilde{\Lambda}$ is the symmetrizer under particles' permutations, and redefining the star mapping (3.5) and the gauge factor appropriately (see Remark 2.12).

We shall now present a formula which describes all the BC_N potentials in the classification in a unified way. In the first place, the spin potential can be written (barring some irrelevant constant terms) as

$$\begin{aligned} V_{\rm spin}(\mathbf{x}) &= a\sum_{i\neq j}\Big[\Big(v(x_{ij}^-) + v(x_{ij}^+ + P_1)\Big)(a+S_{ij}) \\ &\qquad\qquad + \Big(v(x_{ij}^+) + v(x_{ij}^- + P_1)\Big)(a+\tilde{S}_{ij})\Big] \\ &\quad + b\sum_i \big(v(x_i)+v(x_i+P_1)\big)(b+S_i) \\ &\quad + b'\sum_i \big(v(x_i+P_2)+v(x_i+P_1+P_2)\big)(b'+(-1)^m S_i)\,, \end{aligned} \tag{3.17}$$

where v is a (possibly degenerate) elliptic function with primitive half-periods P_1 and P_2 (see Table III). In this formula, expressions like $v(x+P_i)$ or $v(x+P_1+P_2)$ are defined as zero if any of the periods are infinite. Using this notation, in Cases 3 and 5–7 the scalar external potential $U(x)$ can be written as

$$\begin{aligned} U(x) &= \lambda(\lambda-1)\Big[v(x+\tfrac{1}{2}P_1) + v(x-\tfrac{1}{2}P_1)\Big] \\ &\quad + \lambda'(\lambda'-1)\Big[v(x+\tfrac{1}{2}P_1+P_2) + v(x-\tfrac{1}{2}P_1+P_2)\Big], \end{aligned} \tag{3.18}$$

where $\lambda=\beta$, $\lambda'=0$ in Cases 3, $\lambda=-\alpha+\mathrm{i}\beta$, $\lambda'=-\alpha-\mathrm{i}\beta$ in Case 5, and $\lambda=\beta$, $\lambda'=\beta'$ in Cases 6–7. In Cases 1 and 4 Eq. (3.18) cannot be directly applied, since in these cases all the terms in (3.18) are either indeterminate or zero. However, the potentials in Cases 4a and 4b can be obtained from that of Case 5 in the limits $\theta\to 0$ and $\theta\to\pi/2$, respectively. Likewise, applying the rescaling $x_i\mapsto \nu x_i$ ($i=1,\dots,N$, $\nu>0$) to the potential of type 3a or 3b one obtains the potential of type 1 by replacing β by $\beta/(4\nu^2)$ and letting $\nu\to 0$.

In Cases 5–7, the function v that determines the potential $V(\mathbf{x})$ can be expressed up to an irrelevant additive constant in terms of the Weierstrass $\wp$ function as

$$v(x)=\wp(x;P_1,P_2)\,, \tag{3.19}$$

where the primitive half-periods P_1 and P_2 are listed in Table III. Substituting Eq. (3.19) into Eqs. (3.17) and (3.18), and applying a modular transformation of the Weierstrass function [45] to the one-particle terms we obtain the following remarkable formula for the potential $V(\mathbf{x})$ in Cases 5–7:

$$\begin{aligned}
V(\mathbf{x}) &= a\sum_{i\neq j}\Big[\Big(\wp(x_{ij}^-;P_1,P_2)+\wp(x_{ij}^++P_1;P_1,P_2)\Big)\,(a+S_{ij})\\
&\qquad\qquad +\Big(\wp(x_{ij}^+;P_1,P_2)+\wp(x_{ij}^-+P_1;P_1,P_2)\Big)\,(a+\tilde S_{ij})\Big]\\
&\quad +4b\sum_i \wp(2x_i;P_1,2P_2)(b+S_i)\\
&\quad +4b'\sum_i \wp(2x_i+2P_2;P_1,2P_2)(b'+(-1)^m S_i)\\
&\quad +4\sum_i\Big[\lambda(\lambda-1)\,\wp(2x_i+P_1;P_1,2P_2)\\
&\qquad\qquad +\lambda'(\lambda'-1)\,\wp(2x_i+P_1+2P_2;P_1,2P_2)\Big]\,.
\end{aligned} \tag{3.20}$$

One of the main results of Ref. [20] is the fact that the potential (3.20) is QES provided that the *ordered* pair (P_1,P_2) is chosen from Cases 5–7 in Table III. In fact, the remaining BC_N-type (Q)ES spin potentials in Cases 1–4 can be obtained from the potentials in Eq. (3.20) by sending one or both of the half-periods of the Weierstrass function to infinity. This is of course reminiscent of the analogous formula for integrable scalar Calogero–Sutherland models associated to root systems [48].

4. Conclusions

In this article we have given an overview of a novel systematic method for constructing (Q)ES scalar or spin Calogero–Sutherland models using

Dunkl operators. Our approach is close in spirit to the hidden symmetry algebra approach to scalar (Q)ES models described in Section 1. The quasi-exact solvability of the Hamiltonian is a consequence of the invariance of the polynomial space $\mathcal{R}_m$ under all three families of Dunkl operators (of a given type). We emphasize that all the previously known ES spin Calogero–Sutherland models are obtained as particular cases, as well as several new families of (Q)ES spin CS models, some of them of elliptic type. Although in this review we have focused on the construction of new (Q)ES spin CS models, the spectrum and the integrability of these models can also be studied using the properties of Dunkl operators.

Another important development not covered in this review is the connection between Calogero–Sutherland models and integrable spin chains with long-range interactions of Haldane–Shastry type [33, 58]. In fact, every spin Calogero–Sutherland model can be related to a spin chain with long-range interactions by taking the strong coupling limit, in which the particles are "frozen" in their classical equilibrium positions, and their interaction are restricted to the spin degrees of freedom [53, 63]. In this limit, the constants of motion of the dynamical CS model yield those of the spin chain, thereby proving its integrability [69, 21]. Moreover, in the exactly solvable cases the spectrum and partition function of the spin chain can be derived in principle from those of the corresponding scalar and spin dynamical models [54]. An interesting open problem in this respect is the generalization of the above results to the new (Q)ES spin CS models presented in Sections 2 and 3.

The potentials presented in Section 2 are all invariant under the realization of the A_N Weyl group spanned by the total permutation operators Π_{ij}. A remarkable feature of the potentials in Cases 2–4 and 7–9 is their additional invariance under a change of sign of the spatial coordinate of any particle. Therefore, although these potentials are not invariant under the full B_N Weyl group spanned by total permutation and sign-reversal operators Π_{ij} and Π_i, they are invariant under the restriction of the action of this group to the spatial coordinates. These models thus occupy an intermediate position between the usual spin CS models of A_N type and the fully B_N-invariant spin CS models discussed in Section 3.

We note finally that in the A_N case, the rational, trigonometric and hyperbolic models with both $V^+ = 0$ and $V^+ = V^-$ (see Eq. (2.26)) are obtained in the classification (in Cases 1–6). In contrast, in the elliptic Cases 7–9 the interaction potential is always of the form $V^+ = V^-$. It remains an open question to determine whether the purely A_N (with $V^+ = 0$) elliptic potentials are also QES; see Refs. [14, 8] for partial results in this direction.

References

1. H. Azuma and S. Iso, *Explicit relation of the quantum Hall effect and the Calogero–Sutherland model*, Phys. Lett. B **331** (1994), 107–113.
2. T.H. Baker and P.J. Forrester, *The Calogero–Sutherland model and generalized classical polynomials*, Commun. Math. Phys. **188** (1997), 175–216.
3. B. Basu-Mallick, *Spin-dependent extension of Calogero–Sutherland model through anyon-like representations of permutation operators*, Nucl. Phys. B **482** (1996), 713–730.
4. D. Bernard, M. Gaudin, F.D.M. Haldane and V. Pasquier, *Yang–Baxter equation in long-range interacting systems*, J. Phys. A: Math. Gen. **26** (1993), 5219–5236.
5. L. Brink, T.H. Hansson, S. Konstein, and M.A. Vasiliev, *The Calogero model—anyonic representation, fermionic extension and supersymmetry*, Nucl. Phys. B **401** (1993), 591–612.
6. F. Calogero, *Solution of the one-dimensional N-body problems with quadratic and/or inversely quadratic pair potentials*, J. Math. Phys. **12** (1971), 419–436.
7. A.L. Carey and E. Langmann, *Loop groups, anyons and the Calogero–Sutherland model*, Commun. Math. Phys. **201** (1999), 1–34.
8. O. Chalykh, P. Etingof and A. Oblomkov, *Generalized Lamé operators*, preprint math.QA/0212029.
9. I. Cherednik, *A unification of Knizhnik–Zamolodchikov and Dunkl operators via affine Hecke algebras*, Invent. math. **106** (1991), 411–431.
10. I. Cherednik, *Integration of quantum many-body problems by affine Knizhnik–Zamolodchikov equations*, Adv. Math. **106** (1994), 65–95.
11. É. Cotton, *Sur les invariants différentiels de quelques équations linéaires aux dérivées partielles du second ordre*, Ann. École Norm. **17** (1900), 211–244.
12. E. D'Hoker and D.H. Phong, *Calogero–Moser systems in* SU(N) *Seiberg–Witten theory*, Nucl. Phys. B **513** (1998), 405–444.
13. J.F. van Diejen, *Confluent hypergeometric orthogonal polynomials related to the rational quantum Calogero system with harmonic confinement*, Commun. Math. Phys. **188** (1997), 467–497.
14. J. Dittrich and V.I. Inozemtsev, *On the structure of eigenvectors of the multidimensional Lamé operator*, J. Phys. A: Math. Gen. **26** (1993), L753–L756.
15. C.F. Dunkl, *Differential-difference operators associated to reflection groups*, Trans. Am. Math. Soc. **311** (1989), 167–183.
16. C.F. Dunkl, *Orthogonal polynomials of types A and B and related Calogero models*, Commun. Math. Phys. **197** (1998), 451–487.
17. F. Finkel and N. Kamran, *On the equivalence of matrix differential operators to Schrödinger form*, Nonlin. Math. Phys. **4** (1997), N 3–4, 278–286.
18. F. Finkel and N. Kamran, *The Lie algebraic structure of differential operators admitting invariant spaces of polynomials*, Adv. Appl. Math. **20** (1998), 300–322.
19. F. Finkel, D. Gómez-Ullate, A. González-López, M.A. Rodríguez and R. Zhdanov, *A_N-type Dunkl operators and new spin Calogero–Sutherland models*, Commun. Math. Phys. **221** (2001), 477–497.
20. F. Finkel, D. Gómez-Ullate, A. González-López, M.A. Rodríguez and R. Zhdanov, *New spin Calogero–Sutherland models related to B_N-type Dunkl operators*, Nucl. Phys. B **613** (2001), 472–496.
21. F. Finkel, D. Gómez-Ullate, A. González-López, M.A. Rodríguez and R. Zhdanov, *On the Sutherland model of B_N type and its associated spin chain*, Commun. Math. Phys. **233** (2003), 191–209.

22. D. Gómez-Ullate, A. González-López and M.A. Rodríguez, *New algebraic quantum many body problems*, J. Phys. A: Math. Gen. **33** (2000), 7305–7335.
23. D. Gómez-Ullate, A. González-López and M.A. Rodríguez, *Exact solutions of a new elliptic Calogero–Sutherland model*, Phys. Lett. B **511** (2001), 112–118.
24. A. González-López, N. Kamran and P.J. Olver, *Quasi-exactly solvable Lie algebras of first order differential operators in two complex variables*, J. Phys. A: Math. Gen. **24** (1991), 3995–4008.
25. A. González-López, N. Kamran and P.J. Olver, *Normalizability of one-dimensional quasi-exactly solvable Schrödinger operators*, Commun. Math. Phys. **153** (1993), 117–146.
26. A. González-López, N. Kamran and P.J. Olver, *New quasi-exactly solvable Hamiltonians in two dimensions*, Commun. Math. Phys. **159** (1994), 503–537.
27. A. González-López, N. Kamran and P.J. Olver, *Quasi-exact solvability*, Contemp. Math. **160** (1994), 113–140.
28. A. González-López, N. Kamran and P.J. Olver, *Real Lie algebras of differential operators and quasi-exactly solvable potentials*, Phil. Trans. R. Soc. Lond. A **354** (1996), 1165–1193.
29. A. Gorsky and N. Nekrasov, *Hamiltonian systems of Calogero type, and 2-dimensional Yang–Mills theory*, Nucl. Phys. B **414** (1994), 213–238.
30. Z.N.C. Ha, *Exact dynamical correlation functions of Calogero–Sutherland model and one-dimensional fractional statistics*, Phys. Rev. Lett. **73** (1994), 1574–1577.
31. Z.N.C. Ha, *Fractional statistics in one dimension: view from an exactly solvable model*, Nucl. Phys. B **435** (1995), 604–636.
32. Z.N.C. Ha and F.D.M. Haldane, *Models with inverse-square exchange*, Phys. Rev. B **46** (1992), 9359–9368.
33. F.D.M. Haldane, *Exact Jastrow–Gutzwiller resonating-valence-bond ground state of the spin-1/2 antiferromagnetic Heisenberg chain with $1/r^2$ exchange*, Phys. Rev. Lett. **60** (1988), 635–638.
34. K. Hikami and M. Wadati, *Integrability of Calogero–Moser spin system*, J. Phys. Soc. Jap. **62** (1993), 469–472.
35. K. Hikami and M. Wadati, *Integrable spin-$\frac{1}{2}$ particle systems with long-range interactions*, Phys. Lett. A **173** (1993), 263–266.
36. X. Hou and M. Shifman, *A quasi-exactly solvable N-body problem with the* $\mathrm{sl}(N+1)$ *algebraic structure*, Int. J. Mod. Phys. A **14** (1999), 2993–3003.
37. T. Guhr, A. Müller-Groeling and H.A. Weidenmüller, *Random-matrix theories in quantum physics: common concepts*, Phys. Rep. **299** (1998), 189–425.
38. V.I. Inozemtsev, *Lax representation with spectral parameter on a torus for integrable particle systems*, Lett. Math. Phys. **17** (1989), 11–17.
39. V.I. Inozemtsev, *Integrable model of interacting fermions confined by the Morse potential*, Int. J. Mod. Phys. A **12** (1997), 195–200.
40. V.I. Inozemtsev and R. Sasaki, *Universal Lax pairs for spin Calogero–Moser models and spin exchange models*, J. Phys. A: Math. Gen. **34** (2001), 7621–7632.
41. S.B. Isakov, G. Lozano and S. Ouvry, *Non-abelian Chern–Simons particles in an external magnetic field*, Nucl. Phys. B **552** (1999), 677–706.
42. N. Kamran and P.J. Olver, *Lie algebras of differential operators and Lie-algebraic potentials*, J. Math. Anal. Appl. **145** (1990), 342–356.
43. A. Kasman, *Bispectral KP solutions and linearization of Calogero–Moser particle systems*, Commun. Math. Phys. **172** (1995), 427–448.
44. L. Lapointe and L. Vinet, *Exact operator solution of the Calogero–Sutherland model*, Commun. Math. Phys. **178** (1996), 425–452.

45. D.F. Lawden, *Elliptic Functions and Applications*, Springer-Verlag, Berlin, 1989.
46. J.A. Minahan and A.P. Polychronakos, *Integrable systems for particles with internal degrees of freedom*, Phys. Lett. B **302** (1993), 265–270.
47. A. Minzoni, M. Rosenbaum and A. Turbiner, *Quasi-exactly solvable many-body problems*, Mod. Phys. Lett. A **11** (1996), 1977–1984.
48. M.A. Olshanetsky and A.M. Perelomov, *Quantum integrable systems related to Lie algebras*, Phys. Rep. **94** (1983), 313–404.
49. P.J. Olver, *A quasi-exactly solvable travel guide*, in H.-D. Doebner et al., eds., GROUP21: Physical Applications and Mathematical Aspects of Geometry, vol. 1, World Scientific, Singapore, 1997, pp. 285–95.
50. P.J. Olver, *Classical Invariant Theory*, Cambridge University Press, Cambridge, 1999.
51. A.P. Polychronakos, *Non-relativistic bosonization and fractional statistics*, Nucl. Phys. B **324** (1989), 597–622.
52. A.P. Polychronakos, *Exchange operator formalism for integrable systems of particles*, Phys. Rev. Lett. **69** (1992), 703–705.
53. A.P. Polychronakos, *Lattice integrable systems of Haldane–Shastry type*, Phys. Rev. Lett. **70** (1993), 2329–2331.
54. A.P. Polychronakos, *Exact spectrum of* SU(n) *spin chain with inverse-square exchange*, Nucl. Phys. B **419** (1994), 553–566.
55. A.P. Polychronakos, *Waves and solitons in the continuum limit of the Calogero–Sutherland model*, Phys. Rev. Lett. **74** (1995), 5153–5157.
56. A.P. Polychronakos, *Generalized Calogero models through reductions by discrete symmetries*, Nucl. Phys. B **543** (1999), 485–498.
57. W. Rühl and A. Turbiner, *Exact solvability of the Calogero and Sutherland models*, Mod. Phys. Lett. A **10** (1995), 2213–2221.
58. B.S. Shastry, *Exact solution of an $S = 1/2$ Heisenberg antiferromagnetic chain with long-ranged interactions*, Phys. Rev. Lett. **60** (1988), 639–642.
59. M.A. Shifman, *New findings in quantum mechanics (partial algebraization of the spectral problem)*, Int. J. Mod. Phys. A **4** (1989), 2897–2952.
60. M.A. Shifman and A.V. Turbiner, *Quantal problems with partial algebraization of the spectrum*, Commun. Math. Phys. **126** (1989), 347–365.
61. B. Sutherland, *Exact results for a quantum many-body problem in one dimension*, Phys. Rev. A **4** (1971), 2019–2021.
62. B. Sutherland, *Exact results for a quantum many-body problem in one dimension. II*, Phys. Rev. A **5** (1972), 1372–1376.
63. B. Sutherland and B.S. Shastry, *Solution of some integrable one-dimensional quantum systems*, Phys. Rev. Lett. **71** (1993), 5–8.
64. N. Taniguchi, B.S. Shastry and B.L. Altshuler, *Random matrix model and the Calogero–Sutherland model: A novel current-density mapping*, Phys. Rev. Lett. **75** (1995), 3724–3727.
65. A.V. Turbiner, *Quasi-exactly solvable problems and* $\mathfrak{sl}(2)$ *algebra*, Commun. Math. Phys. **118** (1988), 467–474.
66. A. Turbiner, *Lie algebras and polynomials in one variable*, J. Phys. A: Math. Gen. **25** (1992), L1087–L1093.
67. A.G. Ushveridze, *Quasi-Exactly Solvable Models in Quantum Mechanics*, Institute of Physics Publishing, Bristol, 1994.
68. T. Yamamoto, *Multicomponent Calogero model of B_N-type confined in a harmonic potential*, Phys. Lett. A **208** (1995), 293–302.

69. T. Yamamoto and O. Tsuchiya, *Integrable $1/r^2$ spin chain with reflecting end*, J. Phys. A: Math. Gen. **29** (1996), 3977–3984.
70. O.B. Zaslavskii, *Two-dimensional quasi-exactly solvable models and classical orthogonal polynomials*, J. Phys. A: Math. Gen. **27** (1994), 323–328.

Table I. Coefficients A, B, C, D of the external potential of A_N type in Cases 1–9. Here $\alpha_\epsilon = \frac{c_\epsilon}{4}$, with $\epsilon = \pm, 0$, and the parameter ℓ is given in Eq. (2.14).

Case	A	B	C	D
1	α_+^2	$4\alpha_0\alpha_+$	$4(\alpha_0^2 + 2\alpha_-\alpha_+)$	$2(8\alpha_-\alpha_0 + \ell\alpha_+)$
2	α_+^2	$2\alpha_0\alpha_+$	$\alpha_0^2 + \alpha_+(3\ell + 2\alpha_-)$	$\frac{1}{4}\left((2\alpha_- - \ell)^2 - 1\right)$
3	α_+^2	$2\alpha_+(\alpha_0 + \ell)$	$2(\alpha_+ - \alpha_-)(\alpha_0 - \ell)$	$(\alpha_+ + \alpha_-)^2 + (\alpha_0 - \ell)^2 - 1$
4	$-\alpha_+^2$	$2\alpha_+(\ell - \alpha_0)$	$2(\alpha_+ - \alpha_-)(\ell + \alpha_0)$	$(\alpha_+ + \alpha_-)^2 + (\ell + \alpha_0)^2 - 1$
5	α_+^2	$2\alpha_+(\alpha_0 + \ell)$	$2\alpha_-(\alpha_0 - \ell)$	α_-^2
6	$\frac{1}{2}\left((\alpha_+ - \alpha_-)^2 - \alpha_0^2\right)$	$2\left(\alpha_-^2 - \alpha_+^2 + \ell\alpha_0\right)$	$\alpha_0(\alpha_- - \alpha_+)$	$2\big(\alpha_0(\alpha_+ + \alpha_-) + \ell(\alpha_+ - \alpha_-)\big)$
7	$k^2(\ell^2 - 1) + \frac{k^2\alpha_0}{k'^2}\left(\frac{\alpha_0}{k'^2} - 2\ell\right)$ $+\frac{1}{k'^4}(\alpha_+ + k^2\alpha_-)^2$	$\frac{2}{k'^2}(\alpha_+ + k^2\alpha_-)\left(\ell - \frac{\alpha_0}{k'^2}\right)$	$\ell^2 - 1 + \frac{\alpha_0}{k'^2}\left(\frac{\alpha_0}{k'^2} + 2\ell\right)$ $+\frac{1}{k'^4}(\alpha_+ + \alpha_-)^2$	$\frac{2}{k'^2}(\alpha_+ + \alpha_-)\left(\ell + \frac{\alpha_0}{k'^2}\right)$
8	$\ell^2 - 1 + \frac{\alpha_0}{k^2}\left(\frac{\alpha_0}{k^2} - 2\ell\right)$ $+\frac{1}{k^4}(\alpha_+ + \alpha_-)^2$	$\frac{2}{k^2}(\alpha_+ + \alpha_-)\left(\frac{\alpha_0}{k^2} - \ell\right)$	$k'^2(\ell^2 - 1) + \frac{k'^2\alpha_0}{k^2}\left(\frac{\alpha_0}{k^2} + 2\ell\right)$ $+\frac{1}{k^4}(\alpha_- + k'^2\alpha_-)^2$	$\frac{2}{k^2}(\alpha_+ + k'^2\alpha_-)\left(\frac{\alpha_0}{k^2} + \ell\right)$
9	$k'^2(1 - \ell^2) + k'^2\alpha_0(2\ell - \alpha_0)$ $+\frac{1}{k^2}\left(k'^2\alpha_+ - k^2\alpha_-\right)^2$	$2(\ell - \alpha_0)(k'^2\alpha_+ - k^2\alpha_-)$	$(\ell + \alpha_0)^2 + (\alpha_+ + \alpha_-)^2 - 1$	$2(\ell + \alpha_0)(\alpha_+ + \alpha_-)$

TABLE II. Parameters β and β' of the scalar external potential of BC_N type (none of them appear in Cases 2a-2b). The parameter α is given in Eq. (3.14).

Case	β	β'
1	$-\frac{1}{2}c_0$	–
3a	$\frac{c_0}{8} - \left(a(N-1) + b + \frac{m}{2}\right)$	–
3b	$-\left(\frac{c_0}{8} + a(N-1) + b + \frac{m}{2}\right)$	–
4a	$\frac{1}{8}\left(c_0 + 2(b - b')\right)$	–
4b	$-\frac{1}{8}\left(c_0 + 2(b' - b)\right)$	–
5	$-\frac{1}{8kk'}\left(c_0 + 2(k^2 - k'^2)(b - b')\right)$	–
6a	$-\frac{1}{8k'^2}\left(c_0 + 4(1 + k^2)(b - b')\right) - \alpha$	$\frac{1}{8k'^2}\left(c_0 + 4(1 + k^2)(b - b')\right) - \alpha$
6b	$\frac{1}{8k^2}\left(c_0 + 4(1 + k'^2)(b' - b)\right) - \alpha$	$-\frac{1}{8k^2}\left(c_0 + 4(1 + k'^2)(b' - b)\right) - \alpha$
7	$-\left(\frac{c_0}{8} + \frac{1}{2}(k^2 - k'^2)(b' - b) + \alpha\right)$	$\frac{c_0}{8} + \frac{1}{2}(k^2 - k'^2)(b' - b) - \alpha$

TABLE III. Function $v(x)$ and its primitive half-periods P_i (see Eq. (3.17)) for each of the BC_N-type potentials. The constant $K \equiv K(k)$ is the complete elliptic integral of the first kind, $K' = K(k')$, k is the modulus of the elliptic functions, and $k' = \sqrt{1-k^2}$.

Case	$v(x)$	P_1	P_2
1	x^{-2}	∞	∞
3a	$\sinh^{-2} x$	$\frac{\mathrm{i}\pi}{2}$	∞
3b	$\sin^{-2} x$	$\frac{\pi}{2}$	∞
4a	$\sinh^{-2} x$	∞	$\frac{\mathrm{i}\pi}{2}$
4b	$\sin^{-2} x$	∞	$\frac{\pi}{2}$
5	$\frac{\mathrm{dn}^2 x}{\mathrm{sn}^2 x}$	$K + \mathrm{i}K'$	K
6a	$\frac{\mathrm{cn}^2 x\, \mathrm{dn}^2 x}{\mathrm{sn}^2 x}$	K	$\frac{\mathrm{i}K'}{2}$
6b	$\frac{\mathrm{dn}^2 x}{\mathrm{sn}^2 x\, \mathrm{cn}^2 x}$	$\mathrm{i}K'$	$\frac{K}{2}$
7	$\frac{\mathrm{cn}^2 x}{\mathrm{sn}^2 x\, \mathrm{dn}^2 x}$	K	$\frac{1}{2}(K + \mathrm{i}K')$

YANG–BAXTER MAPS AND MATRIX SOLITONS

V.M. GONCHARENKO (mike-gonchar@mtu-net.ru)
Chair of Mathematics and Financial Applications, Financial Academy, Leningradsky prospect, 49, Moscow, Russia

A.P. VESELOV (a.p.veselov@lboro.ac.uk)
Department of Mathematical Sciences, Loughborough University, Loughborough, Leicestershire, LE 11 3TU, UK
Landau Institute for Theoretical Physics, Kosygina 2, Moscow, 117940, Russia

Abstract. New examples of the Yang–Baxter maps (or set-theoretical solutions to the quantum Yang–Baxter equation) on the Grassmannians, arising from the theory of the matrix KdV equation are discussed. The Lax pairs for these maps are produced using the relations with the inverse scattering problem for the matrix Schrödinger operator

1. Introduction.

The problem of studying the set-theoretical solutions to the quantum Yang–Baxter equation was suggested by V.G. Drinfeld [1]. This stimulated research in this direction, mainly from the algebraic point of view (see e.g. [2],[3]). The dynamical aspects of this problem were discussed in the paper [4] where also a shorter term "Yang–Baxter map" for such solutions was suggested.

In this paper we present some new examples of the Yang–Baxter maps appeared in relation with the theory of solitons. In the case when the solitons have internal degrees of freedom described by some manifold X their pairwise interaction gives a map from $X \times X$ into itself which satisfies the Yang–Baxter relation, which means that the final result of multiparticle interaction is independent of the order of collisions (see Kulish's paper [5] which is the first one we know containing such a statement).

As an example of the equation with the soliton solutions having non-trivial internal parameters we consider the matrix KdV equation

$$U_t = 3UU_x + 3U_xU - U_{xxx}\,, \tag{1.1}$$

A.B. Shabat et al. (eds.), New Trends in Integrability and Partial Solvability, 191–197.

where U is $n \times n$ matrix. This equation was introduced in the famous P. Lax's paper [6] and was the subject of investigations in several papers including [7–9]. The related inverse scattering problem for the matrix Schrödinger operator was investigated by Martinez Alonso and Olmedilla [10, 11].

We will show that the formulas from [9] for two matrix KdV soliton interaction can be generalized to determine some Yang–Baxter maps on the Grassmannians $G(k,n)$ and products of two Grassmannians $G(k,n)\times G(n-k,n)$. We produce also the Lax pairs for these maps using the relations with the inverse scattering problem for the matrix Schrödinger operator [10, 11].

2. Two-soliton interaction as Yang–Baxter map.

Let us start with the definition of the Yang–Baxter map (cf. [1], [4]). Let X be any set and R be a map: $X \times X \to X \times X$. Let $R_{ij} : X^n \to X^n, \quad X^n = X \times X \times \ldots\ldots \times X$ be the maps which acts as R on i-th and j-th factors and identically on the others. If $P : X^2 \to X^2$ is the permutation: $P(x,y) = (y,x)$, then

$$R_{21} = PRP.$$

The map R is called *Yang–Baxter map* if it satisfies the Yang–Baxter relation

$$R_{12}R_{13}R_{23} = R_{23}R_{13}R_{12}, \tag{2.1}$$

considered as the equality of the maps of $X \times X \times X$ into itself. If additionally R satisfies the relation

$$R_{21}R = Id, \tag{2.2}$$

we will call it *reversible Yang–Baxter map.*

We will actually consider the *parameter-dependent Yang–Baxter maps* $R(\lambda,\mu)$ $(\lambda,\mu \in \mathbf{C})$ satisfying the corresponding version of Yang–Baxter relation

$$R_{12}(\lambda_1,\lambda_2)R_{13}(\lambda_1,\lambda_3)R_{23}(\lambda_2,\lambda_3) = R_{23}(\lambda_2,\lambda_3)R_{13}(\lambda_1,\lambda_3)R_{12}(\lambda_1,\lambda_2) \tag{2.3}$$

and reversibility condition

$$R_{21}(\mu,\lambda)R(\lambda,\mu) = Id. \tag{2.4}$$

Although this case can be considered as a particular case of the previous one by introducing $\tilde{X} = X \times \mathbf{C}$ and $\tilde{R}(x,\lambda;y,\mu) = R(\lambda,\mu)(x,y)$ it is more convenient for us to keep the parameter separately.

To construct examples of such maps consider the two-soliton interaction in the matrix KdV equation (1.1). At the beginning let U be a general $n \times n$ complex matrix, no symmetry conditions are assumed.

It is easy to check that the matrix KdV equation has the soliton solution of the form

$$U = 2\lambda^2 P \operatorname{sech}^2(\lambda x - 4\lambda^3 t),$$

where P must be a projector: $P^2 = P$. If we assume that P has rank 1 then P should have the form $P = \frac{\xi \otimes \eta}{(\xi, \eta)}$. Here ξ is a vector in a complex vector space V of dimension d, η is a vector from the dual space V^* (covector) and bracket (ξ, η) means the canonical pairing between V and V^*.

To find the two-soliton solutions one can use the inverse scattering problem for the general matrix Schrödinger operator developed in [10, 11]. The corresponding formulas have been found and analyzed in [9]. In particular, it was shown that the change of the matrix amplitudes P ("polarizations") of two solitons with velocities λ_1 and λ_2 after their interaction is described by the following map:

$$R(\lambda_1, \lambda_2) : (\xi_1, \eta_1; \xi_2, \eta_2) \to (\tilde{\xi_1}, \tilde{\eta_1}; \tilde{\xi_2}, \tilde{\eta_2})$$

$$\tilde{\xi_1} = \xi_1 + \frac{2\lambda_2(\xi_1, \eta_2)}{(\lambda_1 - \lambda_2)(\xi_2, \eta_2)}\xi_2, \qquad \tilde{\eta_1} = \eta_1 + \frac{2\lambda_2(\xi_2, \eta_1)}{(\lambda_1 - \lambda_2)(\xi_2, \eta_2)}\eta_2, \tag{2.5}$$

$$\tilde{\xi_2} = \xi_2 + \frac{2\lambda_1(\xi_2, \eta_1)}{(\lambda_2 - \lambda_1)(\xi_1, \eta_1)}\xi_1, \qquad \tilde{\eta_2} = \eta_2 + \frac{2\lambda_1(\xi_1, \eta_2)}{(\lambda_2 - \lambda_1)(\xi_1, \eta_1)}\eta_1. \tag{2.6}$$

We claim that this map is a reversible parameter-dependent Yang–Baxter map. This can be checked directly although the calculations are quite long.

A better way is explained in the next section.

3. Matrix factorizations and Lax pairs.

Suppose we have a matrix $A(x, \lambda; \zeta)$ depending on the point $x \in X$, a parameter λ and an additional parameter $\zeta \in \mathbf{C}$, which we will call *spectral parameter.* We assume that A depends on ζ polynomially or rationally. The case of elliptic dependence is also very interesting (see [13]) but we will not consider it here.

Consider the product $L = A(y, \mu; \zeta)A(x, \lambda; \zeta)$, then change the order of the factors $L \to \tilde{L} = A(x, \lambda; \zeta)A(y, \mu; \zeta)$ and re-factorize it as: $\tilde{L} = A(\tilde{y}, \mu; \zeta)A(\tilde{x}, \lambda; \zeta)$. Suppose that this re-factorization relation

$$A(x, \lambda; \zeta)A(y, \mu; \zeta) = A(\tilde{y}, \mu; \zeta)A(\tilde{x}, \lambda; \zeta) \tag{3.1}$$

uniquely determines $\tilde{x}, \tilde{y}$.

It is easy to see that the map

$$R(\lambda,\mu)(x,y) = (\tilde{x},\tilde{y}) \tag{3.2}$$

determined by (3.1) satisfies the Yang–Baxter relation. Indeed if we consider the product $A(x_1)A(x_2)A(x_3)$ (we omit here the parameters λ_i and ζ for shortness) then applying the left hand side of (2.1) to this product we have

$$\begin{aligned} A(x_1)A(x_2)A(x_3) &= A(x_1^{(1)})A(x_3^{(1)})A(x_2^{(1)}) \\ &= A(x_3^{(2)})A(x_1^{(2)})A(x_2^{(2)}) \\ &= A(x_3^{(3)})A(x_2^{(3)})A(x_1^{(3)}). \end{aligned}$$

Similarly the right hand side corresponds to the relations

$$\begin{aligned} A(x_1)A(x_2)A(x_3) &= A(\tilde{x}_2^{(1)})A(\tilde{x}_1^{(1)})A(\tilde{x}_3^{(1)}) \\ &= A(\tilde{x}_2^{(2)})A(\tilde{x}_3^{(2)})A(\tilde{x}_1^{(2)}) \\ &= A(\tilde{x}_3^{(3)})A(\tilde{x}_2^{(3)})A(\tilde{x}_1^{(3)}). \end{aligned}$$

If the factorization is unique we have $x_i^{(3)} = \tilde{x}_i^{(3)}$, which is exactly the Yang–Baxter relation.

If a parameter-dependent Yang–Baxter map $R(\lambda,\mu)$ can be described in such a way we will say that $A(x,\lambda;\zeta)$ is a *Lax pair for R*. As it was shown in [4] such a Lax pair allows to produce the integrals for the dynamics of the related transfer-maps.

Let us come back now to matrix solitons. We claim that the map described by the formulas (2.5),(2.6) has the Lax pair of the following form (Yuri Suris suggested a simple explanation of this form which works also for a wide class of the Yang–Baxter maps (see [15])) motivated by the inverse spectral problem for the matrix Schrödinger operator [11]:

$$A(\xi,\eta,\lambda;\zeta) = I + \frac{2\lambda}{\zeta-\lambda}\frac{\xi\otimes\eta}{(\xi,\eta)} \tag{3.3}$$

In the soliton theory this type of matrices were first used by Zakharov and Shabat [14].

One can check directly that re-factorization relation for this matrix leads to the map (2.5, 2.6) but we would prefer to do this in a more general situation.

4. Generalization: Yang–Baxter maps on the Grassmannians.

Let V be an n-dimensional real (or complex) vector space, $P : V \to V$ be a projector of rank k: $P^2 = P$. Any such projector is uniquely determined by its kernel $K = \mathrm{Ker}P$ and image $L = \mathrm{Im}P$, which are two subspaces of V of dimensions k and $n-k$ complementary to each other: $K \oplus L = V$. The space of all projectors X of rank k is an open set in the product of two Grassmannians $G(k, n) \times G(n-k, n)$.

Consider the following matrix

$$A(P, \lambda; \zeta) = I + \frac{2\lambda}{\zeta - \lambda} P \tag{4.1}$$

and the related re-factorization relation

$$\Big(I + \frac{2\lambda_1}{\zeta - \lambda_1} P_1\Big)\Big(I + \frac{2\lambda_2}{\zeta - \lambda_2} P_2\Big) = \Big(I + \frac{2\lambda_2}{\zeta - \lambda_2} \tilde{P}_2\Big)\Big(I + \frac{2\lambda_1}{\zeta - \lambda_1} \tilde{P}_1\Big) \tag{4.2}$$

which we can rewrite in the polynomial form as

$$\begin{aligned}((\zeta - \lambda_1)I + 2\lambda_1 P_1)&((\zeta - \lambda_2)I + 2\lambda_2 P_2) \\ &= ((\zeta - \lambda_2)I + 2\lambda_2 \tilde{P}_2)((\zeta - \lambda_1)I + 2\lambda_1 \tilde{P}_1).\end{aligned} \tag{4.3}$$

We claim that if $\lambda_1 \neq \pm\lambda_2$ it has a unique solution. This follows from the general theory of matrix polynomials (see e.g. [12]) but in this case we can see this directly.

Indeed let us compare the kernels of both sides of the relation (4.3) when the spectral parameter $\zeta = \lambda_1$. In the right hand side we obviously have $\tilde{K}_1$ while the left hand side gives

$$((\lambda_1 - \lambda_2)I + 2\lambda_2 P_2)^{-1} K_1 = \Big(I + \frac{2\lambda_2}{\lambda_1 - \lambda_2} P_2\Big)^{-1} K_1.$$

Now we use the following property of the matrix (4.1):

$$A(P, -\lambda; \zeta) = A(P, \lambda; \zeta)^{-1} \tag{4.4}$$

to have

$$\tilde{K}_1 = \Big(I - \frac{2\lambda_2}{\lambda_1 + \lambda_2} P_2\Big) K_1. \tag{4.5}$$

Similarly taking the image of both sides of (4.3) at $\zeta = \lambda_2$ we will have

$$\tilde{L}_2 = \Big(I + \frac{2\lambda_1}{\lambda_2 - \lambda_1} P_1\Big) L_2. \tag{4.6}$$

To find $\tilde{K}_2$ and $\tilde{L}_1$ one should take first the inverse of both sides of (4.2), use the property (4.4) and then repeat the procedure. This will lead us to the formulas:

$$\tilde{K}_2 = \Big(I - \frac{2\lambda_1}{\lambda_1 + \lambda_2} P_1\Big) K_2 \tag{4.7}$$

and

$$\tilde{L}_1 = \Big(I + \frac{2\lambda_2}{\lambda_1 - \lambda_2} P_2\Big) L_1. \tag{4.8}$$

The formulas (4.5,4.6,4.7,4.8) determine a parameter-dependent Yang–Baxter map on the set of projectors. One can easily check that for $k = 1$ one has the formulas (2.5, 2.6) for two matrix soliton interaction.

If we supply now our vector space V with the Euclidean (Hermitian) structure and consider the self-adjoint projectors P of rank k then the corresponding space X will coincide with the Grassmannian $G(k, n)$ such a projector is completely determined by its image L (which is a k-dimensional subspace in V and thus a point in $G(k, n)$) since the kernel K in this case is the orthogonal complement to L.

The corresponding Yang–Baxter map R on the Grassmannian is determined by the formulas

$$\tilde{L}_1 = \Big(I + \frac{2\lambda_2}{\lambda_1 - \lambda_2} P_2\Big) L_1, \tag{4.9}$$

$$\tilde{L}_2 = \Big(I + \frac{2\lambda_1}{\lambda_2 - \lambda_1} P_1\Big) L_2. \tag{4.10}$$

It would be very interesting to investigate the dynamics of the corresponding transfer-maps [4]. As we have shown here the Lax pair for them is given by (4.1).

Acknowledgements

The second author (A.P.V.) is grateful to the organizers and participants of the NEEDS conference and NATO Advanced Research Workshop on "New Trends in Integrability and Partial Solvability" (Cádiz, 10-15 June 2002) and SIDE-V conference (Giens, 21-26 June 2002) where these results were first presented and especially to P.P. Kulish, A.B. Shabat and Yu. Suris for stimulating and helpful discussions.

References

1. V.G. Drinfeld, *On some unsolved problems in quantum group theory.* In "Quantum groups" (Leningrad, 1990), Lecture Notes in Math., 1510, Springer, 1992, p. 1–8.

2. A. Weinstein, P. Xu, *Classical solutions of the quantum Yang–Baxter equation.* Comm. Math. Phys. **148**, 309–343 (1992).
3. P. Etingof, T. Schedler, A. Soloviev, *Set-theoretical solutions to the quantum Yang–Baxter equation.* Duke Math. J. **100**, (1999).
4. A.P. Veselov, *Yang–Baxter maps and integrable dynamics.* math.QA/0205335.
5. P.P. Kulish *Factorization of the classical and quantum S-matrix and conservation laws.* Theor. Math. Phys. **26**, 132–137 (1976).
6. P.D. Lax., *Integrals of nonlinear equations of evolution and solitary waves.* Comm. Pure Appl. Math. **21**, 467–490 (1968).
7. M. Wadati, T. Kamijo, *On the extension of inverse scattering method.* Prog. Theor. Physics **52**, 397–414 (1974).
8. F. Calogero, A. Degasperis, *Nonlinear evolution equations solvable by the inverse spectral transform. II.* Nuovo Cimento. **39B**, 1 (1977).
9. V.M. Goncharenko, *Multisoliton solutions of the matrix KdV equation.* Theor. Math. Phys. **126**, 81–91 (2001).
10. L. Martinez Alonso, E. Olmedilla, *Trace identities in the inverse scattering transform method associated with matrix Schrdinger operators.* J. Math. Phys. **23**, 2116 (1982).
11. E. Olmedilla, *Inverse scattering transform for general matrix Schrdinger operators and the related symplectic structure.* Inverse Problems **1**, 219–236 (1985).
12. I. Gohberg, P. Lancaster, L. Rodman, *Matrix polynomials.* New York: Academic Press, 1982.
13. A. Odesskii, *Set-theoretical solutions to the Yang–Baxter relation from factorization of matrix polynomials and theta-functions.* math.QA/0205051.
14. V.E. Zakharov, A.B.Shabat, *Integration of the nonlinear equations of mathematical physics by the method of the inverse scattering problem. II.* Funct. Anal. Appl. **13**, 13–22 (1979).
15. Yu.B. Suris, A.P. Veselov *Lax pairs for Yang–Baxter maps.* In preparation.

NONLOCAL SYMMETRIES AND GHOSTS

PETER J.OLVER (olver.math.umn.edu)*
Department of Mathematics, University of Minnesota, Minneapolis, MN, USA 55455

1. Introduction

The local theory of symmetries of differential equations has been well-established since the days of Sophus Lie. Generalized, or higher order symmetries can be traced back to the original paper of Noether, [32], but were not exploited until the discovery that they play a critical role in integrable (soliton) partial differential equations, cf. [30, 33, 35].

While the local theory is very well developed, the theory of nonlocal symmetries of nonlocal differential equations remains incomplete. Particular results on certain classes of nonlocal symmetries and nonlocal differential equations have been developed by several groups, including Abraham–Shrauner et. al., [1–3, 13], Bluman et. al., [5–7], Chen et. al., [8–10], Fushchich et. al., [17], Guthrie and Hickman, [20–22], Ibragimov et. al., [4], [23, Chapter 7], and many others, [11, 12, 16, 18, 19, 24, 28, 29, 31, 37]. Perhaps the most promising proposed foundation for a general theory of nonlocal symmetries is the Krasilshchik-Vinogradov theory of coverings, [25–27, 38, 39]. However, their construction relies on the a priori specification of the underlying differential equation, and so, unlike local jet space, does not form a universally valid foundation for the theory.

One of the reasons for the lack of a proper foundation is a continuing lack of understanding of the calculus of nonlocal vector fields. Recently, [34], during an attempt to systematically investigate the symmetry properties of the Kadomtsev–Petviashvili (KP) equation, Sanders and Wang made a surprising discovery that the Jacobi identity for nonlocal vector fields appears to fail! The observed violation of the naïve version of the Jacobi

* Research supported in part by NSF Grant DMS 98–03154.

A.B. Shabat et al. (eds.), New Trends in Integrability and Partial Solvability, 199–215.

identity applies to all of the preceding nonlocal symmetry calculi, and, consequently, many statements about the "Lie algebra" of nonlocal symmetries of differential equations are, by and large, not valid as stated. This indicates the need for a comprehensive re-evaluation of all earlier results on nonlocal symmetry algebras.

In this paper, I propose a new theoretical and computational basis for a nonlocal theory which, like the original jet bundle construction, does not rely on a specific differential equation, but applies equally well to a wide variety of nonlocal systems. I will also review the concept of a ghost symmetry, introduced in [34], that resolves the apparent Jacobi paradox. Applications to the classification of symmetries of the KP equation appear in [34]. Similar issues appear in the study of recursion operators by Sanders and Wang, [36].

2. Generalized symmetries

Let us recall the basic theory of generalized symmetries in the local jet bundle framework as presented in [33]. We specify p independent variables $x = (x^1, \ldots, x^p)$ and q dependent variables $u = (u^1, \ldots, u^q)$. The induced jet space coordinates are denoted by

$$u_J^\alpha = \frac{\partial^{\#J} u^\alpha}{(\partial x^1)^{j_1} \cdots (\partial x^p)^{j_p}},$$

in which $1 \leq \alpha \leq q$, and $J = (j_1, \ldots, j_p) \in \mathbb{N}^p$ is a (nonnegative) multi-index, so $j_\nu \geq 0$, of order $\#J = j_1 + \cdots + j_p$. We let $u^{(\infty)} = (\ldots\ u_J^\alpha\ \ldots)$ denote the collection of all such local jet variables. A *differential function* is a smooth function $P[u] = P(x, u^{(\infty)})$ depending on finitely many jet variables. If $u = f(x)$ is any smooth function, we let $P[f]$ denote the evaluation of the differential function P on f.

The total derivatives $D_1, \ldots, D_p$ are defined so that $D_i P[f] = \partial_i(P[f])$ where $\partial_i = \partial/\partial x^i$. They act on the space of differential functions as derivations, and so are completely determined by their action

$$D_i(x^j) = \delta_i^j, \qquad D_i(u_J^\alpha) = u_{J+e_i}^\alpha, \tag{2.1}$$

on the coordinate functions. Here $e_i \in \mathbb{N}^p$ denotes the ith basis multi-index having a 1 in the ith position and zeros elsewhere. If J is a multi-index, we let $D^J = D_1^{j_1} \cdots D_p^{j_p}$ denote the corresponding higher order total derivative; in particular, $u_J^\alpha = D^J u^\alpha$.

We consider *generalized vector fields* in *evolutionary form*

$$\mathbf{v} = \mathbf{v}_Q = \sum_{\alpha=1}^{q} \sum_{J \geq 0} D^J Q^\alpha \frac{\partial}{\partial u_J^\alpha}, \tag{2.2}$$

where $Q = (Q^1, \ldots, Q^q)$ is the *characteristic*, and serves to uniquely specify $\mathbf{v}$. We note the basic formula

$$\mathbf{v}_Q(P) = \mathrm{D}_P(Q) \tag{2.3}$$

where D_P denotes the Fréchet derivative of the differential function P, [33], which is a total differential operator with components

$$\mathrm{D}_P^\alpha = \sum_J \frac{\partial P}{\partial u_J^\alpha} D^J, \qquad \alpha = 1, \ldots, q. \tag{2.4}$$

The *Lie bracket* or *commutator* between two evolutionary vector fields is again an evolutionary vector field

$$[\mathbf{v}_P, \mathbf{v}_Q] = \mathbf{v}_{[P,Q]},$$

with characteristic

$$[P, Q] = \mathbf{v}_P(Q) - \mathbf{v}_Q(P) = \mathrm{D}_Q(P) - \mathrm{D}_P(Q). \tag{2.5}$$

The Lie bracket satisfies the Jacobi identity, and hence endows the space of evolutionary vector fields with the structure of a Lie algebra.

3. Counterexamples to the Jacobi identity?

Attempting to generalize the algebra of evolutionary vector fields to nonlocal variables runs into some immediate, unexpected difficulties. Intuitively, the nonlocal variables should be given by iterating the inverse total derivatives D_i^{-1}, applied to either the jet coordinates u_J^α, or, more generally, to differential functions. In particular, we allow nonlocal variables $u_J^\alpha = D^J u^\alpha$ in which $J \in \mathbb{Z}^p$ is an arbitrary multi-index. Even more generally, one might allow inversion of arbitrary total differential operators $\mathcal{D}^{-1}$, where $\mathcal{D} = \sum_K P_K[u] D^K$, whose coefficients P_K can be either constants, or even general differential functions.

However, the following fairly simple computation appears to indicate that the Jacobi identity does *not* hold between nonlocal vector fields.

Example 3.1. Let $p = q = 1$, with independent variable x and dependent variable u. Consider the vector fields $\mathbf{v}, \mathbf{w}$, and $\mathbf{z}$ with respective characteristics 1, u_x and $D_x^{-1}u$. The first two are local vector fields, and, in fact, correspond to the infinitesimal generators of the translation group

$$(x, u) \longmapsto (x + \delta, u + \varepsilon).$$

The Jacobi identity for these three vector fields has the form

$$[1, [u_x, D_x^{-1}u]] + [u_x, [D_x^{-1}u, 1]] + [D_x^{-1}u, [1, u_x]] = 0, \tag{3.1}$$

where we work on the level of the characteristics, using the induced commutator bracket (2.5). Since

$$[1, u_x] = \mathrm{D}_{u_x}(1) - \mathrm{D}_1(u_x) = D_x(1) = 0, \tag{3.2}$$

reflecting the fact that the group of translations is abelian, we only need to compute the first two terms in (3.1). First, using the definition of the Fréchet derivative, we compute

$$\begin{aligned} [u_x, D_x^{-1} u] &= \mathrm{D}_{D_x^{-1} u}(u_x) - \mathrm{D}_{u_x}(D_x^{-1} u) \\ &= D_x^{-1} u_x - D_x(D_x^{-1} u) = u + c - u = c, \end{aligned}$$

where c is an arbitrary constant representing the ambiguity in the antiderivative D_x^{-1}. Thus,

$$[1, [u_x, D_x^{-1} u]] = [1, c] = 0,$$

irrespective of the integration constant c. On the other hand,

$$[D_x^{-1} u, 1] = -\,D_x^{-1}(1) = -\,x + d,$$

where d is another arbitrary constant, and so

$$[u_x, [D_x^{-1} u, 1]] = [u_x, -x + d] = -D_x(-x + d) = 1.$$

Therefore, no matter how we choose the integration "constants" c, d, the left hand side of (3.1) equals 1, *not* zero, and so the Jacobi identity appears to be invalid!

This example is, in fact, the simplest of a wide variety of apparent nonlocal counterexamples to the Jacobi identity. Similar problems arise in the structure of the Lie algebra of nonlocal symmetries of the KP equation, [34], and the theory of recursion operators, [36].

4. Nonlocal differential algebra

In order to keep the constructions reasonably simple, we will work entirely within the polynomial category throughout. Thus, we only consider differential polynomials with polynomial coefficients. Also we work, without any significant loss of generality, with real-valued polynomials, the complex version being an easy adaptation.

By a *multi-index* we mean a p-tuple $J = (j_1, \ldots, j_p) \in \mathbb{Z}^p$ with integer entries. The *order* of J is $\#J = j_1 + \cdots + j_p$. The multi-index is *positive*, written $J \geq 0$, if *all* its entries are positive: $j_\nu \geq 0$. We impose a partial order on the space of multi-indices with $J \leq K$ if and only if $K - J \geq 0$. We will also impose a total ordering $J \prec K$ on the multi-indices that respects degree, so if $\#J < \#K$, then $J \prec K$. In particular, degree lexicographic ordering is a convenient choice of total order, [14].

Let $\mathcal{A} = \mathbb{R}[x] = \mathbb{R}\{x^1, \ldots, x^p\}$ denote the algebra of polynomial functions $f(x)$ depending upon p variables. The derivatives $\partial_1, \ldots, \partial_p$ make $\mathcal{A}$ into a (partial) differential algebra. Given a possibly infinite set of dependent variables $\mathcal{U} = \{\ldots u^\alpha \ldots\}$, let $\mathcal{A}\{\mathcal{U}\}$ denote the differential algebra consisting of all polynomials in their local derivatives $u^\alpha_J = D^J u^\alpha$, $J \geq 0$, whose coefficients are polynomial functions in $\mathcal{A}$. We write $P[u] = P(\ldots x^i \ldots u^\alpha_J \ldots)$ for a differential polynomial in $\mathcal{A}\{\mathcal{U}\}$. Even though $\mathcal{U}$ may contain infinitely many variables, any differential polynomial $P \in \mathcal{A}\{\mathcal{U}\}$ has only finitely many summands and hence depends on only finitely many variables u^α_J. The set of polynomials that only depend on x can be identified with $\mathcal{A}$ itself, and there is a natural decomposition

$$\mathcal{A}\{\mathcal{U}\} = \mathcal{A} \oplus \mathcal{A}_\star\{\mathcal{U}\},$$

where $\mathcal{A}_\star\{\mathcal{U}\}$ consists of all differential polynomials that vanish whenever we set all $u^\alpha_J = 0$. Any ordering $u^\alpha \prec u^\beta$ of $\mathcal{U}$ induces an ordering of the derivatives, so $u^\alpha_J \prec u^\beta_K$ whenever $u^\alpha \prec u^\beta$ or $\alpha = \beta$ and $J \prec K$. This ordering in turn induces the degree lexicographic ordering on the differential monomials in $\mathcal{A}_\star\{\mathcal{U}\}$.

The total derivatives $D_1, \ldots, D_p$ act on $\mathcal{A}\{\mathcal{U}\}$ as derivations. Their kernels are well-known:

Lemma 4.1 *The kernel of the* i^{th} *total derivative is*

$$\ker D_i = \mathcal{A}_i = \{f(x^1, \ldots, x^{i-1}, x^{i+1}, \ldots, x^p) \mid f \in \mathcal{A}\}. \tag{4.1}$$

In particular, the restriction of $D_i \colon \mathcal{A}_\star\{\mathcal{U}\} \to \mathcal{A}_\star\{\mathcal{U}\}$ *has trivial kernel.*

We begin our construction with the algebra of *local differential polynomials*

$$\mathcal{B}_{\text{loc}} = \mathcal{B}^{(0)} = \mathcal{A} \oplus \mathcal{B}^{(0)}_\star = \mathcal{A}\{u^1, \ldots, u^q\}$$

Our goal is to construct a nonlocal differential algebra

$$\mathcal{B}^{(\infty)} = \mathcal{A} \oplus \mathcal{B}^{(\infty)}_\star \supset \mathcal{B}_{\text{loc}}$$

such that each total derivative $D_i \colon \mathcal{B}^\infty_\star \longrightarrow \mathcal{B}^\infty_*$ defines an invertible map everywhere except on the ordinary polynomials $f(x) \in \mathcal{A}$. The polynomials in $\mathcal{B}^\infty$ will, therefore, be polynomials involving expressions of the form $D_i^{-1} P$ where P is any local or nonlocal differential polynomial, e.g. $D_i^{-1} u$, $D_i^{-1}(u^2 u_j)$, or even $D_i^{-1}(u_{ij} D_j^{-1}(u^2) D_k^{-1}(u_i^2))$, and so on. Our construction will accomplished by inductively implementing the following construction.

At each step, we are given an infinite[1] collection of dependent variables

$$\mathcal{U}^{(m)} = \mathcal{U}_0 \cup \mathcal{U}_1 \cup \mathcal{U}_2 \cup \cdots \cup \mathcal{U}_m,$$

which is the disjoint union of the subsets

$$\mathcal{U}_k = \{u^\alpha \mid \operatorname{depth} u^\alpha = k\}.$$

consisting of all dependent variables of a given *depth*. Roughly speaking, the depth of a variable will measure its "depth of nonlocality". In particular, all the original variables in our local differential algebra $\mathcal{U}^{(0)} = \mathcal{U}_0 = \{u^1, \ldots, u^q\}$ have depth 0. We also assign a weight $w_\alpha = \operatorname{wt} u^\alpha$ to each variable in $\mathcal{U}^{(m)}$. For simplicity, the original dependent variables $u^\alpha \in \mathcal{U}_0$ can have weight 1, although the initial weighting can be adapted to particular applications, as in [35].

We let $\mathcal{B}^{(m)} = \mathcal{A} \oplus \mathcal{B}_\star^{(m)} = \mathcal{A}\{\mathcal{U}^{(m)}\}$ denote the algebra of polynomials in the variables $u_J^\alpha = D^J u^\alpha$ for all $u^\alpha \in \mathcal{U}^{(m)}$ and $J \in \mathbb{Z}^p$. We define

$$\operatorname{depth} u_J^\alpha = \operatorname{depth} u^\alpha, \qquad \operatorname{wt} u_J^\alpha = \operatorname{wt} u^\alpha.$$

Note that linearly nonlocal variables u_J^α for $u^\alpha \in \mathcal{U}_0$ will continue to have depth 0. The total derivatives act on $\mathcal{B}^{(m)}$ as derivations subject to the same rules (2.1).

Let $\mathcal{M}^{(m)}$ denote the set of x-independent monomials in $\mathcal{B}_\star^{(m)}$, i.e., products of the form $M = u_{J_1}^{\alpha_1} \cdots u_{J_k}^{\alpha_k}$. Therefore, $\mathcal{B}^{(m)} = \mathcal{A}[\mathcal{M}^{(m)}]$ consists of finite linear combinations of monomials with coefficients in $\mathcal{A}$. We extend the notion of depth and weight to monomials in $\mathcal{M}^{(m)}$ by setting

$$\begin{aligned} \operatorname{depth}(M\,N) &= \max\{\operatorname{depth} M, \operatorname{depth} N\}, \\ \operatorname{wt}(M\,N) &= \operatorname{wt} M + \operatorname{wt} N, \end{aligned} \tag{4.2}$$

whenever $M, N \in \mathcal{M}^{(m)}$. Thus, we write

$$\mathcal{M}^{(m)} = \mathcal{M}_0 \cup \mathcal{M}_1 \cup \cdots \cup \mathcal{M}_m,$$

where $\mathcal{M}_k$ denotes the set of monomials of depth k.

We describe the induction step. For each monomial $M_\gamma \in \mathcal{M}_m$, and each $1 \le i \le p$, we introduce a new dependent variable $u^{\gamma,i} \in \mathcal{U}_{m+1}$ of depth $m+1$ such that

$$D_i u^{\gamma,i} = M_\gamma. \tag{4.3}$$

Formally, we can write

$$u^{\gamma,i} = D_i^{-1} M_\gamma, \tag{4.4}$$

[1] Except the initial step, where we start with only finitely many dependent variables $u^1, \ldots, u^q$.

but it is better, for the time being, to regard each of these as a completely new dependent variable. Once the induction procedure is complete, we shall impose the relations implied by (4.3). The weight and depth of each new variable is

$$\operatorname{wt} u^{\gamma,i} = \operatorname{wt} M_\gamma, \qquad \operatorname{depth} u^{\gamma,i} = m+1 > \operatorname{depth} M_\gamma.$$

The inductive step sets

$$\mathcal{U}_{m+1} = \{u^{\gamma,i}\}, \qquad \mathcal{U}^{(m+1)} = \mathcal{U}^{(m)} \cup \mathcal{U}_{m+1}, \qquad \mathcal{B}^{(m+1)} = \mathcal{A}\{\mathcal{U}^{(m+1)}\}.$$

Finally, we let

$$\mathcal{U}^{(\infty)} = \bigcup_{m=0}^{\infty} \mathcal{U}_m, \qquad \mathcal{B}^{(\infty)} = \mathcal{A} \oplus \mathcal{B}_\star^{(\infty)} = \mathcal{A}\{\mathcal{U}^{(\infty)}\}. \tag{4.5}$$

We can identify $\mathcal{B}^{(\infty)}$ as the injective limit $\mathcal{B}^{(0)} \hookrightarrow \mathcal{B}^{(1)} \hookrightarrow \mathcal{B}^{(2)} \hookrightarrow \cdots$ of subalgebras of progressively higher and higher depth. Note that $D_i^{-1} P \in \mathcal{B}_\star^{(\infty)}$ is well-defined for any $P \in \mathcal{B}_\star^{(\infty)}$. A nonlocal differential polynomial is said to be *homogeneous* of weight k if all its constituent monomials have weight k. We write $\mathcal{B}_k^{(\infty)}$ for the set of all homogeneous differential polynomials of weight k, and so $\mathcal{B}^{(\infty)} = \mathcal{A} \oplus \bigoplus_{k=1}^{\infty} \mathcal{B}_k^{(\infty)}$.

Of course, the differential algebra $\mathcal{B}^{(\infty)}$ contains a huge number of redundancies, since we have not yet taken into account the defining relations (4.3) of our nonlocal variables. Thus, we need to determine which of these nonlocal expressions are trivial, meaning that they vanish when evaluated upon any smooth function. In local differential algebra, one can prove triviality by evaluating the differential polynomial on all polynomial functions $u = p(x)$. In the nonlocal case, the class of polynomial functions is not appropriate because the inverse derivatives ∂_i^{-1} include a possible integration constant, and so are not uniquely defined on the space of polynomial functions. To check the vanishing of a nonlocal differential polynomial, one needs to keep track of a consistent choice of integration constants used to evaluate the nonlocal terms, and this rapidly becomes a difficult, if not intractable issue. A more enlightened approach is to introduce the following functions.

Definition 1. A function of the form $f(x) = p(x)\, e^{n\cdot x}$ in which $p(x)$ is a polynomial and $n = \{n_1, \ldots, n_p\} \in \mathbb{Z}^p$ will be called a *polynomial–exponential function.* It will be called *positive* if $n > 0$, meaning $n_i > 0$ for $i = 1, \ldots, p$. Let $\mathcal{F} = \{p(x)\, e^{n\cdot x} \mid n > 0\}$ denote the algebra of all positive polynomial–exponential functions.

The key property is that, in contrast to the space of polynomials, derivatives are invertible when restricted to the positive polynomial–exponential space $\mathcal{F}$.

Lemma 4.2 *The derivative $\partial_i : \mathcal{F} \to \mathcal{F}$ is a one-to-one linear map, and hence its inverse $\partial_i^{-1} : \mathcal{F} \to \mathcal{F}$ is uniquely defined. Consequently, if $P \in \mathcal{B}_k^{(\infty)}$ is any homogeneous nonlocal differential polynomial, its evaluation on $f \in \mathcal{F}$ gives a uniquely defined polynomial–exponential function $P[f] \in \mathcal{F}$. Moreover, evaluation commutes with (anti-) differentiation, so $D^J P[f] = \partial^J(P[f])$ for any $J \in \mathbb{Z}^p$.*

Proof: We recall the well-known formula

$$D_i^{-1}(P\,Q) = \sum_{j\geq 0}(-1)^j\,(D_i^j P)\,(D_i^{-j-1}Q). \tag{4.6}$$

If $P = p(x)$ is a polynomial, then the sum terminates, and gives an explicit formula for $D_i^{-1}(p(x)\,e^{n\cdot x})$. *Q.E.D.*

Definition 2. A homogeneous nonlocal differential polynomial $P \in \mathcal{B}_k^{(\infty)}$ is *trivial* $P[f] = 0$ for all $f = (f^1, \ldots, f^q) \in \mathcal{F}^q$.

The fact that testing a nonlocal differential polynomial on all polynomial–exponentials is sufficient to detect triviality is a consequence of the fact that polynomial–exponential functions are sufficiently extensive to match any finite nonlocal jet. We let

$$\mathcal{I}^{(\infty)} = \bigoplus_{k=1}^{\infty} \mathcal{I}_k^{(\infty)}, \tag{4.7}$$

where

$$\mathcal{I}_k^{(\infty)} = \{P \in \mathcal{B}_\star^{\infty} \mid P[f] = 0, \text{ for all } f \in \mathcal{F}\},$$

denote the ideal of all trivial nonlocal differential polynomials. If $P \in \mathcal{I}^{(\infty)}$ then, by the last remark in Lemma 4.2, $D^J P \in \mathcal{I}^{(\infty)}$ for any $J \in \mathbb{Z}^p$, and hence $\mathcal{I}^{(\infty)} \subset \mathcal{B}_\star^{(\infty)}$ is a homogeneous nonlocal differential ideal. In particular, the defining relations (4.3) of our nonlocal variables belong to the ideal, meaning $D_i u^{\gamma,i} - M_\gamma \in \mathcal{I}^{(\infty)}$.

Finally, we define our *nonlocal differential algebra* to be the quotient algebra

$$\pi: \mathcal{B}^{(\infty)} \longrightarrow \mathcal{Q}^{(\infty)} = \mathcal{A} \oplus \mathcal{Q}_\star^{(\infty)} = \mathcal{B}^{(\infty)} / \mathcal{I}^{(\infty)}. \tag{4.8}$$

This algebra incorporates all the relations implied by (4.3) and their (anti-) derivatives. We easily check that D_i^{-1} is uniquely defined on all of $\mathcal{Q}^{(\infty)}$, and moreover forms an inverse to D_i when restricted to $\mathcal{Q}_\star^{(\infty)}$.

Theorem 4.3 *If $P \in \mathcal{Q}_\star^{(\infty)}$ and $1 \leq i \leq p$, then there exists a unique $S_i \in \mathcal{Q}_\star^{(\infty)}$ such that $P = D_i S_i$. We write $S_i = D_i^{-1} P$.*

In practical applications, the key issue is whether we can perform effective computations in the nonlocal differential algebra $\mathcal{Q}^{(\infty)}$. The main question is how to recognize whether a given differential polynomial $P \in \mathcal{B}^{(\infty)}$ lies in the differential ideal $\mathcal{I}^{(\infty)}$. We assume, without loss of generality, that $P \in \mathcal{I}_k^{(\infty)}$ is homogeneous. Roughly speaking, differentiating the polynomial P sufficiently often will eventually (and in a finite number of steps) produce a purely local differential polynomial $P^\star \in \mathcal{B}_{\mathrm{loc}}$ with the property that $P \in \mathcal{I}^{(\infty)}$ if and only if $P^\star \in \mathcal{I}^{(\infty)}$. However, the latter will occur if and only if $P^\star = 0$ in $\mathcal{B}_{\mathrm{loc}}$, which is trivial to check.

In order to implement an algorithm, we extend our original term ordering to include all the nonlocal variables $\mathcal{U}^{(\infty)}$. We set $u^{\alpha,i} \prec u^{\beta,j}$ if and only if $\operatorname{depth} u^{\alpha,i} < \operatorname{depth} u^{\beta,j}$, or if $\operatorname{depth} u^{\alpha,i} = \operatorname{depth} u^{\beta,j} = m \geq 1$ and the corresponding monomials satisfy $M_\alpha \prec M_\beta$ in the induced term ordering on $\mathcal{M}^{(m)}$.

Although the full differential algebra $\mathcal{B}^{(\infty)}$ contains a gigantic number of different variables, any given polynomial

$$P(\ldots x^i \ldots u_J^\alpha \ldots u_K^{\gamma,i} \ldots) \in \mathcal{B}_k^{(\infty)}$$

only depends on finitely many of them, and so all computations are finite in extent. Let $u_K^{\gamma,i} = D_K u^{\gamma,i}$ be the highest order variable occurring in P. We can assume $k_i > 0$, since otherwise we replace $u_K^{\gamma,i} \longmapsto D_{K-e_i} M_\gamma$, in accordance with (4.3), which has smaller depth and hence appears earlier in the term ordering. We write out

$$P = P_n\,(u_K^{\gamma,i})^n + \sum_{\ell=0}^{n-1} P_\ell\,(u_K^{\gamma,i})^\ell, \tag{4.9}$$

where each coefficient $P_\ell \in \mathcal{B}^{(\infty)}$ depends on lower order variables $u_J^{\alpha,l} \prec u_K^{\gamma,i}$ and we assume $P_n \notin \mathcal{I}^{(\infty)}$. Since P_n is of lower order than P, the latter condition can be checked by the same algorithm. The derivative

$$D_i P = D_i P_n\,(u_K^{\gamma,i})^n + \sum_{\ell=0}^{n-1} [D_i P_\ell + (\ell+1) P_{\ell+1} D_K M_\gamma](u_K^{\gamma,i})^\ell. \tag{4.10}$$

does not appear earlier in the term ordering. However, the combination

$$P_n\, D_i P - P\, D_i P_n = Q = \sum_{\ell=0}^{n-1} Q_\ell (u_K^{\gamma,i})^\ell \tag{4.11}$$

is of lower order than P in $u_K^{\gamma,i}$. The induction step claims that $P \in \mathcal{I}^{(\infty)}$ if and only if $Q \in \mathcal{I}^{(\infty)}$, and hence the same algorithm can be used on Q.

Since Q is of lower order, we use the same algorithm on Q, and so eventually — but in a finite number of steps — reducing to a purely local differential polynomial, as desired.

To prove the claim, equation (4.11) implies that, for any $f \in \mathcal{F}$,

$$P_n[f]\,\partial_i P[f] - P[f]\,\partial_i P_n[f] = Q[f] = 0.$$

Therefore,

$$\frac{\partial}{\partial x^i}\left(\frac{P[f]}{P_n[f]}\right) = 0\,,$$

and hence $P[f] = 0$ or $P_n[f] = 0$. Now, if $P_n \notin \mathcal{I}^{(\infty)}$ is nontrivial, then the jets of polynomial–exponential functions that solve the nonlocal differential equation $P_n[f] = 0$ forms a proper subvariety, and $P[f] = 0$ everywhere outside this subvariety, which, by continuity, implies $P[f] = 0$ for all $f \in \mathcal{F}$, and proves the claim. Of course, the implementation of this algorithm might be quite lengthy, and so developing more efficient algorithms would be an interesting research topic.

5. Evolutionary vector fields and symmetries

In this section we extend the space of evolutionary vector fields to our nonlocal differential algebra. Since they are defined as commutators, the Jacobi Identity will be automatically valid.

Definition 3. A *evolutionary vector field* $\mathbf{v}$ on a differential algebra $\mathcal{B} = \mathcal{A} \oplus \mathcal{B}^{(0)}$ is a derivation $\mathbf{v}\colon \mathcal{B} \to \mathcal{B}$, with $\mathcal{A} \subset \ker \mathbf{v}$, while $[\mathbf{v}, D_i] = 0$ commutes with all total derivatives.

Remark: If we drop the hypothesis $\mathcal{A} \subset \ker \mathbf{v}$ then the only additional derivations that commute with the total derivatives are the partial derivatives $\partial/\partial x^j$; see [33].

Therefore, an evolutionary vector field $\mathbf{v}$ must satisfy

$$\begin{aligned} \mathbf{v}(P+Q) &= \mathbf{v}(P) + \mathbf{v}(Q), \quad & \mathbf{v}(x^i) &= 0 \\ \mathbf{v}(P\,Q) &= \mathbf{v}(P)\,Q + P\,\mathbf{v}(Q), \quad & [\mathbf{v}, D_i] &= 0, \end{aligned} \tag{5.1}$$

for all $P, Q \in \mathcal{B}$ and $i = 1, \ldots, p$. Each evolutionary vector field is uniquely specified by its action $\mathbf{v}(u^\alpha_J)$ on the coordinate variables. We denote the space of evolutionary vector fields by $\mathcal{V} = \mathcal{V}(\mathcal{B})$. The commutator bracket

$$[\mathbf{v}, \mathbf{w}](P) = \mathbf{v}(\mathbf{w}(P)) - \mathbf{w}(\mathbf{v}(P)), \qquad P \in \mathcal{B},$$

between two evolutionary vector fields endows $\mathcal{V}$ with the structure of a Lie algebra, satisfying the usual skew symmetry and Jacobi identities. The proof of the latter is elementary.

Warning: The space of evolutionary vector fields is *not* a $\mathcal{B}$ module. The product $P\mathbf{v}$ of $P \in \mathcal{B}$ and $\mathbf{v} \in \mathcal{V}$ does not commute with total differentiation.

Given an evolutionary vector field $\mathbf{v}$, we define its *characteristic* $Q \in \mathcal{B}^q$ to have components

$$\mathbf{v}(u^\alpha) = Q^\alpha, \quad \alpha = 1, \dots, q.$$

The commutation condition implies

$$\mathbf{v}(u^\alpha_J) = \mathbf{v}(D_J u^\alpha) = D_J \mathbf{v}(u^\alpha) = D_J Q^\alpha$$

for all positive multi-indices $J \geq 0$. Thus, in the local situation, an evolutionary vector field is uniquely determined by its characteristic. This basic fact is *not* true in nonlocal differential algebras — there are nonzero evolutionary vector fields with zero characteristic — and this observation motivates the following key definition.

Definition 4. An evolutionary vector field γ is called a a K-*ghost* for some $K \in \mathbb{Z}^p$ if $\gamma(u^\alpha_L) = 0$ for all $L \geq K$ and $\alpha = 1, \dots, q$.

There are no ghost vector fields in a local differential algebra $\mathcal{B}_{\text{loc}}$ because each evolutionary vector field is uniquely determined by its characteristic Q. There are, however, positive ghost vector fields; for example the vector field with characteristic $Q = 1$ is a K-ghost for any positive multi-index $K > 0$.

Example 5.1. Let us see how the existence of ghost vector fields serves to resolve the Jacobi identity paradox in (3.1). Surprisingly, the problem is *not* with the nonlocal vector field $\mathbf{z}$ with characteristic $D_x^{-1}u$, but rather the local commutator $[\mathbf{v}, \mathbf{w}]$ corresponding to the vector fields with characteristics 1 and u_x, respectively. While $[\mathbf{v}, \mathbf{w}] = 0$ on the local differential algebra $\mathcal{B}_{\text{loc}}$, it is, in fact, a ghost vector field on a nonlocal differential algebra. Thus, surprisingly, in a nonlocal setting, the group of translations is not abelian!

The action of the vector fields on the local variables does not uniquely specify their action on the nonlocal variables, due to the presence of possible integration constants. However, as we have seen, the integration constants do not play a significant role in the resolution of the Jacobi identity paradox, and so we shall fix all the integration constants to be zero by default. Therefore,

$$\begin{aligned} \mathbf{v}(u_k) &= D_x^k(1) = \chi_k(x) \equiv \begin{cases} 0, & k > 0, \\ \dfrac{x^{-k}}{(-k)!}, & k \leq 0, \end{cases} \\ \mathbf{w}(u_k) &= D_x(u_k) = u_{k+1}. \end{aligned} \tag{5.2}$$

Since $\mathbf{v}(u_k)$ only depends on x, we have $\mathbf{w}(\mathbf{v}(u_k)) = 0$, and so

$$[\mathbf{v}, \mathbf{w}](u_k) = \mathbf{v}(u_{k+1}) = \chi_{k+1}(x).$$

Therefore, $[\mathbf{v}, \mathbf{w}] = \gamma$ is a ghost vector field that satisfies

$$\gamma(u_k) = \chi_{k+1}(x) = \begin{cases} 0 & k \geq 0, \\ \dfrac{x^{-k-1}}{(-k-1)!} & k < 0. \end{cases}$$

This ghost provides the missing term in the Jacobi identity (3.1). Indeed,

$$[\mathbf{z}, \gamma](u) = -\gamma(\mathbf{z}(u)) = -\gamma(D_x^{-1} u) = -1.$$

In [34], we introduced a "ghost calculus" for general nonlocal evolutionary vector fields. The first remark is that only evolutionary vector fields that depend on the independent variables can be ghosts. Indeed, if γ is a K-ghost, then

$$\gamma(u_J) = D^{J-K}\gamma(u_K) = 0, \qquad J \geq K.$$

Therefore, if $\gamma(u_I) = P_I$ and $J \geq 0$ is any positive multi-index such that $J + I \geq K$, then

$$0 = \gamma(u_{J+I}) = D^J \gamma(u_I) = D^J P_I,$$

and we know that $\ker D^J \subset \mathcal{A}$, so that P_I is a function of x only.

Lemma 5.2 *An evolutionary vector field γ is a K-ghost for some $K \in \mathbb{Z}^p$ if and only if $\gamma(u_J^\alpha) = p_J^\alpha(x)$ is a polynomial function of $x^1, \ldots, x^p$.*

Definition 5. Given a multi-index $K \in \mathbb{Z}^p$, define

$$\chi_K = D^K(1) = \begin{cases} \dfrac{x^{-K}}{(-K)!}, & K \leq 0, \\ 0 & \text{otherwise}, \end{cases} \qquad \text{where} \quad (-K)! = \prod_{\nu=1}^{p} (-k_\nu)!\,. \tag{5.3}$$

Definition 6. Given a multi-index $J \in \mathbb{Z}^p$, define the basis ghost vector field γ_J so that $\gamma_J(u_K) = \chi_{J+K}$, which is a K-ghost for any $K + J > 0$.

Proposition 5.3 *Every ghost vector field is a linear combination of the basis ghosts, $\gamma = \sum_J c_J \gamma_J$, where the $c_J \in \mathbb{R}$ are constants.*

The summation in Proposition 5.3 can be infinite. However, only certain "configurations" of the nonzero coefficients c_J are allowed in order that γ map $\mathcal{Q}^{(\infty)}$ to $\mathcal{Q}^{(\infty)}$.

Let us formulate the results in the one dependent variable case where $u \in \mathbb{R}$, and so $q = 1$. (The multi-variable case can be found in [34].) Let us split the space of evolutionary vector fields $\mathcal{V} = \mathcal{V}_x \oplus \mathcal{V}_\star$ where $\mathcal{V}_x$ denotes the space of purely x-dependent vector fields, so $\mathbf{v}(u_K) = p_K(x) \in \mathcal{A}$. In the polynomial category, every $\mathbf{v} \in \mathcal{V}_x$ is a ghost vector field. The remainder, $\mathcal{V}_\star$, consists of u-dependent vector fields, where $\mathbf{v}(u_K) = D^K Q \in \mathcal{B}_\star^{(\infty)}$. Since $\ker D^K = \{0\}$ on $\mathcal{B}_\star^{(\infty)}$, the evolutionary vector fields in $\mathcal{V}_\star$ are uniquely determined by their characteristics $Q = \mathbf{v}(u)$, and we write $\mathbf{v} = \mathbf{v}_Q$ as in the local category. Thus, to re-emphasize: only the x-dependent vector fields can be ghosts and hence cause any difficulty in the non-local category.

Corollary 5.4 *If $\mathcal{B}^{(\infty)}$ is a polynomial differential algebra, then any evolutionary vector field $\mathbf{v} \in \mathcal{V}$ can be written a linear combination of basis ghosts and a u-dependent vector field $\mathbf{v}_Q \in \mathcal{V}_\star$:*

$$\mathbf{v} = \mathbf{v}_Q + \sum_J c_J \gamma_J, \quad \text{whereby} \quad \mathbf{v}(u_K) = D^K Q + \sum_J c_J \chi_{K+J}. \tag{5.4}$$

To implement a calculus of evolutionary vector fields, we identify a vector field with its "characteristic". The characteristic of the evolutionary vector field $\mathbf{v}_Q$ is, as usual, Q. The characteristic of the ghost vector field γ_J will be formally written as χ_J. In this manner, every nonlocal vector field (5.4) has a unique characteristic

$$S = Q + \sum_J c_J \chi_J. \tag{5.5}$$

In particular, a local vector field with polynomial characteristic x^K becomes a ghost characteristic $K!\,\chi_{-K}$. Indeed, one can, again in the one dependent variable case, replace all polynomials $x^K \longmapsto K!\,\chi_{-K}$ wherever they appear in the characteristic (5.5). The only place true ghosts appear, i.e., χ_J with $J \not\leq 0$, is in the u-independent terms in the summation. Only when the vector field has been evaluated on a nonlocal differential polynomial are we allowed to replace the ghost functions χ_J by their actual formulas (5.3).

In this calculus, the product rule $x^J x^K = x^{J+K}$ becomes the ghost product rule

$$\chi_K \chi_J = \binom{-K-J}{-K} \chi_{K+J}, \qquad J \geq 0. \tag{5.6}$$

The product makes sense as long as one of the multi-indices is non-negative, provided we adopt the Pochhammer definition

$$\binom{J+K}{J} = \frac{1}{J!} \prod_{\nu=1}^{p} \prod_{i=0}^{j_\nu - 1} (j_\nu + k_\nu - i) \tag{5.7}$$

for the multinomial symbol. And, indeed, only such products will appear when we evaluate commutators and apply vector fields to nonlocal differential polynomials.

The precise ghost calculus rules for computing the commutators of ghost characteristics will now be described. The commutators of ordinary characteristics $[Q, R]$ for $Q, R \in \mathcal{B}_\star^{(\infty)}$ follow the same rules (2.5) as in the local case, where we replace the multiplication of monomials by the ghost multiplication rule (5.6). Secondly, since ghosts do not involve the dependent variables, they mutually commute:

$$[\chi_J, \chi_K] = 0. \tag{5.8}$$

Finally, the ghost characteristics χ_J act as derivations on the ordinary characteristics:

$$[\chi_K, Q\, R] = Q\, \chi_K(R) + R\, \chi_K(Q).$$

Thus, we only need to know how to commute ghosts and derivative coordinates,

$$[\chi_J, u_K] = \chi_{J+K} \tag{5.9}$$

in order to compute in the ghost characteristic space.

Example 5.5. Let us revisit Example 5.1. The three ghost characteristics are

$$1 = \chi_0, \quad u_x = u_1, \quad D_x^{-1} u = u_{-1}.$$

Then the three terms are

$$[\chi_0, [u_1, u_{-1}]] = 0, \qquad [u_1, [u_{-1}, \chi_0]] = -[u_1, \chi_{-1}] = \chi_0,$$
$$[u_{-1}, [\chi_0, u_1]] = [u_{-1}, \chi_1] = -\chi_0.$$

The sum of these three terms is 0, and so the Jacobi paradox is resolved.

Example 5.6. The first Jacobi identity paradox that was found in [34], while working on the symmetry algebra of the KP equation, [9, 10, 15, 29], was more complicated than (3.1). Here $p = 2$, with independent variables x, y, and $q = 1$, with dependent variable u. Consider the vector fields with characteristics y, $y\, u_x$ and $u_x D_x^{-1} u_y$. As before, without the introduction of ghost terms, the Jacobi sum

$$[y, [u_x D_x^{-1} u_y, y u_x]] + [y u_x, [y, u_x D_x^{-1} u_y]] + [u_x D_x^{-1} u_y, [y u_x, y]] \tag{5.10}$$

equals $-2\,y\,u_x$, not zero. In this case, the three ghost characteristics are

$$y = \chi_{0,-1}, \qquad y u_x = \chi_{0,-1}\, u_{1,0}, \qquad u_x D_x^{-1} u_y = u_{1,0}\, u_{-1,1}.$$

Then,

$$\begin{aligned}
[\chi_{0,-1}, \chi_{0,-1}\, u_{1,0}] &= 2\,\chi_{1,-2}, \\
[\chi_{0,-1}, u_{0,0} u_{1,0}] &= \chi_{0,-1}\, u_{1,0}, \\
[\chi_{0,-1}\, u_{1,0}, u_{1,0}\, u_{-1,1}] &= D_{u_{1,0}\, u_{-1,1}}(\chi_{0,-1}\, u_{1,0}) - \chi_{0,-1}\, D_x(u_{1,0}\, u_{-1,1}) \\
&= u_{0,0} u_{1,0},
\end{aligned}$$

and so,

$$\begin{aligned}
[u_{1,0}\, u_{-1,1}, 2\,\chi_{1,-2}] &= -2\,\chi_{0,-1}\, u_{1,0}, \\
[u_{1,0}\, u_{-1,1}, \chi_{0,-1}] &= -\chi_{-1,0}\, u_{1,0}, \\
[\chi_{0,-1}\, u_{1,0}, -\chi_{-1,0}\, u_{1,0}] &= \chi_{0,-1}\, u_{1,0}.
\end{aligned}$$

The latter three terms add up to 0, and so the Jacobi identity is valid in the ghost framework.

6. Conclusions

In this paper, I have introduced a general framework for a nonlocal differential algebra that will handle quite general nonlocal polynomial expressions. Several further topics of investigation are now of importance:

(a) A complete re-evaluation of earlier work on nonlocal symmetries of local and non-local partial differential equations is required. A proper understanding of the hitherto undetected ghost terms needs to be properly incorporated into earlier results, including the study of recursion operators and master symmetries, all of which typically involve nonlocal operations.

(b) The framework for the geometric and algebraic study of nonlocal symmetries and nonlocal differential equations requires further development. The establishment of a complete nonlocal variational calculus on the nonlocal differential algebra $\mathcal{Q}^{(\infty)}$, including nonlocal conservation laws, [40] and a nonlocal form of Noether's Theorem, [32, 33], would be a very worthwhile project for both theoretical developments and practical applications.

(c) Implementation of the nonlocal ghost calculus in standard computer algebra packages would help a lot in these investigations.

References

1. Abraham–Shrauner, B., Hidden symmetries and nonlocal group generators for ordinary differential equations, *IMA J. Appl. Math.* **56** (1996), 235–252.
2. Abraham–Shrauner, B., Hidden symmetries, first integrals and reduction of order of nonlinear ordinary differential equations, *J. Nonlinear Math. Phys.* **9** Suppl. 2 (2002), 1–9.
3. Abraham–Shrauner, B., Guo, A., Hidden and nonlocal symmetries of nonlinear differential equations, *in: Modern Group Analysis: Advanced Analytical and Computational Methods in Mathematical Physics*, N.H. Ibragimov, M. Torrisi, and A. Valenti, eds., Kluwer, Dordrecht, The Netherlands, 1993, pp. 1–5.
4. Akhatov, I.S., Gazizov, R.K., Ibragimov, N.H., Nonlocal symmetries. Heuristic approach, *J. Sov. Math.* **55** (1991), 1401–1450
5. Bluman, G.W., Kumei, S., *Symmetries and Differential Equations,* Springer–Verlag, New York, 1989.
6. Bluman, G.W., Kumei, S., Reid, G., New classes of symmetries for partial differential equations, *J. Math. Phys.* **29** (1988), 806–811.
7. Bluman, G.W., Reid, G., New classes of symmetries for ordinary differential equations, *IMA J. Appl. Math.* **40** (1988), 87–94.
8. Chen, H.H., Lee, Y.C., Lin, J.-E., On a new hierarchy of symmetries for the integrable nonlinear evolution equations, *in: Advances in Nonlinear Waves* vol. 2, L. Debnath, ed., Research Notes in Math., vol. 111, Pitman Publ., Marshfield, Mass., 1985, pp. 233–239.
9. Chen, H.H., Lee, Y.C., Lin, J.-E., On a new hierarchy of symmetries for the Kadomtsev–Petviashvili equation, *Physica D* **9** (1983), 439–445.
10. Chen, H.H., Lin, J.E. On the infinite hierarchies of symmetries and constants of motion for the Kadomtsev–Petviashvili equation, *Physica D* **26** (1987), 171–180.
11. Chetverikov, V.N., and Kudryavtsev, A.G., Modeling integro-differential equations and a method for computing their symmetries and conservation laws, *Amer. Math. Soc. Transl.* **167** (1995), 1–22.
12. Cieśliński, J., Non-local symmetries and a working algorithm to isolate integrable geometries, *J. Phys. A* **26** (1993), L267–L271.
13. Coggeshall, S.V., Abraham–Shrauner, B., Knapp, C., Hidden symmetries of partial differential equations, *in: Proceedings of IMACS World Congress,* W.F. Ames, ed., Georgia Tech., 1994, Vol. 1, 102-107.
14. Cox, D., Little, J., O'Shea, D., *Ideals, Varieties, and Algorithms,* 2nd ed., Springer–Verlag, New York, 1996.
15. David, D., Kamran, N., Levi, D., Winternitz, P., Subalgebras of loop algebras and symmetries of the Kadomtsev-Petviashivili equation, *Phys. Rev. Lett.* **55** (1985), 2111–2113.
16. Dodd, R.K., The general prolongation formulae for nonlocal symmetries, *Phys. Lett. A* **195** (1994), 125–127.
17. Fushchich, W.I., Nikitin, A.G. *Symmetries of Equations of Quantum Mechanics,* Allerton Press, New York, 1994.
18. Galas, F., New non-local symmetries with pseudopotentials, *J. Phys. A* **25** (1992), L981–L986.
19. Govinder, K.S., Leach, P.G.L., On the determination of nonlocal symmetries, *J. Phys. A* **28** (1995), 5349–5359.
20. Guthrie, G.A., More nonlocal symmetries of the KdV equation, *J. Phys. A* **26** (1993), L905–L908.

21. Guthrie, G.A., Recursion operators and non-local symmetries, *Proc. Roy. Soc. London A* **446** (1994), 107–114.
22. Guthrie, G.A., Hickman, M.S., Nonlocal symmetries of the KdV equation, *J. Math. Phys.* **34** (1993), 193–205.
23. Ibragimov, N.H. ed., *CRC Handbook of Lie Group Analysis of Differential Equations*, vol. 1, CRC Press, Boca Raton, Fl., 1994.
24. Kapcov, O.V., Extension of the symmetry of evolution equations, *Sov. Math. Dokl.* **25** (1982), 173–176.
25. Krasil'shchik, I.S., Kersten, P.H.M., *Symmetries and and Recursion Operators for Classical and Supersymmetric Differential Equations*, Kluwer Academic Publ., Dordrecht, The Netherlands, 2000.
26. Krasil'shchik, I.S., Vinogradov, A.M., Nonlocal symmetries and the theory of coverings: an addendum to A.M. Vinogradov's 'Local symmetries and conservation laws', *Acta Appl. Math.* **2** (1984), 79–96.
27. Krasil'shchik, I.S., Vinogradov, A.M. (eds.), *Symmetries and Conservation Laws for Differential Equations of Mathematical Physics*, American Mathematical Society, Providence, R.I., 1998.
28. Leo, M., Leo, R.A., Soliani, G., Tempesta, P., On the relation between Lie symmetries and prolongation structures of nonlinear field equations, *Prog. Theor. Phys.* **105** (2001), 77–97.
29. Lou, S-Y, Hu, X-B., Infinitely many Lax pairs and symmetry constraints of the KP equation, *J. Math. Phys.* **38** (1997), 6401–6427.
30. Mikhailov, A.V., Shabat, A.B., Sokolov, V.V., The symmetry approach to classification of integrable equations, *in What is Integrability?*, V.E. Zakharov, ed., Springer Verlag, New York, 1991, pp. 115–184.
31. Muriel, C., Romero, J.L., C^∞–symmetries and non-solvable symmetry algebras, *IMA J. Appl. Math.* **66** (2001), 477–498.
32. Noether, E., Invariante Variationsprobleme, *Nachr. Konig. Gesell. Wissen. Gottingen, Math.-Phys. Kl.*, (1918) 235–257. (See *Transport Theory and Stat. Phys.* **1** (1971), 186–207 for an English translation.)
33. Olver, P.J., *Applications of Lie Groups to Differential Equations*, Second Edition, Graduate Texts in Mathematics, vol. 107, Springer–Verlag, New York, 1993.
34. Olver, P.J., Sanders, J., Wang, J.P., Ghost symmetries, *J. Nonlinear Math. Phys.* **9** Suppl. 1 (2002), 164–172.
35. Sanders, J.A., Wang, J.P., On the integrability of homogeneous scalar evolution equations, *J. Diff. Eq.* **147** (1998), 410–434.
36. Sanders, J.A., Wang, J.P., On recursion operators, *Physica D* **149** (2001), 1–10.
37. Sluis, W.M., Kersten, P.H.M., Non-local higher-order symmetries for the Federbush model, *J. Phys. A* **23** (1990), 2195–2204.
38. Vinogradov, A.M., Krasil'shchik, I.S., A method of computing higher symmetries of nonlinear evolution equations, and nonlocal symmetries, *Soviet Math. Dokl.* **22** (1980), 235–239.
39. Vinogradov, A.M., Krasil'shchik, I.S., On the theory of nonlocal symmetries of nonlinear partial differential equations, *Soviet Math. Dokl.* **29** (1984), 337–341.
40. Vladimirov, V.S., Volovich, I.V., Local and nonlocal currents for nonlinear equations, *Theor. Math. Phys.* **62** (1985), 1–20.

INTEGRABLE BC_N ANALYTIC DIFFERENCE OPERATORS: HIDDEN PARAMETER SYMMETRIES AND EIGENFUNCTIONS

S.N.M. RUIJSENAARS
Centre for Mathematics and Computer Science,
P.O. Box 94079, 1090 GB Amsterdam, The Netherlands

Abstract. We consider integrable N-particle quantum systems of Calogero-Moser type, focusing on the 'relativistic' BC_N setting, where commuting analytic difference operators arise. We show that the defining operators at the hyperbolic/elliptic levels, which depend on four/eight coupling constants, can be transformed to a manifestly D_4/D_8 symmetric form, resp. We survey various results on special eigenfunctions (including 'ground states') with regard to the latter symmetries and other ones. We also sketch a symmetry scenario for the arbitrary-N eigenfunctions, motivated by the hyperbolic BC_1 case, where our 'relativistic' hypergeometric function has all of the expected properties.

1. Introduction

There exists a wide-spread belief that integrable N-particle quantum Hamiltonians—especially those leading to factorized scattering—are highly exceptional objects. We have recently shown that this is not the case [1], but in the following we focus on the quite special class of Calogero-Moser type integrable quantum systems, and more specifically on those connected to the root system BC_N.

We will detail the most general Hamiltonians of this family in the main text, whereas in this introduction we only present some illuminating special cases. Moreover, we will be quite brief on background material, referring to the survey by Olshanetsky and Perelomov [2] and our lecture notes [3] for extensive information on the 'nonrelativistic' Calogero-Moser systems connected to root systems, and their 'relativistic' generalizations, resp. (To be precise, only for the root system A_{N-1} there is a natural relation to the nonrelativistic/relativistic space-time symmetry groups [3]. For convenience, we use the same terminology for arbitrary root systems.)

A.B. Shabat et al. (eds.), New Trends in Integrability and Partial Solvability, 217–261.

As explained in [3], the relativistic setting gives rise to Hamiltonians that are commuting difference operators, whose step size is proportional to $1/c$, with c the speed of light. In the limit $c \to \infty$, these difference operators reduce to the commuting differential operators of the nonrelativistic case.

We recall that the defining A_{N-1} type differential operator is of the form

$$H_{\mathrm{nr}}^{(A)} = -\sum_{j=1}^{N} \frac{\partial_j^2}{2} + g(g-1) \sum_{1\le j<k\le N} V(x_j - x_k), \tag{1.1}$$

with the most general $V(x)$ being the Weierstrass $\wp$-function. For the remaining classical root systems B_N, C_N, D_N and BC_N, a term of the form

$$g(g-1) \sum_{1\le j<k\le N} V(x_j + x_k) + \sum_{j=1}^{N} U(x_j) \tag{1.2}$$

should be added to $H_{\mathrm{nr}}^{(A)}$. For our purposes, the elliptic BC_N case, where one may choose

$$\begin{aligned} H_{\mathrm{nr}}^{(BC)} &= -\sum_{j=1}^{N} \frac{\partial_j^2}{2} + g(g-1) \sum_{1\le j<k\le N} [V(x_j - x_k) \\ &\quad + V(x_j + x_k)] + \sum_{j=1}^{N}\sum_{t=0}^{3} \frac{g_t(g_t-1)}{2} V(x_j + \omega_t), \end{aligned} \tag{1.3}$$

should be explicitly mentioned. This case was first studied by Inozemtsev [4], and its integrability was shown by Oshima and H. Sekiguchi [5] (cf. also [6]).

The above Hamiltonians have a manifest coupling constant symmetry: The pair potential part is invariant under $g \to 1-g$ and the external field part under $g_t \to 1-g_t$. A principal aim of this contribution is to reveal nonmanifest symmetries of the relativistic/difference operator generalizations of the above Hamiltonians, and to discuss their eventual relevance for the joint eigenfunctions of the commuting difference operators.

The defining difference operator for the relativistic elliptic A_{N-1} case may be chosen as [7]

$$\begin{aligned} H_{\mathrm{rel}}^{(A)} &= \sum_{j=1}^{N} \left(\prod_{k\neq j} \frac{\sigma(x_j - x_k + \mu)}{\sigma(x_j - x_k)} \right)^{\frac{1}{2}} \exp(i\beta\partial_{x_j}) \left(\prod_{l\neq j} \frac{\sigma(x_j - x_l - \mu)}{\sigma(x_j - x_l)} \right)^{\frac{1}{2}} \\ &\quad + (x \to -x), \end{aligned} \tag{1.4}$$

where $\sigma(z)$ is the Weierstrass σ-function. Its BC_N version was first proposed by van Diejen [8], who found some commuting Hamiltonians and conjectured the general form of the desired N commuting difference operators. Subsequently, Hikami and Komori [9], in a tour de force of R-matrix technology, arrived at N commuting operators of the expected form.

The latter cannot easily be written down explicitly (by contrast to the A_{N-1} case [7]). Even in the trigonometric/hyperbolic setting, the N commuting difference operators first found by van Diejen [10] have a quite complicated combinatorial structure. Our concern in this paper is primarily the defining ('simplest') BC_N Hamiltonian, both at the hyperbolic and at the elliptic level.

In the trigonometric regime this Hamiltonian was first introduced by Koornwinder [11], who found and studied its polynomial eigenfunctions, following previous work by Macdonald. Specializing to the rank-one ($N = 1$) case, these polynomials are the well-known Askey-Wilson polynomials [12, 13], which depend on four coupling type parameters. For $N > 1$ the four parameters are present as well, together with an extra one. This fifth parameter plays the role of the coupling g in (1.3) and μ in (1.4), whereas the four parameters may be viewed as external field couplings, just as $g_0, \ldots, g_3$ in (1.3).

In the hyperbolic regime the same number of parameters occurs, whereas in the elliptic regime one can allow eight parameters of external field coupling type. The hidden parameter symmetries alluded to above concern, first and foremost, certain non-manifest invariances w.r.t. the four/eight external field parameters in the hyperbolic/elliptic case. (Other non-obvious symmetries are present as well, as discussed later on.) Specifically, we show that the defining hyperbolic/elliptic difference operator has a D_4/D_8 symmetry in the parameters. That is, it is not only invariant under arbitrary permutations in S_4/S_8, but also under any even number of sign flips of the parameters.

As will be seen, the S_4-invariance property in the hyperbolic case becomes manifest when the four 'physical' couplings are traded for four shifted parameters. (For the rank-one trigonometric case, the resulting Askey-Wilson operator is just the usual one [12, 13]; in this case the S_4-invariance of the suitably normalized polynomial eigenfunctions was already pointed out by Askey and Wilson [12].)

The S_8-invariance at the elliptic level, however, is far from clear from Refs. [8, 9], even for the $N = 1$ case. Indeed, van Diejen [8] works with shifted σ-functions $\sigma(z + \omega_t)$, and Komori/Hikami [9] with theta functions $\theta_s(z)$ (where $t, s - 1 = 0, \ldots, 3$), using eight associated couplings whose permutations give rise to manifestly distinct operators. Even so, also at the elliptic level suitable shifts, alongside with multiplication by a (non-permutation-invariant!) constant, give rise to S_8-invariance. To be sure, this follows only after substantial calculations.

We will therefore proceed differently. We start from difference operators whose S_8-invariance is 'built in', and then show that they are in fact D_8-invariant. In Appendix B we spell out how these operators relate to those of Refs. [8, 9]. The main tool for arriving at the manifestly symmetric versions of the hyperbolic/elliptic operators is the hyperbolic/elliptic gamma function from our paper [14], and therefore we begin by collecting some salient features of these functions and related ones in Section 2.

In Section 3 we study the $N = 1$ Hamiltonians, with Subsections 3.1 and 3.2 treating the hyperbolic and elliptic case, resp. Once the D_4/D_8 symmetry is well-understood for $N = 1$, the general-N case handled in Section 4 presents no new difficulties of principle. Here we also discuss further properties that are relevant to the search for eigenfunctions.

Both in Section 3 and in Section 4 we need a few crucial functional identities that we collect in some lemmas. The proofs of these lemmas are of a calculational rather than a conceptual nature, and so we have relegated them to Appendix A. We would like to stress that in the elliptic case these identities amount to an adaptation of an $N = 1$ result due to van Diejen [8] and an arbitrary-N result due to Komori/Hikami [9], cf. also Appendix B.

In Section 5 we discuss eigenfunctions of the pertinent AΔOs with an eye on the parameter symmetries at issue, and also with regard to their suitability for Hilbert space purposes. Subsection 5.1 may be viewed as an introduction and overview of various aspects. It also contains remarks on a possible scenario for the arbitrary N, arbitrary coupling and arbitrary eigenvalue eigenfunctions that are yet to be constructed/discovered.

In Subsection 5.2 we have tried to sketch in a discursive fashion various results from our papers [15–17], which deal with a 'relativistic' hypergeometric function. This is an eigenfunction of the hyperbolic BC_1 AΔO, and we focus on features that may be illuminating as a guide for the general case. In particular, the character and occurrence of a 'ground state' can be completely controlled in the hyperbolic BC_1 setting. This ground state issue is studied for the general case in Subsection 5.3.

Section 6 contains information regarding the nonrelativistic limit. In particular, we detail how the defining AΔOs reduce to differential operators, and we clarify the fate of the various symmetries under this limit. In the process, we obtain the limits of the eigenfunctions of ground state type. At the end of Section 6 we comment on the connection of the latter functions with recent work on quasi-exactly solvable models by the Madrid school [18, 19].

Appendix B is concerned with the precise relation of our form of the elliptic BC_N operators and the one found in the papers by van Diejen [8] and Komori/Hikami [9].

2. Preliminaries

In this section we collect known results on the hyperbolic and elliptic gamma functions, including some formulas relating to elliptic and allied functions. For more information and background material we refer to [20, 14, 21].

To begin with, throughout this paper we use parameters a_+, a_-, r that are positive. It is convenient to set

$$a \equiv (a_+ + a_-)/2, \quad \alpha \equiv 2\pi/a_+a_-. \tag{2.1}$$

With these conventions, the hyperbolic gamma function can be defined by the integral representation

$$G(a_+, a_-; z) = \exp\left(i\int_0^\infty \frac{dy}{y}\left(\frac{\sin 2yz}{2\sinh(a_+y)\sinh(a_-y)} - \frac{z}{a_+a_-y}\right)\right),$$
$$|\mathrm{Im}\, z| < a. \tag{2.2}$$

It extends to a meromorphic function satisfying the analytic difference equation (AΔE)

$$\frac{G(z + ia_+/2)}{G(z - ia_+/2)} = 2\cosh(\pi z/a_-). \tag{2.3}$$

Up to normalization, it is the unique minimal solution to this AΔE, in a sense explained in [14].

The symmetry under $a_+ \leftrightarrow a_-$ that is manifest from (2.2) entails that $G(a_+, a_-; z)$ also obeys the AΔE

$$\frac{G(z + ia_-/2)}{G(z - ia_-/2)} = 2\cosh(\pi z/a_+). \tag{2.4}$$

The only further properties of G we invoke in this paper are the reflection equation

$$G(a_+, a_-; -z) = 1/G(a_+, a_-; z), \tag{2.5}$$

(which is clear from (2.2)), the duplication formula

$$G(a_+, a_-; 2z) = \prod_{l,m=+,-} G(a_+, a_-; z + i(la_+ + ma_-)/4), \tag{2.6}$$

cf. [14] (3.24)–(3.25), and the limits

$$\lim_{a_-\downarrow 0} \frac{G(a_+, a_-; z - ia_-\kappa)}{G(a_+, a_-; z - ia_-\lambda)} = \exp[(\lambda - \kappa)\ln(2\cosh(\pi z/a_+))], \tag{2.7}$$

where $\kappa, \lambda \in \mathbb{R}$, and z belongs to the cut plane $\mathbb{C}\backslash\{\pm i[a_+/2, \infty)\}$ (cf. [14] (3.91)),

$$\lim_{a_-\downarrow 0} G(a_+, a_-; ia_+/2 - ia_-\kappa)\exp[\kappa\ln(2\pi a_-/a_+)] = \frac{(2\pi)^{1/2}}{\Gamma(\kappa + 1/2)}, \tag{2.8}$$

where $\kappa \in C$, cf. Proposition III.6 in [14].

The elliptic gamma function depends not only on the parameters $a_+, a_- > 0$, but also on a parameter $r > 0$. It can be defined by the infinite product

$$G(r, a_+, a_-; z) = \prod_{m,n=1}^{\infty} \frac{1 - q_+^{2m-1} q_-^{2n-1} e^{-2irz}}{1 - q_+^{2m-1} q_-^{2n-1} e^{2irz}}, \quad q_\delta \equiv \exp(-ra_\delta). \quad (2.9)$$

From this it is clear that G is meromorphic in z, with zeros and poles that can be read off.

Defining the 'right-hand-side function'

$$R(r, a; z) = \prod_{k=1}^{\infty} (1 - \exp[2irz - (2k-1)ra])(z \to -z), \quad a > 0, \quad (2.10)$$

it is readily verified that G solves the AΔEs

$$\frac{G(z + ia_\delta/2)}{G(z - ia_\delta/2)} = R(r, a_{-\delta}; z), \quad \delta = +, -. \quad (2.11)$$

Up to normalization, our elliptic gamma function is the unique minimal solution to each of these AΔEs. (Here we think of R as a right-hand-side function given by (2.10).)

We will have occasion to invoke the reflection equation

$$G(r, a_+, a_-; -z) = 1/G(r, a_+, a_-; z), \quad (2.12)$$

and the duplication formula

$$\begin{aligned} G(r, a_+, a_-; 2z) &= \prod_{l,m=+,-} G\left(r, a_+, a_-; z - \frac{i}{4}(la_+ + ma_-)\right) \\ &\quad \times G\left(r, a_+, a_-; z - \frac{i}{4}(la_+ + ma_-) - \frac{\pi}{2r}\right), \end{aligned} \quad (2.13)$$

cf. [14] (3.106). Another feature that is needed is the alternative representation

$$G(r, a_+, a_-; z) = \exp\left(i \sum_{n=1}^{\infty} \frac{\sin 2nrz}{2n \sinh(nra_+) \sinh(nra_-)}\right), \quad |\mathrm{Im}\, z| < a. \quad (2.14)$$

It entails in particular the hyperbolic limit

$$\lim_{r \downarrow 0} G(r, a_+, a_-; z) \exp(\pi^2 z / 6ira_+a_-) = G(a_+, a_-; z), \quad (2.15)$$

cf. Proposition III.12 in [14]. We also need another $r \downarrow 0$ limit that cannot be found in [14], namely

$$\lim_{r \downarrow 0} G(r, a_+, a_-; z + \pi/2r) \exp(-\pi^2 z / 12ira_+a_-) = 1, \quad z \in C. \quad (2.16)$$

But this limit easily follows from (2.14) and (2.15). (See also the proof of Proposition III.12 in [14].) Finally, we have occasion to invoke the limit

$$\lim_{a_- \downarrow 0} \frac{G(r, a_+, a_-; z - ia_-\kappa)}{G(r, a_+, a_-; z - ia_-\lambda)} = \exp[(\lambda - \kappa)\ln(R(r, a_+; z))], \tag{2.17}$$

with $\kappa, \lambda \in \mathbb{R}$, and z in the cut plane $\mathbb{C} \setminus \cup_{k\in\mathbb{Z}}\{k\pi/r \pm i[a_+/2, \infty)\}$, cf. [14] (3.138).

We now turn to properties of $R(r, a; z)$ and related functions. To start with, it is plain from the product representation (2.10) that $R(r, a; z)$ is an even, π/r-periodic, entire function of z, which satisfies

$$\frac{R(r, a; z + ia/2)}{R(r, a; z - ia/2)} = -\exp(-2irz), \tag{2.18}$$

and which has simple zeros for $z = k\pi/r + i(l + 1/2)a$, with $k, l \in \mathbb{Z}$. Combining (2.13) and (one of) the AΔEs (2.11), we obtain the duplication formula

$$\begin{aligned} R(r, a; 2z) &= R(r, a; z + ia/4)R(r, a; z - ia/4)R(r, a; z + ia/4 - \pi/2r) \\ &\quad \times R(r, a; z - ia/4 - \pi/2r). \end{aligned} \tag{2.19}$$

Likewise, (2.14) entails the representation

$$R(r, a; z) = \exp\left(-\sum_{n=1}^{\infty} \frac{\cos 2nrz}{n\sinh(nra)}\right), \quad |\mathrm{Im}\, z| < a/2, \tag{2.20}$$

and (2.15)–(2.16) and (2.11) yield the limits

$$\lim_{r\downarrow 0} \exp(\pi^2/6ra)R(r, a; z) = 2\cosh(\pi z/a), \tag{2.21}$$

$$\lim_{r\downarrow 0} \exp(-\pi^2/12ra)R(r, a; z + \pi/2r) = 1. \tag{2.22}$$

Another convenient function can be defined by

$$s(r, a; z) \equiv ie^{-irz}R(r, a; z - ia/2)/2r\prod_{k=1}^{\infty}(1 - p^k)^2, \quad p \equiv e^{-2ra}. \tag{2.23}$$

The product representation

$$s(r, a; z) = \frac{\sin rz}{r}\prod_{k=1}^{\infty}\frac{(1 - p^k e^{2irz})(z \to -z)}{(1 - p^k)^2} \tag{2.24}$$

following from (2.10) shows that we have

$$s(r, a; z) = \exp(-\eta z^2 r/\pi)\sigma(z; \pi/2r, ia/2), \tag{2.25}$$

where σ is the Weierstrass σ-function.

We continue by tying in our functions $s(r, a; z)$ and $R(r, a; z)$ with the shifted σ-functions and θ-functions employed by van Diejen [8] and Komori/Hikami [9], resp. (The remainder of this section is invoked only in Appendix B.) To this end we set

$$\omega_0 = 0, \quad \omega_1 = \pi/2r, \quad \omega_2 = ia/2, \quad \omega_3 = -\omega_1 - \omega_2, \tag{2.26}$$

$$\eta_0 = 0, \qquad \eta_t = \zeta(\omega_t), \quad t = 1, 2, 3, \tag{2.27}$$

where ζ is the Weierstrass ζ-function. Now van Diejen works with the four functions

$$\sigma_0(z) \equiv \sigma(z), \quad \sigma_t(z) \equiv \exp(-\eta_t z)\sigma(z + \omega_t)/\sigma(\omega_t), \quad t = 1, 2, 3. \tag{2.28}$$

Their connection to s and R can be made via (2.25) and (2.23).

The θ-functions $\theta_s(z)$ in Eqs. (A.1a)–(A.1d) of [9] are related to the θ-functions $\theta_s^{WW}(z)$ from [20] by

$$\theta_s(z) = \theta_s^{WW}(\pi z), \quad s = 1, \ldots, 4. \tag{2.29}$$

Their period lattice is $\mathbb{Z} + \tau\mathbb{Z}$. Hence our parameters r, a should be chosen as

$$r = \pi, \quad a = -i\tau. \tag{2.30}$$

Putting

$$q \equiv \exp(i\pi\tau) = \exp(-\pi a), \quad G \equiv \prod_{k=1}^{\infty}(1 - q^{2k}), \tag{2.31}$$

the product formula (2.24) with $r = \pi$ and $p = q^2$ can be compared to the product formula for $\theta_1(z)$ (cf. p. 470 in [20]), to obtain the relation

$$\theta_1(z) = 2\pi q^{1/4} G^3 s(\pi, a; z). \tag{2.32}$$

In particular, this entails

$$\theta_1'(0) = 2\pi q^{1/4} G^3. \tag{2.33}$$

Next, comparing (2.10) to p. 469 in [20], we deduce

$$\theta_4(z) = GR(\pi, a; z). \tag{2.34}$$

The relation of our functions s and R to the remaining θ-functions

$$\theta_2(z) \equiv \theta_1(z + 1/2), \quad \theta_3(z) \equiv \theta_4(z + 1/2), \tag{2.35}$$

is then clear. Finally, for later use we note that (2.23) entails

$$s(\pi, a; z) = (-2i\pi G^2)^{-1} \exp(-i\pi z) R(\pi, a; z - ia/2). \tag{2.36}$$

3. The BC_1 Hamiltonians

The hyperbolic/elliptic operators depend on four/eight parameters h_n, with $n = 0, \ldots, 3/7$, resp. Unless specified otherwise, the h_n's are arbitrary complex numbers. A key role in both regimes is played by the c-function

$$c(z) \equiv \frac{1}{G(2z + ia)} \prod_n G(z - h_n), \tag{3.1}$$

where G is the hyperbolic/elliptic gamma function and the product is over $n = 0, \ldots, 3/n = 0, \ldots, 7$ in the hyperbolic/elliptic case. (Here and in the sequel, we often suppress the dependence on r, a_+, a_- and various other parameters, whenever the context is such that no confusion is likely to arise.) Two other functions occurring in both cases are the w-function

$$w(z) \equiv 1/c(z)c(-z), \tag{3.2}$$

and the u-function

$$u(z) \equiv -\exp(-2irz)c(z)/c(-z). \tag{3.3}$$

(Of course, $r = 0$ in the hyperbolic case.) Note that all of these functions are invariant under interchange of a_+ and a_- and under arbitrary permutations of the parameters h_n. Using the reflection equations (2.5) and (2.12), the u-function can be rewritten as

$$u(z) = -\exp(-2irz)\frac{\prod_n G(z - h_n)G(z + h_n)}{G(2z + ia)G(2z - ia)}. \tag{3.4}$$

Hence it is also invariant under arbitrary sign flips of h_n.

Consider now the coefficient function

$$V_a(z) \equiv \exp(-2ra_-)u(z + ia_-)/u(z). \tag{3.5}$$

Obviously, the z-shift breaks the $a_+ \leftrightarrow a_-$ symmetry. But V_a is still invariant under arbitrary permutations and sign flips of h_n. The defining BC_1 analytic difference operator (AΔO) can now be chosen to be of the form

$$\mathcal{A} = \exp(-ia_-\partial_z) + V_a(z)\exp(ia_-\partial_z) + V_b(z). \tag{3.6}$$

As will be detailed shortly, the additive 'potential' $V_b(z)$ is even in z and permutation invariant in the parameters h_n, but no longer invariant under arbitrary sign flips. It is however invariant under sign flips that involve an even number of parameters. Thus we arrive at the above-mentioned D_4/D_8-invariance.

There are two similarity transforms of $\mathcal{A}$ that are also crucial. First, we can transform $\mathcal{A}$ with the c-function, yielding

$$\begin{aligned} A &\equiv c(z)\mathcal{A}c(z)^{-1} \\ &= V(z)\exp(-ia_-\partial_z) + V(-z)\exp(ia_-\partial_z) + V_b(z), \end{aligned} \tag{3.7}$$

with

$$V(z) \equiv c(z)/c(z - ia_-). \tag{3.8}$$

Note this entails the relation

$$V_a(z) = V(-z)V(z + ia_-). \tag{3.9}$$

The second transformation is given by

$$H \equiv (-e^{2irz}u(z))^{1/2}\mathcal{A}(-e^{2irz}u(z))^{-1/2}. \tag{3.10}$$

Thus H can be rewritten as

$$H = V(z)^{1/2}\exp(-ia_-\partial_z)V(-z)^{1/2} + (z \to -z) + V_b(z), \tag{3.11}$$

or as

$$H = V_a(z)^{1/2}\exp(ia_-\partial_z) + (z \to -z) + V_b(z). \tag{3.12}$$

Moreover, we have

$$H = w(z)^{1/2}Aw(z)^{-1/2}, \tag{3.13}$$

as is easily verified.

On first acquaintance, having three choices for a defining operator may seem strange. Therefore we add some brief comments. To begin with, the Hamiltonian H in the form (3.11) arises most naturally from quantization of a classical version, cf. (1.4) and [3]. In the representation (3.11) (together with the definition of $V_b(z)$ detailed below), permutation invariance is plain, but D_4/D_8 invariance is non-obvious. Even so, H is in fact D_4/D_8-symmetric, as follows from the D_4/D_8-invariance of $V_a(z)$ and $V_b(z)$. The operator A, however, is only permutation invariant.

Imposing suitable restrictions on the parameters h_n (specified below), the u-function is a phase for real z, and H is a formally hermitian operator on $L^2(\mathbb{R}, dz)$ that commutes with parity. Thus the operator $\mathcal{A}$ is also formally hermitian on $L^2(\mathbb{R}, dz)$, but it no longer commutes with parity. Furthermore, $w(z)$ is positive for real z and the operator A is formally hermitian on $L^2(\mathbb{R}, w(z)dz)$; since $w(z)$ is even and A commutes with parity, it leaves even and odd functions invariant.

To be sure, this appraisal of the most conspicuous differences between the three versions is quite sketchy even at a formal level. Later on, we will say more about the pros and cons of the three choices, in connection with known and 'expected' Hilbert space eigenfunction transforms. But now we proceed by zooming in on the hyperbolic and elliptic cases.

3.1. THE HYPERBOLIC REGIME: D_4 SYMMETRY

From now on we work with the functions

$$s_\delta(z) \equiv \sinh(\pi z/a_\delta), \quad c_\delta(z) \equiv \cosh(\pi z/a_\delta), \quad \delta = +, -. \tag{3.14}$$

Using the definitions (3.8) and (3.1), and then the AΔE (2.4), we readily calculate

$$V(z) = -\frac{4\prod_n c_+(z - h_n - ia_-/2)}{s_+(2z)s_+(2z - ia_-)}. \tag{3.15}$$

Hence (3.9) yields

$$V_a(z) = \frac{16\prod_n c_+(z + h_n + ia_-/2)c_+(z - h_n + ia_-/2)}{s_+(2z)s_+(2z + ia_-)^2 s_+(2z + 2ia_-)}. \tag{3.16}$$

Clearly, (3.16) exhibits manifest invariance under arbitrary sign flips of h_n. Likewise, (3.15) shows $V(z)$ is not invariant under any flips (for generic $\mathbf{h} = (h_0, \ldots, h_3)$, of course).

In the hyperbolic case the function $V_b(z)$ is defined by

$$V_b(z) \equiv -V(z) - V(-z) - 2c_+(\sum h_n + ia_-). \tag{3.17}$$

Thus $V_b(z)$ is S_4-invariant, just as V and V_a. The additive constant ensures the asymptotics

$$V_b(z) \to 0, \quad |\mathrm{Re}\, z| \to \infty, \tag{3.18}$$

as is easily verified.

The point is now that $V_b(z)$ admits a second representation that reveals it is also invariant under an even number of sign flips; hence it is (by definition) D_4-invariant.

Lemma 3.1 *We have*

$$V_b(z) = \frac{2p_c}{c_+(z - ia)c_+(z + ia)} - \frac{2p_s}{s_+(z - ia)s_+(z + ia)}, \tag{3.19}$$

where

$$p_c \equiv \prod_{n=0}^{3} c_+(h_n), \quad p_s \equiv \prod_{n=0}^{3} s_+(h_n). \tag{3.20}$$

We prove this lemma in Appendix A. Note that p_c is invariant under arbitrary sign flips; in contrast, p_s is only invariant under an even number of flips.

Next, we introduce couplings $\mathbf{c} = (c_0, \dots, c_3)$ by setting

$$h_0 = i(c_0 - a), \quad h_1 = i(c_1 - a_-/2), \quad h_2 = i(c_2 - a_+/2), \quad h_3 = ic_3. \tag{3.21}$$

Then we can rewrite (3.15) as

$$\begin{aligned} V(z) &= \frac{s_+(z - ic_0)}{s_+(z)} \frac{c_+(z - ic_1)}{c_+(z)} \frac{s_+(z - ic_2 - ia_-/2)}{s_+(z - ia_-/2)} \\ &\quad \times \frac{c_+(z - ic_3 - ia_-/2)}{c_+(z - ia_-/2)}. \end{aligned} \tag{3.22}$$

From this the terminology 'couplings' can be understood: for $c_n = 0$ the corresponding 'interaction term' becomes 1. In particular, for $\mathbf{c} = 0$, we have

$$V(z) = 1, \quad V_b(z) = 0, \quad V_a(z) = 1, \quad (\mathbf{c} = 0), \tag{3.23}$$

so that

$$\mathcal{A} = A = H = \exp(-ia_-\partial_z) + \exp(ia_-\partial_z), \quad (\mathbf{c} = 0). \tag{3.24}$$

In words, the three versions of the BC_1 AΔO reduce to the same 'free' AΔO. Note also that the c-function (3.1) reduces to 1 for $\mathbf{c} = 0$, due to the duplication formula (2.6).

On the one hand, the parametrization of the AΔO-coefficients via the couplings $\mathbf{c}$ is natural from a physical viewpoint, as just illustrated. On the other hand, it breaks the above symmetries, in the sense that e.g. $V(z)$ (3.22) is no longer invariant under any permutation of $c_0, \dots, c_3$. In terms of $\mathbf{c}$, therefore, even permutation invariance is 'hidden'. We will return to this issue in Subsection 5.2, where we review our BC_1-eigenfunction.

Finally, we observe that when we choose $\mathbf{c} \in \mathbb{R}^4$ and $z \in \mathbb{R}$, then $u(z)$ is a phase, $w(z)$ is positive, $V_b(z)$ is real-valued, and the complex conjugate of $V(z)$ equals $V(-z)$, cf. (3.22). Hence H and $\mathcal{A}$ are formally self-adjoint on $L^2(\mathbb{R}, dz)$, and A is formally self-adjoint on $L^2(\mathbb{R}, w(z)dz)$.

3.2. THE ELLIPTIC REGIME: D_8 SYMMETRY

Proceeding along the same lines as in the hyperbolic case, we begin by introducing the notation

$$s_\delta(z) \equiv s(r, a_\delta; z), \quad R_\delta(z) \equiv R(r, a_\delta; z), \quad \delta = +, -. \tag{3.25}$$

Now we use (3.8), (3.1) and (2.4) to express $V(z)$ in terms of R_+. This yields

$$V(z) = \frac{\prod_n R_+(z - h_n - ia_-/2)}{R_+(2z + ia_+/2)R_+(2z - ia_- + ia_+/2)}. \tag{3.26}$$

From (3.9) we then get

$$V_a(z) = \frac{\prod_n R_+\left(z + h_n + \frac{i}{2}a_-\right) R_+(z - h_n + \frac{i}{2}a_-)}{R_+\left(2z - \frac{i}{2}a_+\right) \prod_{\delta=+,-} R_+\left(2z + ia_- + \frac{i}{2}\delta a_+\right) \cdot R_+\left(2z + 2ia_- + \frac{i}{2}a_+\right)}. \tag{3.27}$$

Just as in the hyperbolic case, we see that $V_a(z)$ is invariant under arbitrary sign flips of h_n, whereas $V(z)$ is not invariant. The obvious analog of the rhs of (3.17) reads $-V(z) - V(-z) + C$, where C is a suitable constant. But as will become clear shortly, this is not a D_8-invariant function. On the other hand, with a certain restriction on $\mathbf{h} = (h_0, \ldots, h_7)$ in effect, $V_b(z)$ is indeed of this form. For general $\mathbf{h}$, however, the definition of $V_b(z)$ is quite different, and yields a function that *is* D_8-invariant.

Admittedly, at this stage the definition of $V_b(z)$ on which we now embark may appear ad hoc. Its motivation arises from the $N > 1$ generalization, cf. Section 4 and Appendix B. First, we introduce half-periods $\omega_t, t = 0, \ldots, 3$:

$$\omega_0 = 0, \quad \omega_1 = \pi/2r, \quad \omega_2 = ia_+/2, \quad \omega_3 = -\omega_1 - \omega_2. \tag{3.28}$$

Second, we introduce four functions that involve a new parameter $\mu \in C$:

$$\mathcal{E}_t(\mu; z) \equiv \frac{R_+(z + \mu - ia - \omega_t) R_+(z - \mu + ia - \omega_t)}{R_+(z - ia - \omega_t) R_+(z + ia - \omega_t)}, \tag{3.29}$$

$t = 0, \ldots, 3$.

Third, we define the product functions

$$p_0 \equiv \prod_n R_+(h_n), \quad p_1 \equiv \prod_n R_+(h_n - \omega_1), \tag{3.30}$$

$$p_2 \equiv \exp(-2ra_+) \prod_n \exp(-irh_n) R_+(h_n - \omega_2), \tag{3.31}$$

$$p_3 \equiv \exp(-2ra_+) \prod_n \exp(irh_n) R_+(h_n - \omega_3). \tag{3.32}$$

Finally, we introduce

$$V_b(z) \equiv \frac{\sum_{t=0}^{3} p_t [\mathcal{E}_t(\mu; z) - \mathcal{E}_t(\mu; z_t)]}{2R_+(\mu - ia_+/2) R_+(\mu - ia_- - ia_+/2)}, \tag{3.33}$$

with

$$z_0 = z_2 = \omega_1, \qquad z_1 = z_3 = 0. \tag{3.34}$$

In the next lemma we show in particular that $-V_b(z)$ only differs from the function

$$\Sigma(z) \equiv V(z) + V(-z) \tag{3.35}$$

by a constant, provided we have

$$i \sum_n h_n = 4a, \quad (\text{mod } i\pi/r). \tag{3.36}$$

To appreciate this restriction, we first note that by virtue of (2.18) the functions $\mathcal{E}_t(z)$ and $V_b(z)$ are elliptic with periods $\pi/r, ia_+$, whereas

$$V(z + ia_+)/V(z) = \exp(2r[i \sum h_n - 4a]). \tag{3.37}$$

From (3.9) we then get

$$V_a(z + ia_+)/V_a(z) = 1, \tag{3.38}$$

so that $V_a(z)$ is elliptic, too. But $V(z)$ is not ia_+-periodic, unless (3.36) holds true. We are now prepared for the lemma, which is proved in Appendix A.

Lemma 3.2 *The function $V_b(z)$ does not depend on μ. Moreover, with (3.36) in effect, the function*

$$\Sigma(z) + V_b(z) \tag{3.39}$$

does not depend on z.

The presence of the extra parameter μ anticipates the arbitrary-N generalization, cf. the next section. In this connection we note the $\mu = 0$ representation

$$\begin{aligned} V_b(z) &= \frac{\rho}{2R_+(ia_- + ia_+/2)} \\ &\quad \times \sum_{t=0}^{3} p_t([L_+(z - ia - \omega_t) - L_+(z + ia - \omega_t)] - [z \to z_t]), \end{aligned} \tag{3.40}$$

where

$$\rho \equiv \text{Res}(1/R_+(z))|_{z=-ia_+/2}, \tag{3.41}$$

and $L_+(z)$ is the logarithmic derivative

$$L_+(z) \equiv R'_+(z)/R_+(z). \tag{3.42}$$

As in the hyperbolic case, D_8-invariance of the BC_1 AΔO $\mathcal{A}$ (3.6) is now easily established. Indeed, the products $p_0, \ldots, p_3$ are manifestly permutation invariant. Since $R_+(z)$ is even and π/r-periodic, p_0 and p_1 are invariant under arbitrary sign flips. Using also (2.18), we see that p_3 has this feature, too, whereas p_2 is only invariant under an even number of flips.

The upshot is that $p_0, \ldots, p_3$ are D_8-invariant. Since $V_a(z)$ is also invariant, we have now shown that $\mathcal{A}$ is D_8-invariant, as announced.

We continue by defining new parameters $c_0, \ldots, c_7$ of coupling constant type. Specifically, $c_0, \ldots, c_3$ are defined as in the hyperbolic case, cf. (3.21), while $c_4, \ldots, c_7$ are given by

$$\begin{aligned} h_4 &= i(c_4 - a) - \omega_1, & h_5 &= i(c_5 - a_-/2) - \omega_1, \\ h_6 &= i(c_6 - a_+/2) + \omega_1, & h_7 &= ic_7 + \omega_1. \end{aligned} \tag{3.43}$$

Using the duplication formula (2.19) we now obtain

$$\begin{aligned} V(z) = {} & \frac{R_+(z - ic_0 + ia_+/2)}{R_+(z + ia_+/2)} \frac{R_+(z - ic_1)}{R_+(z)} \\ & \times \frac{R_+(z - ic_2 + i(a_+ - a_-)/2)}{R_+(z + i(a_+ - a_-)/2)} \frac{R_+(z - ic_3 - ia_-/2)}{R_+(z - ia_-/2)} \\ & \times \frac{R_+(z - ic_4 + ia_+/2 + \omega_1)}{R_+(z + ia_+/2 + \omega_1)} \frac{R_+(z - ic_5 + \omega_1)}{R_+(z + \omega_1)} \\ & \times \frac{R_+(z - ic_6 + i(a_+ - a_-)/2 + \omega_1)}{R_+(z + i(a_+ - a_-)/2 + \omega_1)} \frac{R_+(z - ic_7 - ia_-/2 + \omega_1)}{R_+(z - ia_-/2 + \omega_1)}, \end{aligned} \tag{3.44}$$

showing the coupling character of c_n. Trading $\mathbf{h}$ for $\mathbf{c}$ in the products $p_0, \ldots, p_3$ leads to unilluminating formulas, however. Notice that the restriction (3.36) is equivalent to

$$\sum_n c_n = 0, \quad (\mathrm{mod}\ \pi/r). \tag{3.45}$$

Obviously, for $\mathbf{c} = 0$ all of the products p_t vanish, so that $V_b(z)$ vanishes for zero coupling. More generally, we obtain once again (3.23) and (3.24), and the c-function reduces to 1 for $\mathbf{c} = 0$ (by the duplication formula (2.13)).

It is illuminating to study the hyperbolic limit $r \downarrow 0$, with the couplings $c_0, \ldots, c_7$ fixed. Using (2.21) and (2.22), we get from (3.44)

$$\begin{aligned} \lim_{r \downarrow 0} V(z) = {} & \frac{c_+(z - ic_0 + ia_+/2)}{c_+(z + ia_+/2)} \frac{c_+(z - ic_1)}{c_+(z)} \\ & \times \frac{c_+(z - ic_2 + i(a_+ - a_-)/2)}{c_+(z + i(a_+ - a_-)/2)} \frac{c_+(z - ic_3 - ia_-/2)}{c_+(z - ia_-/2)}. \end{aligned} \tag{3.46}$$

This equals the hyperbolic coefficient $V(z)$ (3.22). From (2.21) and (2.22) we also deduce that $V_b(z)$ has a finite limit. Specifically, we obtain

$$\lim_{r \downarrow 0} V_b(z) = \frac{2}{s_+(\mu) s_+(ia_- - \mu)} (p_c[Q_c(z) - 1] + p_s[Q_s(z) - 1]), \tag{3.47}$$

where

$$
\begin{aligned}
Q_c(z) &\equiv \frac{c_+(z+\mu-ia)c_+(z-\mu+ia)}{c_+(z-ia)c_+(z+ia)}, \\
Q_s(z) &\equiv \frac{s_+(z+\mu-ia)s_+(z-\mu+ia)}{s_+(z-ia)s_+(z+ia)},
\end{aligned}
\tag{3.48}
$$

and p_c, p_s are given by (3.20).

Since $V_b(z)$ does not depend on μ, the same is true for the function on the rhs of (3.47). Indeed, a straightforward calculation shows that it equals the hyperbolic $V_b(z)$, as given by (3.19).

We conclude this subsection by observing that the choices $\mathbf{c} \in \mathbb{R}^8, z \in \mathbb{R}$, entail that $u(z)$ is a phase, $w(z)$ is positive, $V_b(z)$ is real-valued, and $\overline{V(z)}$ equals $V(-z)$, cf. (3.44). It readily follows that H and $\mathcal{A}$ are formally self-adjoint on $L^2((-\pi/2r, \pi/2r), dz)$ and A on $L^2((-\pi/2r, \pi/2r), w(z)dz)$. (Of course, working formally, we may as well choose $\mathbb{R}$ instead of $(-\pi/2r, \pi/2r)$, but the latter choice is more natural from a physical viewpoint.)

4. The BC_N case

In this section we proceed along the same lines as in the previous one. Thus we begin by specifying the features of the BC_N AΔOs that are shared by the hyperbolic and elliptic versions. To complete the definition of the AΔOs, however, we need to consider the hyperbolic and elliptic regimes separately.

Apart from the 'external field' c-function (3.1), we need a second c-function

$$
c_\mu(z) \equiv \frac{G(z-\mu+ia)}{G(z+ia)}, \quad \mu \in \mathbb{C}, \tag{4.1}
$$

as building block. (As before, G denotes the elliptic or hyperbolic gamma function.) We now define factorized C-functions depending on N complex variables $x = (x_1, \ldots, x_N)$, namely the external field C-function

$$
C_e(x) \equiv \prod_{j=1}^{N} c(x_j), \tag{4.2}
$$

the sum and difference C-functions

$$
C_\pm(x) \equiv \prod_{1 \le j < k \le N} c_\mu(x_j \pm x_k), \tag{4.3}
$$

and the total C-function

$$
C(x) \equiv C_e(x) C_+(x) C_-(x). \tag{4.4}
$$

Finally, we define a weight function

$$W(x) \equiv 1/C(x)C(-x). \tag{4.5}$$

The three versions of the defining BC_N AΔO are now given by

$$H \equiv \sum_{j=1}^{N} (\mathcal{V}_j(x)^{1/2} \exp(-ia_-\partial_{x_j})\mathcal{V}_j(-x)^{1/2} + (x \to -x)) + \mathcal{V}(x), \tag{4.6}$$

$$A \equiv W(x)^{-1/2} H W(x)^{1/2}, \tag{4.7}$$

$$\mathcal{A} \equiv C(x)^{-1} A C(x). \tag{4.8}$$

The coefficients $\mathcal{V}_j(x)$ can be written

$$\mathcal{V}_j(x) \equiv \mathcal{V}_j^{(s)}(x)\mathcal{V}_j^{(A)}(x). \tag{4.9}$$

Here, the 'simple' factor is of the form

$$\mathcal{V}_j^{(s)}(x) \equiv V(x_j)C_+(x)/C_+(x - ia_-e_j), \tag{4.10}$$

where $V(z)$ is given by (3.8), and $e_1, \ldots, e_N$ is the canonical basis of $\mathbb{C}^N$. The remaining (A_{N-1}-type) factor $\mathcal{V}_j^{(A)}$ can also be specified by one formula applying at once to the hyperbolic and elliptic cases, but only at the expense of introducing a function

$$f(z) \equiv \begin{cases} 2c_+(z), & \text{(hyperbolic case)}, \\ R_+(z), & \text{(elliptic case)}. \end{cases} \tag{4.11}$$

Then it reads

$$\mathcal{V}_j^{(A)}(x) \equiv \prod_{k \neq j} \frac{f(x_j - x_k - \mu + \omega_2)}{f(x_j - x_k + \omega_2)}, \quad \omega_2 = ia_+/2. \tag{4.12}$$

We note that in terms of f we also have

$$\mathcal{V}_j^{(s)}(x) = \frac{\prod_n f(x_j - h_n - ia_-/2)}{f(2x_j + \omega_2)f(2x_j - ia_- + \omega_2)} \cdot \prod_{k \neq j} \frac{f(x_j + x_k - \mu + \omega_2)}{f(x_j + x_k + \omega_2)}. \tag{4.13}$$

(Recall (3.15) and (3.26) to see this.)

Before turning to the definition of $\mathcal{V}(x)$, we use the above factorizations to calculate the shift coefficients in A and $\mathcal{A}$. First, we assert that we have

$$A = \sum_{j=1}^{N} (\mathcal{V}_j(x) \exp(-ia_-\partial_j) + \mathcal{V}_j(-x) \exp(ia_-\partial_j)) + \mathcal{V}(x). \tag{4.14}$$

Indeed, this amounts to the identities

$$\mathcal{V}_j(x) = \frac{C(x)C(-x)}{C(x-ia_-e_j)C(-x+ia_-e_j)}\mathcal{V}_j(-x+ia_-e_j), \quad j=1,\ldots,N. \tag{4.15}$$

To prove these, we note that (4.10) and (3.8) entail that (4.15) with $\mathcal{V}_j \to \mathcal{V}_j^{(s)}$ and $C \to C_e C_+$ is clearly true. Thus we need only show

$$\mathcal{V}_j^{(A)}(x) = \frac{C_-(x)C_-(-x)}{C_-(x-ia_-e_j)C_-(-x+ia_-e_j)}\mathcal{V}_j^{(A)}(-x+ia_-e_j). \tag{4.16}$$

To this end we calculate the functions

$$\mathcal{V}_j^-(x) \equiv \mathcal{V}_j^{(A)}(x)C_-(x-ia_-e_j)/C_-(x), \quad j=1,\ldots,N. \tag{4.17}$$

Using (4.3), (4.1) and the G-AΔEs, we obtain

$$\begin{aligned}\frac{C_-(x-ia_-e_j)}{C_-(x)} &= \prod_{k>j}\frac{c_\mu(x_j-x_k-ia_-)}{c_\mu(x_j-x_k)}\cdot\prod_{l<j}\frac{c_\mu(x_l-x_j+ia_-)}{c_\mu(x_l-x_j)}\\ &= \prod_{k>j}\frac{f(x_j-x_k+\omega_2)}{f(x_j-x_k-\mu+\omega_2)}\cdot\prod_{l<j}\frac{f(x_l-x_j-\mu+ia_-+\omega_2)}{f(x_l-x_j+ia_-+\omega_2)}.\end{aligned} \tag{4.18}$$

Multiplying by (4.12), we now get

$$\mathcal{V}_j^-(x) = \prod_{l<j}\frac{f(x_j-x_l-\mu+\omega_2)f(x_j-x_l+\mu-ia_--\omega_2)}{f(x_j-x_l+\omega_2)f(x_j-x_l-ia_--\omega_2)}. \tag{4.19}$$

Finally, we observe that the rhs is invariant under $x \to -x+ia_-e_j$ (since f is even), and so (4.16) follows.

We have now proved the representation (4.14) for A. Our calculations also entail a more explicit formula for $\mathcal{A}$ (4.8), viz.,

$$\mathcal{A} = \sum_{j=1}^{N}(\mathcal{V}_j^-(x)\exp(-ia_-\partial_j) + \mathcal{V}_j^+(x)\exp(ia_-\partial_j)) + \mathcal{V}(x), \tag{4.20}$$

with $\mathcal{V}_j^-$ given by (4.19) and $\mathcal{V}_j^+$ by

$$\begin{aligned}\mathcal{V}_j^+(x) &= V_a(x_j)\prod_{k\neq j}\frac{f(x_j+x_k+\mu-\omega_2)f(x_j+x_k-\mu+ia_-+\omega_2)}{f(x_j+x_k-\omega_2)f(x_j+x_k+ia_-+\omega_2)}\\ &\quad\cdot\prod_{l>j}\frac{f(x_j-x_l+\mu-\omega_2)f(x_j-x_l-\mu+ia_-+\omega_2)}{f(x_j-x_l-\omega_2)f(x_j-x_l+ia_-+\omega_2)}.\end{aligned} \tag{4.21}$$

Before presenting the definition of $\mathcal{V}(x)$, we connect the above developments to the A_{N-1} Hamiltonian $H_{\mathrm{rel}}^{(A)}$ (1.4) with $\beta \to a_-$ and σ-periods

$\pi/r, ia_+$. Consider H (4.6) with $\mathcal{V}(x) = 0$ and $\mathcal{V}_j(x)$ replaced by $\mathcal{V}_j^{(A)}(x)$. Then it readily follows from (4.12) with $f = R_+$ that H reduces to a multiple of (1.4). (Recall (2.18), (2.23) and (2.25) to see this.) Transforming with $C_-(x)$ and $W_-(x) \equiv 1/C_-(x)C_-(-x)$ then yields A_{N-1} AΔOs $A_{\rm rel}^{(A)}$ and $\mathcal{A}_{\rm rel}^{(A)}$ with meromorphic coefficients that are clear from the above.

4.1. THE HYPERBOLIC AΔOS

We now specialize to the hyperbolic case, completing our AΔO definitions by setting

$$\mathcal{V}(x) \equiv \frac{2}{s_+(\mu)s_+(ia_- - \mu)} \left(p_c[\prod_{j=1}^{N} Q_c(x_j) - 1] + p_s[\prod_{j=1}^{N} Q_s(x_j) - 1] \right), \tag{4.22}$$

cf. (3.20), (3.48). As we have seen below (3.48), this equals $V_b(x_1)$ for $N = 1$. But for $N > 1$ we do get dependence on μ. (Take e.g. $\mathrm{Re}\,\mu \to \infty$ to see this.) On the other hand, we have

$$\lim_{\mu \to 0} \mathcal{V}(x) = \sum_{j=1}^{N} V_b(x_j), \tag{4.23}$$

as is easily verified, cf. (3.19). We also obtain the limit

$$\mathcal{V}(x) \to 0, \quad |\mathrm{Re}\,x_1|, \ldots, |\mathrm{Re}\,x_N| \to \infty. \tag{4.24}$$

At first sight, $\mathcal{V}(x)$ seems to have no relation to the shift coefficients $\mathcal{V}_j(\pm x)$. However, introducing

$$\Sigma(x) \equiv \sum_{j=1}^{N} (\mathcal{V}_j(x) + \mathcal{V}_j(-x)), \tag{4.25}$$

we have the following lemma, which we prove in Appendix A.

Lemma 4.1 *The function $\Sigma(x) + \mathcal{V}(x)$ does not depend on x. Explicitly, we have*

$$\mathcal{V}(x) = -\Sigma(x) - 2\frac{s_+(N\mu)}{s_+(\mu)} c_+ \left(\sum_{n=0}^{3} h_n + (N-1)\mu + ia_- \right). \tag{4.26}$$

From the above it is again clear that the AΔO $\mathcal{A}$ is D_4-invariant. Since $C(x)/C(-x)$ is also D_4-invariant, the Hamiltonian H is D_4-invariant as well. It should be noted that when $\mathcal{V}(x)$ is replaced by $-\Sigma(x)$, then we obtain AΔOs that are no longer D_4-invariant. (Indeed, the constant on the rhs of (4.26) is only S_4-invariant.)

Another involutive symmetry of $\mathcal{A}$ is easily checked: it is invariant under

$$\mu \mapsto 2ia - \mu. \tag{4.27}$$

Once more, this invariance is present for H, too, but not for A. (Indeed, the C-functions $C_\pm(x)$ are not invariant under (4.27).) On the other hand, $\mathcal{A}$ is only invariant under permutations of $x_1, \ldots, x_N$, whereas A and H are also invariant under arbitrary sign changes. (Thus the latter operators are invariant under the BC_N Weyl group.)

Just as for $N = 1$, we can introduce couplings $\mathbf{c}$ via (3.21). To obtain a positive weight function $W(x)$ on $\mathbb{R}^N$, we should not only take $\mathbf{c} \in \mathbb{R}^4$, but also choose $\tau \in \mathbb{R}$, where τ is defined by

$$\mu = i\tau. \tag{4.28}$$

With these reality conditions in effect, we readily obtain formal self-adjointness of H and $\mathcal{A}$ on $L^2(\mathbb{R}^N, dx)$, and of A on $L^2(\mathbb{R}^N, W(x)dx)$. In order to take BC_N-invariance into account, it is however more natural to replace $\mathbb{R}^N$ by the 'Weyl chamber'

$$G_N \equiv \{x \in \mathbb{R}^N \mid 0 < x_N < \cdots < x_1\}, \tag{4.29}$$

cf. also the next section.

4.2. THE ELLIPTIC AΔOS

The elliptic additive potential $\mathcal{V}(x)$ is given by

$$\mathcal{V}(x) \equiv \frac{\sum_{t=0}^{3} p_t \left(\prod_{j=1}^{N} \mathcal{E}_t(x_j) - \mathcal{E}_t(z_t)^N \right)}{2R_+(\mu - ia_+/2)R_+(\mu - ia_- - ia_+/2)}. \tag{4.30}$$

Thus we obtain $V_b(x_1)$ for $N = 1$, cf. (3.33). Moreover, the limit (4.23) is readily checked in this case, too (cf. the representation (3.40)–(3.42)).

As before, it now follows that $\mathcal{A}$ and H are D_8-invariant, whereas A is only S_8-invariant. Moreover, $\mathcal{A}$ and H are invariant under the μ-involution (4.27), in contrast to A. Finally, A and H are BC_N-invariant, whereas $\mathcal{A}$ is invariant under arbitrary permutations of $x_1, \ldots, x_N$, but not under sign changes.

We proceed with some comments that apply to the hyperbolic case as well. The symmetry properties just discussed show that the operator H is the 'maximally symmetric' one among the three versions. Its drawback is the occurrence of square-root branch points in its coefficients. As a consequence, there is at first no natural choice for a function space on which it acts. By contrast, the AΔOs A and $\mathcal{A}$ have meromorphic coefficients, so they have an obvious action on the space $\mathcal{M}$ of meromorphic functions. In view of the similarity transformations between $H, \mathcal{A}$ and A, the obvious choice for a function space on which H acts is therefore the space

$$\mathcal{M}_C \equiv (C(x)/C(-x))^{1/2}\mathcal{M} = W(x)^{1/2}\mathcal{M}. \tag{4.31}$$

The additive potential $\mathcal{V}(x)$ is not only meromorphic, but also elliptic in $x_1, \ldots, x_N$ with periods $\pi/r, ia_+$. (Indeed, the functions $\mathcal{E}_0(z), \ldots, \mathcal{E}_3(z)$ are elliptic.) Likewise, the coefficients $\mathcal{V}_j^{\pm}(x)$ of the shifts in $\mathcal{A}$ are readily seen to be elliptic, cf. (4.19)–(4.21). Consider now the AΔOs $\tilde{\mathcal{A}}, \tilde{H}$ and $\tilde{A}$ that result from $\mathcal{A}, H$ and A by interchanging a_+ and a_- in all of the above definitions. (In particular, therefore, all coefficients of the new AΔOs are expressed in terms of the functions c_-, s_- and R_-.) Due to the ellipticity properties of the $\mathcal{A}$- and $\tilde{\mathcal{A}}$-coefficients, it is immediate that we have

$$[\mathcal{A}, \tilde{\mathcal{A}}] = 0. \tag{4.32}$$

That is, $\mathcal{A}$ and $\tilde{\mathcal{A}}$ commute as operators on $\mathcal{M}$. Thus this is also true for the operators A and $\tilde{A}$ on $\mathcal{M}$, and for H and $\tilde{H}$ on $\mathcal{M}_C$ (4.31). (Indeed, $C(x)$ is invariant under $a_+ \leftrightarrow a_-$, since the hyperbolic and elliptic gamma functions are.) Hence it is natural to try and find eigenfunctions in these spaces that are also invariant under interchange of a_+ and a_-; we will return to this issue in the next section.

In contrast to the coefficients of $\mathcal{A}$, the functions $\mathcal{V}_j(\pm x)$ (4.9) appearing in A are generically not ia_+-periodic. Indeed, for ia_+-periodicity we should require

$$i\sum_n h_n + 2i(N-1)\mu = 4a, \quad (\text{mod } i\pi/r). \tag{4.33}$$

(Use (4.11)–(4.13) and (2.18) to check this.) As in the hyperbolic case, we now define a sum function $\Sigma(x)$ by (4.25). Then we have the following lemma, whose proof is relegated to Appendix A.

Lemma 4.2 *With the restriction (4.33) in force, the function $\Sigma(x) + \mathcal{V}(x)$ does not depend on x.*

When we trade $\mathbf{h}$ for couplings $\mathbf{c}$ (cf. (3.21) and (3.43)), we see that for $\mathbf{c}$ and μ fixed the limit $r \downarrow 0$ leads from the elliptic AΔOs to their hyperbolic counterparts. (This follows in the same way as in the $N = 1$ case from the two R_+-limits (2.21), (2.22).)

Keeping $r > 0$ and choosing $\mathbf{c} \in \mathbb{R}^8, \tau \in \mathbb{R}$ (with τ defined by (4.28)), we obtain a positive weight function $W(x)$ on $\mathbb{R}^N$. In view of

BC_N-invariance, the natural Hilbert spaces for $H, \mathcal{A}$ and A are $L^2(G_N, dx)$ and $L^2(G_N, W(x)dx)$, resp., where G_N is the BC_N 'Weyl alcove'

$$G_N \equiv \{x \in \mathbb{R}^N \mid 0 < x_N < \cdots < x_1 < \pi/2r\}. \qquad (4.34)$$

The reality conditions just detailed give rise to formally self-adjoint operators, as in the previous cases.

5. Eigenfunctions vs. parameter symmetries

5.1. INTRODUCTORY REMARKS

Compared to eigenfunctions of differential operators (for which abundant existence and uniqueness results exist), very little is known about eigenfunctions of analytic difference operators. We have commented on the general theory at previous occasions [3, 22, 23]. In this section our emphasis is on special eigenfunctions of the above AΔOs, paying particular attention to their parameter symmetries and quantum mechanical relevance.

More specifically, in Subsection 5.2 we summarize various results from our papers [15–17] that have a bearing on these issues for the hyperbolic BC_1 case. To date, this is the only case where detailed results are known for arbitrary couplings and eigenvalues. As such, it can serve as a source of expectations/conjectures for the elliptic and general-N cases, to which we return shortly.

Subsection 5.3 starts from a quite special type of eigenfunction ('ground state') that arises for arbitrary N, both on the hyperbolic and elliptic levels. Its existence is an immediate consequence of the above lemmas, and accordingly we focus on non-obvious consequences, in particular as regards parameter symmetries and its role in eventual Hilbert space reformulations of the pertinent AΔO actions.

We continue with a general appraisal of the latter context. First, we recall that even when one is dealing with a specific AΔO A that is formally self-adjoint on a specific Hilbert space $\mathcal{H}$, it need not be true that A can be naturally associated with a bona fide self-adjoint operator $\dot{A}$ acting on a dense subspace of $\mathcal{H}$. (Since AΔOs are unbounded, $\dot{A}$ cannot act on all of $\mathcal{H}$.) Furthermore, the association need not be unique: even for free AΔOs one can easily arrive at an infinity of distinct Hilbert space operators $\dot{A}$. (See [24, 23] for explicit examples.) Here and below, we are using a dot for Hilbert space operators, since the more obvious hat would give rise to confusion with dual objects.

In all of the concrete cases known to us, the map

$$A \mapsto \dot{A}, \qquad (5.1)$$

from AΔOs to self-adjoint operators is defined by using explicit A-eigenfunctions with very special features. The ambiguity in the map reflects the huge ambiguity in A-eigenfunctions, once their existence has been established.

In the special cases at issue in this paper, however, the ambiguity can be greatly reduced by extra requirements. Consider first the $N = 1$ setting, as detailed in Section 3. Suppose we restrict attention to eigenfunctions that are real-analytic and symmetric in a_+ and a_-, cf. our discussion in the paragraph containing (4.32). There is considerable evidence (but no complete proof) that there exists at most one meromorphic eigenfunction with this property (up to multiplication by $(a_+ \leftrightarrow a_-)$-symmetric constants, of course). For the hyperbolic case, we detail such an eigenfunction in Subsection 5.2. For the elliptic case no general coupling eigenfunctions are known at all, but in a special case with only one variable coupling, there are partial results that are consistent with the required symmetry in a_+, a_- [25].

The BC_1 eigenfunctions just mentioned are also invariant under the remaining parameter symmetries discussed above. More precisely, the H- and $\mathcal{A}$-eigenfunctions are D_4-invariant in the hyperbolic case [16], and $\mathbb{Z}_2$-invariant with respect to the single coupling in the elliptic case [25]. We should stress that these invariances do not necessarily follow from the AΔOs being invariant.

Turning to the arbitrary-N case, there is an obvious extra requirement (in addition to BC_N and parameter invariances): one should search for *joint* eigenfunctions of the N commuting AΔOs. This is the first point in this paper where the latter are explicitly mentioned, and indeed, we have not shown that they share the parameter symmetries of the defining AΔO. Even so, it is quite plausible that this is true, and we also believe there exist BC_N-invariant joint eigenfunctions that generalize our BC_1 hyperbolic eigenfunction, with the parameter symmetries discussed above.

To date, there is admittedly precious little evidence for the existence of joint eigenfunctions with all of these features. Koornwinder's BC_N polynomials [11] should arise as discretizations of the hyperbolic BC_N joint eigenfunction, just as the Askey-Wilson polynomials arise by discretizing our BC_1 eigenfunction, cf. Subsection 5.2. But these polynomial discretizations only yield fragmentary tests of the expected scenario.

Considerably stronger results that are relevant in this connection were recently obtained by Chalykh [26]. He presents eigenfunctions for the hyperbolic BC_N case, which interpolate the Koornwinder polynomials. However, his arbitrary-eigenvalue elementary eigenfunctions are tied to a lattice of coupling constants, which should in turn be interpolated. Even so, inasmuch as his eigenfunctions allow tests of the above scenario, these tests seem to be passed.

Chalykh's results also exhibit a further symmetry feature that is not at all visible from the defining hyperbolic AΔO. (This is why we have not had occasion to mention this property in previous sections.) Roughly speaking, his eigenfunctions are symmetric under interchange of the variables $x_1, \ldots, x_N$ occurring in the AΔO and suitable spectral variables $p_1, \ldots, p_N$. Therefore, they are also eigenfunctions of an AΔO of the same structure as the defining one, which depends on the variables $p_1, \ldots, p_N$.

We will make this *self-duality* property more precise in Subsection 5.2. Again, one may expect that it will persist for the $N > 1$ and arbitrary-coupling hyperbolic eigenfunctions that are yet to be found/invented, in accordance with self-duality properties of the arbitrary-coupling Koornwinder polynomials [27].

5.2. THE HYPERBOLIC BC_1 CASE

Although the AΔOs H and $\mathcal{A}$ from Subsection 3.1 admit a manifestly D_4-invariant representation, this does not imply that a given eigenfunction has this property. The eigenfunctions at issue in this subsection do have D_4-invariance, but as will become apparent shortly, this symmetry property is far from manifest.

More precisely, we are not aware of any representation for which D_4-invariance is clear by inspection. The representation we are about to detail is the only one known to date, and one of its key features is the special character of one parameter. (Permutation invariance with respect to the three remaining parameters *is* manifest.)

Turning to the details, we are going to introduce an A-eigenfunction $R(a_+, a_-, \mathbf{c}; z, \hat{z})$ with eigenvalue $2c_+(2\hat{z}), \hat{z} \in \mathbb{C}$. Here, $\mathbf{c} \in \mathbb{C}^4$ is the coupling vector that is related to the parameter vector $\mathbf{h}$ by (3.21). It is expedient to introduce dual couplings and parameters by setting

$$\hat{\mathbf{c}} \equiv J\mathbf{c}, \quad \hat{\mathbf{h}} \equiv J\mathbf{h}, \quad J \equiv \frac{1}{2} \begin{pmatrix} 1 & 1 & 1 & 1 \\ 1 & 1 & -1 & -1 \\ 1 & -1 & 1 & -1 \\ 1 & -1 & -1 & 1 \end{pmatrix}. \tag{5.2}$$

Since the shift vector $(a, a_-/2, a_+/2, 0)$ in (3.21) is J-invariant, $\hat{\mathbf{h}}$ and $\hat{\mathbf{c}}$ are once more related via (3.21).

To define the R-function, we also need quantities

$$s_1 \equiv c_0 + c_1 - a_-/2, \quad s_2 \equiv c_0 + c_2 - a_+/2, \quad s_3 \equiv c_0 + c_3, \tag{5.3}$$

and functions

$$F(b; y, z) \equiv \frac{G(z + y + ib - ia)}{G(y + ib - ia)} \frac{G(z - y + ib - ia)}{G(-y + ib - ia)}, \tag{5.4}$$

$$K(\mathbf{c}; z) \equiv \frac{1}{G(z+ia)} \prod_{j=1}^{3} \frac{G(is_j)}{G(z+is_j)}, \tag{5.5}$$

where $G(a_+, a_-; z)$ is the hyperbolic gamma function, cf. Section 2. Choosing now at first

$$\mathbf{c} \in I\!R^4, \quad \mathrm{Re}\, z, \mathrm{Re}\, \hat{z} > 0, \quad s_1, s_2, s_3 \in (-a, a), \tag{5.6}$$

the R-function is defined by the contour integral

$$R(\mathbf{c}; z, \hat{z}) = \frac{1}{(a_+a_-)^{1/2}} \int_{\mathcal{C}} F(c_0; z, z') K(\mathbf{c}; z') F(\hat{c}_0; \hat{z}, z') dz'. \tag{5.7}$$

The choice of the contour $\mathcal{C}$ is determined by the location of the poles in the integrand. The function $K(\mathbf{c}; z')$ gives rise to four upward pole sequences on the imaginary axis, beginning at $z' = 0, i(a - s_j), j = 1, 2, 3,$ whereas $F(b; y, z')$ yields two downward sequences, beginning at $z' = \pm y - ib$. The contour is given by a horizontal line $\mathrm{Im}\, z' = d$, indented (if necessary) so that it passes above the points $-z - ic_0, -\hat{z} - i\hat{c}_0$ in the left half plane and $z - ic_0, \hat{z} - i\hat{c}_0$ in the right half plane, and so that it passes below 0. Thus the four upward pole sequences of the integrand are above $\mathcal{C}$ and the four downward ones are below $\mathcal{C}$. The integrand has exponential decay as $|\mathrm{Re}\, z'| \to \infty$, so that the integral does not depend on d.

Since we have

$$is_j = h_0 + h_j + ia, \quad j = 1, 2, 3, \quad i\hat{c}_0 = \frac{1}{2} \sum_{n=0}^{3} h_n + ia, \tag{5.8}$$

the function $R(a_+, a_-, \mathbf{c}(\mathbf{h}); z, \hat{z})$ is manifestly invariant under permutations of h_1, h_2, h_3, whereas h_0 plays a very different role. Viewed as a function of a_+, a_- and $\mathbf{h}$, it is moreover manifestly invariant under $a_+ \leftrightarrow a_-$. Taking for granted that it satisfies the announced eigenvalue AΔE

$$AR(a_+, a_-, \mathbf{c}(\mathbf{h}); z, \hat{z}) = 2c_+(2\hat{z}) R(a_+, a_-, \mathbf{c}(\mathbf{h}); z, \hat{z}), \tag{5.9}$$

it is therefore immediate that we also have

$$\tilde{A}R(a_+, a_-, \mathbf{c}(\mathbf{h}); z, \hat{z}) = 2c_-(2\hat{z}) R(a_+, a_-, \mathbf{c}(\mathbf{h}); z, \hat{z}), \tag{5.10}$$

where $\tilde{A}$ is obtained from A by interchanging a_+ and a_-. Furthermore, noting

$$is_j = \hat{h}_0 + \hat{h}_j + ia, \quad j = 1, 2, 3, \tag{5.11}$$

it follows that R is an eigenfunction with eigenvalues $2c_\pm(2z)$ of two AΔOs acting on $\hat{z}$; the latter are obtained from A and $\tilde{A}$ by taking $\mathbf{h} \to \hat{\mathbf{h}}$ and $z \to \hat{z}$. (See [15] for the details of the results mentioned thus far.)

To retain the latter self-duality property, the similarity transformation leading from A to $\mathcal{A}$ (cf. (3.7)) should be performed on the dual variable $\hat{z}$ as well. Moreover, a different normalization turns out to be expedient for the $\mathcal{A}$-eigenfunction. Specifically, following [16] we work with the function

$$\mathcal{E}(\mathbf{h}; z, \hat{z}) \equiv \chi(\mathbf{h})[c(\mathbf{h}; z)c(\hat{\mathbf{h}}; \hat{z}) \prod_{j=1}^{3} G(is_j)]^{-1} R(\mathbf{c}(\mathbf{h}); z, \hat{z}), \qquad (5.12)$$

where $c(\mathbf{h}; z)$ is defined by (3.1) and χ is the phase

$$\chi(\mathbf{h}) \equiv \exp(-i\alpha[\sum_{n=0}^{3} h_n^2/4 + (a_+^2 + a_-^2 + a_+a_-)/8]). \qquad (5.13)$$

With these conventions, the $\mathcal{E}$-function has plane wave asymptotics

$$\mathcal{E}(\mathbf{h}; z, \hat{z}) \sim \exp(i\alpha z\hat{z}) - u(\hat{\mathbf{h}}; -\hat{z}) \exp(-i\alpha z\hat{z}), \quad z \to \infty, \qquad (5.14)$$

where the u-function is given by (3.4) (with $r = 0$, of course). Moreover, denoting the D_4 Weyl group by W, we have

$$\mathcal{E}(w(\mathbf{h}); z, \hat{z}) = \mathcal{E}(\mathbf{h}; z, \hat{z}), \quad \forall w \in W. \qquad (5.15)$$

In words, the $\mathcal{E}$-function is D_4-invariant.

The latter invariance property and (a far stronger version of) the asymptotics (5.14) are the principal results of [16]. Even though these two results are quite distinct at face value, their proofs in [16] are intimately related. Briefly, the asymptotics is obtained first for a restricted parameter region that is W-invariant. Next, the W-invariance of the asymptotics is used to prove W-invariance of $\mathcal{E}$ for this region. By analytic continuation, W-invariance for arbitrary parameters then follows. The asymptotics for arbitrary parameters can now be established by exploiting W-invariance.

As we have already seen at the end of Subsection 3.1, we need to choose the couplings real to obtain formally self-adjoint AΔOs. In [17] we have studied the question whether these formal insights can be improved to a rigorous Hilbert space formulation. In view of the meromorphy, plane wave asymptotics, and D_4-invariance of the $\mathcal{E}$-function, this problem is best studied by viewing $\mathcal{E}(a_+, a_-, \mathbf{h}; z, \hat{z})$ (for $i\mathbf{h} \in \mathbb{R}^4$) as the kernel of a map between Hilbert spaces

$$\hat{\mathcal{H}} \equiv L^2((0, \infty), d\hat{z}), \quad \mathcal{H} \equiv L^2((0, \infty), dz). \qquad (5.16)$$

More specifically, our starting point in [17] is the linear operator

$$\begin{aligned} \mathcal{F} \ &: \ C_0^\infty((0, \infty)) \subset \hat{\mathcal{H}} \to \mathcal{H}, \\ \phi(\hat{z}) \ &\mapsto \ \left(\frac{\alpha}{2\pi}\right)^{1/2} \int_0^\infty \mathcal{E}(z, \hat{z})\phi(\hat{z})d\hat{z}. \end{aligned} \qquad (5.17)$$

This eigenfunction transform (generalized Fourier transformation) is well defined for generic parameters in $i\mathbb{R}^4$. (The integral converges for all $\mathbf{h} \in i\mathbb{R}^4$ and $z > 0$, but for special parameters the $\mathcal{E}$-function may have a pole at $z = 0$, precluding an image in $\mathcal{H}$.)

In spite of the normalized plane wave asymptotics (5.14) (which is formally unitary) and the formal self-adjointness of $\mathcal{A}$ on $\mathcal{H}$, the operator $\mathcal{F}$ is not isometric in general, and (accordingly) the operator $\dot{\mathcal{A}}$ defined on $\mathcal{F}C_0^\infty((0,\infty))$ by setting

$$\dot{\mathcal{A}}\mathcal{F} = \mathcal{F}M, \tag{5.18}$$

where M is the operator of multiplication by $2c_+(2\hat{z})$, is not symmetric (hermitian) in general. But when the parameters $(a_+, a_-, i\mathbf{h})$ are required to belong to the polytope

$$P \equiv \{(a_+, a_-, p) \in (0,\infty)^2 \times \mathbb{R}^4 \mid |p_n| < a, \quad n = 0, \ldots, 3\}, \tag{5.19}$$

then $\mathcal{F}$ is isometric and (hence) $\dot{\mathcal{A}}$ is symmetric.

Requiring not only $(a_+, a_-, i\mathbf{h}) \in P$, but also $(a_+, a_-, i\hat{\mathbf{h}}) \in P$, the operator $\mathcal{F}$ extends to an isometry from $\hat{\mathcal{H}}$ onto $\mathcal{H}$, and $\dot{\mathcal{A}}$ extends to a self-adjoint operator on (a certain dense subspace of) $\mathcal{H}$. As a consequence, for this parameter set we then obtain D_4-invariant operators $\mathcal{F}$ and $\dot{\mathcal{A}}$.

The case where one has $(a_+, a_-, i\mathbf{h}) \in P$, but $(a_+, a_-, i\hat{\mathbf{h}}) \notin P$, is far richer, since bound states appear on the stage. We proceed by taking a closer look at this situation. As a bonus, this prepares the ground for the next subsection.

To begin with, the bound states take their simplest form in terms of the R-function (as compared to the $\mathcal{E}$-function). Specifically, assuming from now on $a_- \leq a_+$, they arise from the discrete $\hat{z}$-choices $i\hat{c}_0 + ina_-, n \in \mathbb{N}$, where we have

$$R(a_+, a_-, \mathbf{c}; z, i\hat{c}_0 + ina_-) = P_n(c_+(2z)), \quad n \in \mathbb{N}, \tag{5.20}$$

for generic parameters. Here, $P_n(u)$ is a degree-n polynomial in u; these polynomials are basically the Askey-Wilson polynomials [12, 13].

In particular, for $n = 0$ we obtain the constant function

$$P_0(u) = 1, \tag{5.21}$$

and the remaining polynomials can be obtained via the AΔO dual to A (it yields the three-term recurrence). Even though these eigenfunctions are singularity-free for generic parameters, they do not enter the above Hilbert space formulation whenever we have

$$(a_+, a_-, i\mathbf{h}) \in P, \quad (a_+, a_-, i\hat{\mathbf{h}}) \in P. \tag{5.22}$$

Indeed, as already mentioned, the eigenfunction transform $\mathcal{F}$ is unitary in this case, and its kernel consists of 'scattering states', cf. (5.17), (5.14).

This critical dependence on the parameters can already be appreciated by focusing on the 'ground state' (5.21). To start with, note that by (3.7) and (3.17) we have

$$\begin{aligned} A &= V(z)[\exp(-ia_-\partial_z) - 1] \\ &\quad + V(-z)[\exp(ia_-\partial_z) - 1] - 2c_+(\sum h_n + ia_-). \end{aligned} \tag{5.23}$$

It is therefore immediate that the constant functions are A-eigenfunctions with eigenvalue $-2c_+(\sum h_n + ia_-)$. Recalling (3.21), one sees that this is consistent with what we asserted in (5.20)–(5.21), cf. (5.9). (Note however that a priori the R-function might be a quite different eigenfunction for $\hat{z} = i\hat{c}_0$.)

Let us now pass to the transformed eigenvalue equation

$$\mathcal{A}(1/c(\mathbf{h}; z)) = -2c_+(\sum h_n + ia_-)/c(\mathbf{h}; z). \tag{5.24}$$

The first important point is that the $\mathcal{A}$-eigenfunction $1/c(\mathbf{h}; z)$ does not belong to $\mathcal{H}$ for parameters satisfying (5.22). To be sure, this is not clear by inspection. But the known asymptotics of the hyperbolic gamma function for $|\mathrm{Re}\, z| \to \infty$ can be invoked to obtain (cf. (3.5) in [16])

$$c(\mathbf{h}; z)^{-1} \sim \chi(\mathbf{h})^{-1} \exp(\alpha z(a - i\hat{h}_0)), \quad z \to \infty. \tag{5.25}$$

Accepting this, it *is* clear by inspection that the eigenfunction diverges exponentially for $i\hat{h}_0 < a$.

As a second key point, (5.25) entails that $1/c(\mathbf{h}; z)$ is indeed square-integrable at ∞ whenever $i\hat{h}_0 > a$. Now for parameters in P one readily checks that $i\hat{h}_0$ can vary over all of $(-2a, 2a)$. Furthermore, for $i\hat{h}_0 \in (a, 2a)$ the function $1/c(\mathbf{h}; z)$ is not only in $\mathcal{H}$, but also *orthogonal* to the range of $\mathcal{F}$. Defining the action of $\dot{\mathcal{A}}$ on the $\mathcal{H}$-vector

$$\psi_0(z) \equiv 1/c(\mathbf{h}; z), \quad (a_+, a_-, i\mathbf{h}) \in P, \quad i\hat{h}_0 \in (a, 2a), \tag{5.26}$$

to be equal to that of $\mathcal{A}$, we obtain therefore a bound state for $\dot{\mathcal{A}}$ (namely, its ground state), orthogonal to the 'scattering states' in the range of $\mathcal{F}$.

Depending on the precise relation between $i\hat{h}_0$ and the parameters a_+ and a_-, there are further bound states $P_n(c_+(2z))/c(\mathbf{h}; z)$. In view of the asymptotics (5.25), it is actually easy to see for which degrees $n \in \mathbb{N}$ one obtains exponential decay for $z \to \infty$, hence square-integrability: one clearly needs

$$na_- + a - i\hat{h}_0 < 0. \tag{5.27}$$

For far more information on these bound states we refer to [17]; our remaining discussion in this subsection concerns the issue of D_4-invariance vs. the ground state (5.26).

We begin by noting that the requirement

$$i(h_0 + h_1 + h_2 + h_3) \in (2a, 4a) \tag{5.28}$$

is S_4-invariant, but when we flip an even number of signs of $ih_0, \ldots, ih_3$, then it is no longer satisfied. At first sight, this seems to be in conflict with the D_4-invariance we emphasized above. But in fact there is no contradiction here. Indeed, since $\mathcal{A}$ is D_4-invariant, we conclude from (5.24) that for all $w \in W$ the function $1/c(w(\mathbf{h}); z)$ is an $\mathcal{A}$-eigenfunction with eigenvalue $-2c_+(\sum w(\mathbf{h})_n + ia_-)$. Since there are 8 distinct even sign flips, we obtain 8 distinct eigenfunctions with 8 distinct eigenvalues (for generic parameters). Whenever $(a_+, a_-, i\mathbf{h}) \in P$ satisfies

$$\eta \equiv \max(|\hat{h}_0|, |\hat{h}_1|, |\hat{h}_2|, |\hat{h}_3|) > a, \tag{5.29}$$

there is a unique n such that $\eta = |\hat{h}_n|$, cf. (5.2). A suitable W-transformation then ensures

$$i(Jw(\mathbf{h}))_0 \in (a, 2a), \tag{5.30}$$

and so the above R-function properties hold true for $w(\mathbf{h})$.

Therefore, a W-invariant definition of the Hilbert space ground state reads as follows. First, we require $(a_+, a_-, i\mathbf{h})$ to belong to the (W-invariant) polytope P (5.19). Provided the parameters also satisfy the (W-invariant) restriction (5.29), we choose $w \in W$ such that (5.30) holds true, and define the ground state by

$$\psi_0(z) \equiv 1/c(w(\mathbf{h}); z). \tag{5.31}$$

Then we obtain the W-invariant eigenvalue equation

$$\dot{A}\psi_0(z) = 2\cos(2\pi(\eta - a)/a_+)\psi_0(z), \quad \eta \in (a, 2a). \tag{5.32}$$

Switching to the W-invariant AΔO $H(a_+, a_-, \mathbf{h}; z)$ with $(a_+, a_-, \mathbf{h}) \in P$ (cf. (3.10)–(3.13)), we see from the above discussion that the positive even function $w(\mathbf{h}; z)^{1/2}$, $z \in \mathbb{R}$, should only be viewed as a ground state when it is square-integrable over $\mathbb{R}$. Among the 8 (generically) distinct functions $w(w(\mathbf{h}); z)$, $w \in W$, there is at most one with this property.

5.3. GROUND STATES

From Lemma 4.1 we obtain the arbitrary-N version of the A-representation (5.23), namely (cf. (4.14), (4.26))

$$A = \sum_{j=1}^{N} (\mathcal{V}_j(x)[\exp(-ia_-\partial_j) - 1] + \mathcal{V}_j(-x)[\exp(ia_-\partial_j) - 1]) + E(\mathbf{h}, \mu), \tag{5.33}$$

$$E(\mathbf{h}, \mu) \equiv -2\frac{s_+(N\mu)}{s_+(\mu)}c_+ \left(\sum_{n=0}^{3} h_n + (N-1)\mu + ia_-\right). \tag{5.34}$$

Just as for $N = 1$, it is plain that the constant functions are A-eigenfunctions with eigenvalue $E(\mathbf{h}, \mu)$. Since $H(\mathbf{h}, \mu)$ is D_4-invariant in $\mathbf{h}$ and $\mathbb{Z}_2$-invariant in μ (recall the paragraph containing (4.27)), the function $W(w(\mathbf{h}), \sigma(\mu); x)^{1/2}$ is an $H(\mathbf{h}, \mu)$-eigenfunction with eigenvalue $E(w(\mathbf{h}), \sigma(\mu))$ for all $(w, \sigma) \in W \times \mathbb{Z}_2$. Thus we obtain 16 (generically) distinct eigenfunctions.

Requiring from now on $i\mathbf{h} \in \mathbb{R}^4, i\mu \in \mathbb{R}$, the functions $W(w(\mathbf{h}), \sigma(\mu); x)$ are positive (recall (4.1)–(4.5)). For $\mu = 0$ we obtain a product of BC_1 weight functions, and we can use the results of the previous subsection. Thus, choosing $(a_+, a_-, i\mathbf{h}) \in P$, we need $i\hat{h}_0 \in (a, 2a)$ for $W(a_+, a_-, \mathbf{h}, 0; x)$ to have a finite integral over the BC_N Weyl chamber G_N (4.29), whereas the 7 other W-transforms are not integrable over G_N. Clearly, this remains true when we let $\mu = i\tau$ with $\tau \in (-\epsilon, \epsilon)$ and $\epsilon \in (0, a)$ small enough; for the transformed τ-values $2a - \tau$ the W-integral over G_N diverges, however.

More generally, it is probably true that among the 16 distinct weight functions related by $W \times \mathbb{Z}_2$ transformations, at most one is integrable over G_N. In the absence of eigenfunctions for arbitrary N, arbitrary parameters and arbitrary eigenvalues, it is however not clear under what restrictions on the parameters the function $W(\mathbf{h}, \mu; x)^{1/2}$ serves as the ground state of a self-adjoint operator $\dot{H}$ on $L^2(G_N, dx)$, whose action coincides with that of the AΔO H on a dense subspace. (Note that due to the similarities connecting H with $\mathcal{A}$ and A, this is equivalent to $1/C(\mathbf{h}, \mu; x)$ being the ground state of $\dot{\mathcal{A}}$ on $L^2(G_N, dx)$, and the constant function being the ground state of $\dot{A}$ on $L^2(G_N, W(x)dx)$.)

Turning to the elliptic case, we need to impose the restriction (4.33) on $(\mathbf{h}, \mu)$ for the AΔO A to take the form (5.33), with

$$E(\mathbf{h}, \mu) = \sum_{j=1}^{N} (\mathcal{V}_j(\mathbf{h}, \mu; x) + \mathcal{V}_j(\mathbf{h}, \mu; -x)) + \mathcal{V}(\mathbf{h}, \mu; x), \tag{5.35}$$

cf. Lemma 4.2. Doing so from now on, it is once again obvious that the constant function is an A-eigenfunction with eigenvalue $E(\mathbf{h}, \mu)$. Hence $W(\mathbf{h}, \mu; x)^{1/2}$ is an $H(\mathbf{h}, \mu)$-eigenfunction with eigenvalue $E(\mathbf{h}, \mu)$.

Below Lemma 4.2 we have already seen one way to obtain a positive weight function, namely via real couplings $\mathbf{c}$ and τ. But this is not the only possibility: we can also add $i\pi/2r$ to each of these couplings. For all of these choices, the function $W(\mathbf{h}, \mu; x)$ is integrable over the Weyl alcove G_N (4.34). (As before, this is true for generic parameters—taking for example $N = 1$, one can choose special parameters so that $w(\mathbf{h}; z)$ has a pole of arbitrary order at $z = 0$, precluding integrability.) Moreover, choices for which (4.33) holds true yield $H(\mathbf{h}, \mu)$-eigenfunctions $W(\mathbf{h}, \mu; x)^{1/2}$.

Admittedly, in the elliptic case it is not even clear for $N = 1$ which of the above positive H-eigenfunctions admit a reinterpretation as the ground state of a self-adjoint Hilbert space operator $\dot{H}$. A natural expectation is however that the choice of $(\mathbf{h}, \mu)$ should be such that the hyperbolic limit not only exists, but also yields a weight function that is integrable over $\mathbb{R}^N$. In this connection we point out that the limit $r \downarrow 0$ need not exist for arbitrary shifts of the couplings by $i\pi/2r$. For instance, taking $i\mathbf{h} \in \mathbb{R}^8$, the $N = 1$ weight function $w(r, a_+, a_-, \mathbf{h}; z)$ has no $r \downarrow 0$ limit.

Fixing $\mathbf{c} \in \mathbb{R}^8$ and $\tau \in \mathbb{R}$, it is easily verified that the limit does exist, yielding the hyperbolic weight function corresponding to (c_0, c_1, c_2, c_3) and τ. Indeed, using the limits (2.15) and (2.16), one deduces that the $r \downarrow 0$ limit exists already for the c-functions (3.1) and (4.1).

6. Nonrelativistic limits

In this section we complete our account by studying various aspects of the nonrelativistic limit. More specifically, we obtain the limits of the above AΔOs and eigenfunctions, paying special attention to their parameter symmetries.

To begin with, the symmetry under interchange of a_+ and a_- (which is evidently present for the eigenfunctions $1, W(x)^{1/2}, 1/C(x)$ of the AΔOs A, H and $\mathcal{A}$, resp.) disappears, since the limit involves sending one of a_+, a_- to 0. In the hyperbolic $N = 1$ case, the relevant limit is defined by setting

$$\mathbf{c} = a_-(d_0, \ldots, d_3), \tag{6.1}$$

(with $\mathbf{c}$ related to $\mathbf{h}$ via (3.21)), and taking a_- to 0 for fixed $\mathbf{d}$. It is convenient to introduce new parameters

$$\nu = \pi/a_+, \quad g_{02} = d_0 + d_2, \quad g_{13} = d_1 + d_3. \tag{6.2}$$

For the functions V, V_a and V_b we now calculate the expansions

$$\begin{aligned} F(z) &= F^{(0)} + a_- F^{(1)}(z) + a_-^2 F^{(2)}(z) + O(a_-^3), \quad a_- \downarrow 0, \\ F &= V, V_a, V_b. \end{aligned} \tag{6.3}$$

The results read

$$V^{(0)} = 1, \quad V^{(1)}(z) = -i\nu(g_{02}\coth(\nu z) + g_{13}\tanh(\nu z)), \tag{6.4}$$

$$2V^{(2)}(z) = -\nu^2 \left(\frac{2d_0 d_2 - d_2}{\sinh^2(\nu z)} - \frac{2d_1 d_3 - d_3}{\cosh^2(\nu z)} + (g_{02} + g_{13})^2 \right), \tag{6.5}$$

$$V_a^{(0)} = 1, \quad V_a^{(1)} = 0, \tag{6.6}$$

$$V_a^{(2)}(z) = \nu^2 \left(\frac{d_0^2 + d_2^2 - d_0}{\sinh^2(\nu z)} - \frac{d_1^2 + d_3^2 - d_1}{\cosh^2(\nu z)} \right), \tag{6.7}$$

$$V_b^{(0)} = 0, \quad V_b^{(1)} = 0, \tag{6.8}$$

$$V_b^{(2)}(z) = \nu^2 \left(\frac{2d_0 d_2 - d_2}{\sinh^2(\nu z)} - \frac{2d_1 d_3 - d_3}{\cosh^2(\nu z)} \right). \tag{6.9}$$

For the AΔOs A, H and $\mathcal{A}$ we then obtain

$$D = 2 + a_-^2 D_{\rm nr} + O(a_-^3), \quad a_- \downarrow 0, \quad D = A, H, \mathcal{A}, \tag{6.10}$$

where

$$A_{\rm nr} = -\frac{d^2}{dz^2} - 2iV^{(1)}(z)\frac{d}{dz} - \nu^2(g_{02} + g_{13})^2, \tag{6.11}$$

$$H_{\rm nr}, \mathcal{A}_{\rm nr} = -\frac{d^2}{dz^2} + V_h(z), \tag{6.12}$$

with the hyperbolic potential

$$V_h(z) = \frac{\nu^2 g_{02}(g_{02} - 1)}{\sinh^2(\nu z)} - \frac{\nu^2 g_{13}(g_{13} - 1)}{\cosh^2(\nu z)}. \tag{6.13}$$

Next, we deduce from the limit (2.7) that we have

$$\lim_{a_- \downarrow 0} w(a_+, a_-, \mathbf{h}(\mathbf{c}); z) = [2\sinh(\nu z)]^{2g_{02}}[2\cosh(\nu z)]^{2g_{13}} \equiv w_{\rm nr}(z),$$
$$z \in (0, \infty). \tag{6.14}$$

As a check, note that we do get the expected relation

$$H_{\rm nr} = w_{\rm nr}(z)^{1/2} A_{\rm nr} w_{\rm nr}(z)^{-1/2}, \tag{6.15}$$

cf. (3.13). (The equality of $H_{\rm nr}$ and $\mathcal{A}_{\rm nr}$ can be understood from $u_{\rm nr}$ being z-independent.)

The obvious expectation is that the manifest symmetries under $g_{02} \to 1 - g_{02}$ and $g_{13} \to 1 - g_{13}$ of $H_{\rm nr}$ are remnants of the hidden D_4-symmetries of $\mathcal{A}$ and H. In point of fact, this is only true for the second involution; the first one is a new symmetry. To appreciate this, one need only rewrite (6.1) in terms of $\mathbf{h}$ via (3.21) and reinspect the limiting behavior: the $g_{13} \to 1 - g_{13}$ map arises from the sign flip $(h_1, h_3) \to (-h_1, -h_3)$, whereas the $g_{02} \to 1 - g_{02}$ map has no $\mathbf{h}$-analog.

This different ancestry of the two involutions can also be recognized from their different effect on the nonrelativistic limit of the D_4-invariant BC_1 eigenfunction $\mathcal{E}(a_+, a_-, \mathbf{h}; z, \hat{z})$: the limit function is invariant under $g_{13} \to 1 - g_{13}$, but not under $g_{02} \to 1 - g_{02}$.

To substantiate this assertion, we recall from [15] that for the R-function to converge to ${}_2F_1$ we should not only rescale the couplings by a_- (cf. (6.1)), but also the dual variable $\hat{z}$. Here it is convenient to set

$$\hat{z} = a_- k, \tag{6.16}$$

obtaining [15]

$$\lim_{a_- \downarrow 0} R(a_+, a_-, a_- \mathbf{d}; z, a_- k) = {}_2F_1(\alpha, \beta, \gamma; -\sinh^2(\nu z)),$$
$$z \in (0, \infty), \quad k \in \mathbb{R}, \tag{6.17}$$

with

$$\alpha = (g_{02} + g_{13})/2 - ik, \quad \beta = (g_{02} + g_{13})/2 + ik, \quad \gamma = g_{02} + 1/2. \tag{6.18}$$

Using (2.7), we calculate

$$\lim_{a_- \downarrow 0} \chi(\mathbf{h})/c(\mathbf{h}; z) = w_{\mathrm{nr}}(z)^{1/2}, \quad z \in (0, \infty), \tag{6.19}$$

and using also (2.8), we get

$$\lim_{a_- \downarrow 0} c(\hat{\mathbf{h}}; a_- k) \prod_{j=1}^{3} G(is_j) = \hat{c}_{\mathrm{nr}}(k) \Gamma(g_{02} + 1/2), \tag{6.20}$$

$$\hat{c}_{\mathrm{nr}}(k) \equiv \Gamma(ik)\Gamma(ik + 1/2)/\Gamma(ik + (g_{02} + g_{13})/2)\Gamma(ik + (g_{02} + 1 - g_{13})/2). \tag{6.21}$$

Recalling (5.12), we can put the pieces together, obtaining

$$\lim_{a_- \downarrow 0} \mathcal{E}(z, a_- k) = w_{\mathrm{nr}}(z)^{1/2} \hat{c}_{\mathrm{nr}}(k)^{-1} {}_2F_1(\alpha, \beta, \gamma; -\sinh^2(\nu z)),$$
$$z \in (0, \infty), \quad k \in \mathbb{R}. \tag{6.22}$$

Now $\hat{c}_{\mathrm{nr}}(k)$ is obviously invariant under $g_{13} \to 1 - g_{13}$. The asserted invariance of the $\mathcal{E}$-limit (6.22) therefore amounts to the identity

$${}_2F_1(\alpha, \beta, \gamma; x) = (1 - x)^{\gamma - \alpha - \beta} {}_2F_1(\gamma - \beta, \gamma - \alpha, \gamma; x), \tag{6.23}$$

which was already known to Gauss (see e.g. 9.131 in [28]).

As announced, the limit (6.22) is not invariant under $g_{02} \to 1 - g_{02}$. To be sure, the transformed function is also an eigenfunction of the differential operator H_{nr}. For $g_{02} \in (-1/2, 3/2)$ and $g_{13} \in \mathbb{R}$, these two eigenfunctions are associated to distinct self-adjoint operators $\dot{H}_{\mathrm{nr}}^{\pm}$ on $L^2((0, \infty), dz)$, with $\dot{H}_{\mathrm{nr}}^{+}$ defined for $g_{02} \in [1/2, \infty)$ and $\dot{H}_{\mathrm{nr}}^{-}$ for $g_{02} \in (-1/2, 1/2]$. (Of course, for $g_{02} = 1/2$ the two operators coincide.)

For the arbitrary-N hyperbolic case we should also rescale the coupling τ by a_-. That is, we should set

$$\tau = a_- g, \tag{6.24}$$

before taking a_- to 0. Since the parameter symmetry $\tau \to 2a-\tau$ (cf. (4.27)) amounts to a map $g \to 1-g+a_+/a_-$ that has no limit for $a_- \downarrow 0$, this symmetry does not persist on the nonrelativistic level. But a new invariance under $g \to 1-g$ appears, as will be clear from the formulas that follow. This is a manifest invariance property of $H_{\rm nr}$, but in cases where this can be checked, the analytic continuation from $g \in [1/2, \infty)$ to $g < 1/2$ of the eigenfunction that is relevant for Hilbert space purposes is not invariant. (As before, it is associated to a distinct self-adjoint extension of the differential operator $H_{\rm nr}$ restricted to smooth functions with compact support in G_N.)

To expand the coefficients of the AΔOs, we can use our previous calculations (6.3)–(6.9), not only for the external field terms $V(\pm x_j)$, but also for the extra μ-dependent terms. (This is because they give rise to functions of the form $V(x_j \pm x_k), V_a(x_j \pm x_k)$, with $d_0 = g, d_j = 0, j > 0$.) Moreover, it is not hard to check

$$\mathcal{V}(x) = a_-^2 \sum_{j=1}^{N} V_b^{(2)}(x_j) + O(a_-^3), \quad a_- \downarrow 0. \tag{6.25}$$

The result is that we obtain expansions

$$D = 2N + a_-^2 D_{\rm nr} + O(a_-^3), \quad a_- \downarrow 0, \quad D = A, H, \mathcal{A}, \tag{6.26}$$

where

$$\begin{aligned} A_{\rm nr} &= \sum_{j=1}^{N} \Big(-\partial_j^2 - 2\Big(iV^{(1)}(x_j) \\ &\quad + \nu g \sum_{k \neq j} [\coth \nu(x_j - x_k) + \coth \nu(x_j + x_k)] \Big) \partial_j \Big) + E_0, \end{aligned} \tag{6.27}$$

$$\begin{aligned} E_0 &= -N\nu^2 \Big((g_{02} + g_{13})^2 + 2(N-1)g(g_{02} + g_{13}) \\ &\quad + \frac{2}{3}(2N-1)(N-1)g^2 \Big), \end{aligned} \tag{6.28}$$

$$\begin{aligned} \mathcal{A}_{\rm nr}, H_{\rm nr} &= \sum_{j=1}^{N} \Big(-\partial_j^2 + V_h(x_j) \\ &\quad + \sum_{k \neq j} \Big(\frac{\nu^2 g(g-1)}{\sinh^2 \nu(x_j - x_k)} + \frac{\nu^2 g(g-1)}{\sinh^2 \nu(x_j + x_k)} \Big) \Big), \end{aligned} \tag{6.29}$$

with $V^{(1)}(z)$ and $V_h(z)$ given by (6.4) and (6.13), resp. We used (5.33)–(5.34) to simplify the expansion of A.

From the limit (2.7) we deduce

$$\begin{aligned}\lim_{a_-\downarrow 0} W(x) &= \prod_{j=1}^{N} w_{\rm nr}(x_j)\cdot \prod_{1\le j<k\le N} [4\sinh\nu(x_j-x_k)\sinh\nu(x_j+x_k)]^{2g} \\ &\equiv W_{\rm nr}(x), \qquad x\in G_N. \end{aligned}\tag{6.30}$$

The constant function is manifestly an $A_{\rm nr}$-eigenfunction with eigenvalue E_0, but the 'ground state' equation

$$H_{\rm nr}W_{\rm nr}(x)^{1/2} = E_0 W_{\rm nr}(x)^{1/2}\tag{6.31}$$

is not obvious. It amounts to a functional equation that can be obtained from the $a_-\downarrow 0$ limit of Lemma 4.1. The latter also yields a direct proof of the similarity

$$A_{\rm nr} = W_{\rm nr}(x)^{-1/2}H_{\rm nr}W_{\rm nr}(x)^{1/2},\tag{6.32}$$

which is the limit of (4.7).

Next, we consider the elliptic $N=1$ case. As in the hyperbolic case, we should first switch from the parameter vector $\mathbf{h}$ to the coupling vector $\mathbf{c}$ (now given by (3.21) and (3.43)), and set

$$\mathbf{c} = a_-(d_0,\dots,d_7).\tag{6.33}$$

Taking a_- to 0 for fixed $\mathbf{d}$, we can obtain the pertinent expansions via (2.23) and (2.25) (with $a\to a_+$). It is convenient to introduce

$$g_0=d_0+d_2,\quad g_1=d_4+d_6,\quad g_2=d_1+d_3,\quad g_3=d_5+d_7,\quad \gamma=r(g_0+g_1),\tag{6.34}$$

the shifted $\wp$-function

$$\tilde{\wp}(z)=\wp(z)+\eta/\omega_1,\tag{6.35}$$

and the sum of logarithmic derivatives

$$\mathcal{L}_+(z)=g_0\frac{s_+'(z)}{s_+(z)}+g_1\frac{s_+'(z+\omega_1)}{s_+(z+\omega_1)}+g_2\frac{R_+'(z)}{R_+(z)}+g_3\frac{R_+'(z+\omega_1)}{R_+(z+\omega_1)}.\tag{6.36}$$

Now we obtain by long, but straightforward calculations the elliptic generalization of (6.4)–(6.9):

$$V^{(0)}=1,\quad V^{(1)}(z)=-i\mathcal{L}_+(z)-\gamma,\tag{6.37}$$

$$\begin{aligned}2V^{(2)}(z) &= V^{(1)}(z)^2+(d_0^2+d_2^2+d_2)\tilde{\wp}(z)+(d_4^2+d_6^2+d_6)\tilde{\wp}(z+\omega_1)\\ &\quad +(d_1^2+d_3^2+d_3)\tilde{\wp}(z+\omega_2)\\ &\quad +(d_5^2+d_7^2+d_7)\tilde{\wp}(z+\omega_3),\end{aligned}\tag{6.38}$$

$$V_a^{(0)} = 1, \quad V_a^{(1)} = -2\gamma, \tag{6.39}$$

$$\begin{aligned} V_a^{(2)}(z) &= 2\gamma^2 + (d_0^2 + d_2^2 - d_0)\tilde{\wp}(z) + (d_4^2 + d_6^2 - d_4)\tilde{\wp}(z+\omega_1) \\ &\quad + (d_1^2 + d_3^2 - d_1)\tilde{\wp}(z+\omega_2) \\ &\quad + (d_5^2 + d_7^2 - d_5)\tilde{\wp}(z+\omega_3), \end{aligned} \tag{6.40}$$

$$V_b^{(0)} = 0, \quad V_b^{(1)} = 0, \tag{6.41}$$

$$\begin{aligned} V_b^{(2)}(z) &= (2d_0d_2 - d_2)\wp(z) + (2d_4d_6 - d_6)\wp(z+\omega_1) \\ &\quad + (2d_1d_3 - d_3)\wp(z+\omega_2) \\ &\quad + (2d_5d_7 - d_7)\wp(z+\omega_3) + c_b, \end{aligned} \tag{6.42}$$

$$\begin{aligned} c_b &= -e_1(2d_0d_2 - d_2 + 2d_4d_6 - d_6) - e_3(2d_1d_3 - d_3 + 2d_5d_7 - d_7), \\ e_j &= \wp(\omega_j). \end{aligned} \tag{6.43}$$

Therefore, the AΔOs have an expansion of the form

$$D = 2 - 2a_-\gamma + a_-^2 D_{\rm nr} + O(a_-^3), \quad a_- \downarrow 0, \quad D = A, H, \mathcal{A}, \tag{6.44}$$

where we have

$$A_{\rm nr} = -\frac{d^2}{dz^2} - 2\mathcal{L}_+(z)\frac{d}{dz} - \mathcal{L}_+(z)^2 + \sum_{t=0}^{3} g_t^2 \wp(z+\omega_t) + \chi_A, \tag{6.45}$$

$$H_{\rm nr} = -\frac{d^2}{dz^2} + V_e(z) + \chi_H, \tag{6.46}$$

$$\mathcal{A}_{\rm nr} = -\frac{d^2}{dz^2} - 2i\gamma\frac{d}{dz} + V_e(z) + \chi_{\mathcal{A}}, \tag{6.47}$$

with the elliptic potential

$$V_e(z) = \sum_{t=0}^{3} g_t(g_t - 1)\wp(z+\omega_t), \tag{6.48}$$

and the constants

$$\chi_H = \gamma^2 + c_b + \left(\sum_{n=0}^{7} d_n^2 - d_0 - d_1 - d_4 - d_5\right)\eta/\omega_1, \tag{6.49}$$

$$\chi_{\mathcal{A}} = \chi_H + \gamma^2, \tag{6.50}$$

$$\chi_A = \chi_H + 2\left(\sum_{n=0}^{7} d_n\right)\eta/\omega_1. \tag{6.51}$$

From the limit (2.17) we see that the c-, u- and w-functions have limits for $z \in (0, \omega_1)$ given by

$$c_{\mathrm{nr}}(z) = R_+(z+\omega_2)^{-g_0} R_+(z-\omega_3)^{-g_1} R_+(z)^{-g_2} R_+(z+\omega_1)^{-g_3}, \quad (6.52)$$

$$u_{\mathrm{nr}}(z) = \exp((g_0+g_1-1)(2irz - i\pi)), \quad (6.53)$$

$$w_{\mathrm{nr}}(z) = p_w s_+(z)^{2g_0} s_+(z+\omega_1)^{2g_1} R_+(z)^{2g_2} R_+(z+\omega_1)^{2g_3}, \quad (6.54)$$

with p_w a positive constant. Once again, (6.15) can be checked directly, just as the relation

$$H_{\mathrm{nr}} = (-e^{2irz} u_{\mathrm{nr}}(z))^{1/2} \mathcal{A}_{\mathrm{nr}} (-e^{2irz} u_{\mathrm{nr}}(z))^{-1/2}. \quad (6.55)$$

Rewriting (6.33) in terms of $\mathbf{h}$, we see that only the sign flips on h_1, h_3 and on h_5, h_7 survive in the limit. (Indeed, H_{nr} and $\mathcal{A}_{\mathrm{nr}}$ are invariant under $d_1, d_3 \to 1-d_1, -d_3$ and under $d_5, d_7 \to 1-d_5, -d_7$.) When we omit the constants in (6.46) and (6.47), these involutions amount to $g_2 \to 1-g_2$ and $g_3 \to 1-g_3$, cf. (6.34). But the invariances under $g_0 \to 1-g_0$ and $g_1 \to 1-g_1$ are new. Since general coupling eigenfunctions are neither known for the relativistic nor for the nonrelativistic case, we cannot study these parameter symmetries any further.

As we have seen above, the relation (3.45) guarantees that $w(\mathbf{h}(\mathbf{c}); z)^{1/2}$ is an H-eigenfunction. In terms of $\mathbf{d}$ (6.33), this says $\sum d_j = 0$. Therefore, requiring

$$\sum_{t=0}^{3} g_t = 0, \quad (6.56)$$

we deduce that the functions $w_{\mathrm{nr}}(z)^{1/2}, 1/c_{\mathrm{nr}}(z)$ and the constant function are eigenfunctions of $H_{\mathrm{nr}}, \mathcal{A}_{\mathrm{nr}}$ and A_{nr}, resp. This is by no means clear by inspection, but can also be checked directly, as follows.

First, the condition (6.56) ensures that $\mathcal{L}_+(z)$ (6.36) is elliptic. Second, it clearly implies

$$\mathcal{L}'_+(z) = -\sum_{t=0}^{3} g_t \wp(z+\omega_t), \quad \Big(\sum g_t = 0\Big). \quad (6.57)$$

Third, $\mathcal{L}_+(z)$ is odd, so $\mathcal{L}_+(z)^2$ is even. Comparing principal parts at poles, one deduces that $\mathcal{L}_+(z)^2 - \sum g_t^2 \wp(z+\omega_t)$ is constant:

$$\mathcal{L}_+(z)^2 = \sum_t g_t^2 \wp(z+\omega_t) + C(\mathbf{g}), \quad \Big(\sum g_t = 0\Big). \quad (6.58)$$

A moment's thought now reveals that (6.57)–(6.58) entail the relevant eigenfunction properties.

Next, we consider the arbitrary-N elliptic case. Proceeding as in the hyperbolic case, we note that with the reparametrization (6.24) in effect, the $N=1$ calculations (6.37)–(6.43) can also be applied to the μ-dependent terms. Furthermore, the expansion (6.25) is still valid on the elliptic level. As a result, we obtain expansions of the form

$$D = 2N - 2Na_-\zeta + a_-^2 D_{\rm nr} + O(a_-^3), \qquad a_- \downarrow 0, \quad D = A, H, \mathcal{A}, \tag{6.59}$$

where

$$\zeta = \gamma + 2(N-1)rg. \tag{6.60}$$

In particular, we get

$$H_{\rm nr} = 2H_{\rm nr}^{(BC)} + \chi, \tag{6.61}$$

with $H_{\rm nr}^{(BC)}$ the Inozemtsev Hamiltonian (1.3) and χ a constant. For brevity, we omit the explicit form of χ and of $A_{\rm nr}$ and $\mathcal{A}_{\rm nr}$.

From the limit (2.17) we deduce

$$\begin{aligned}\lim_{a_-\downarrow 0} W(x) &= p\prod_{j=1}^N w_{\rm nr}(x_j)\cdot \prod_{1\le j<k\le N}[s_+(x_j-x_k)s_+(x_j+x_k)]^{2g} \\ &\equiv W_{\rm nr}(x), \qquad x\in G_N,\end{aligned} \tag{6.62}$$

with p a positive constant, $w_{\rm nr}(z)$ given by (6.54), and G_N by (4.34). The explicit form of the differential operator $A_{\rm nr}$ in (6.59) can also be calculated by using (6.61), (6.62) and (6.32).

The constraint (4.33) turns into

$$\sum_{t=0}^3 g_t + 2(N-1)g = 0. \tag{6.63}$$

With this restriction of the couplings, we obtain a factorized 'ground state' eigenfunction $W_{\rm nr}(x)^{1/2}$ of the Inozemtsev Hamiltonian (1.3). The associated functional equation follows from the $a_- \downarrow 0$ limit of Lemma 4.2.

We proceed with a remark about the expansions (6.59). Suppose we work instead of D with the renormalized AΔO

$$D_{\rm ren} = \rho_{\rm ren} D, \qquad \rho_{\rm ren} = \exp[r(c_0+c_2+c_4+c_6+2(N-1)\tau)]. \tag{6.64}$$

The multiplier $\rho_{\rm ren}$ destroys the D_8-invariance of H and $\mathcal{A}$, but it entails the simpler expansions

$$D_{\rm ren} = 2N + a_-^2(D_{\rm nr} - N\zeta^2) + O(a_-^3), \quad a_- \downarrow 0, \quad D = A, H, \mathcal{A}. \tag{6.65}$$

This observation is also relevant for the relation between our AΔO H (4.6) and the AΔO $\mathcal{H}_D$ introduced by Hikami and Komori [9]. Indeed, as we detail in Appendix B, these two operators are connected by a multiplicative constant that amounts to ρ_{ren}, cf. (B.15)–(B.16).

We conclude this section with additional comments on related literature. First, we note that Hikami and Komori [9] connected their versions of the elliptic AΔOs A and H by a similarity transformation that does not amount to our manifestly $(a_+ \leftrightarrow a_-)$-invariant transformation (4.7) involving elliptic gamma functions. The difference consists in multipliers that are elliptic with periods π/r and ia_-, so that there is no contradiction with our similarity transformation. But their elliptic multipliers do not have a limit for $a_- \downarrow 0$, so that their different weight function has no limit either.

Another item of special note is the connection of the nonrelativistic limits of the above (candidates for) ground states to a recent series of papers by Gómez-Ullate et al., cf. in particular [18, 19]. They are focusing on more general Calogero-Moser type systems, since their work pertains to particles with internal degrees of freedom. Even so, their results do admit a specialization to the context of this section, since one can restrict attention to 'scalar' wave functions with suitable symmetries. Doing so, one obtains in particular finite-dimensional invariant spaces for the Inozemtsev operator (1.3) (hence for H_{nr} (6.61)), provided the couplings are suitably restricted. In case the invariant subspace is one-dimensional, the restriction amounts to our restriction (6.63). At face value, the positive H_{nr}-eigenfunctions they obtain for this case look quite different from our H_{nr}-eigenfunctions $W_{\text{nr}}(x)^{1/2}$. But in fact it can be verified that one is dealing with distinct representations for the same functions.

Quite recently, the connection of the Spanish work to the Inozemtsev Hamiltonian was also pointed out by Takemura [29]. He shows more generally that the finite-dimensional invariant subspaces admit a bijective correspondence with subspaces of multi-dimensional theta functions that are invariant under the AΔO H (4.6). The existence of the latter subspaces (for restricted parameters) was previously shown by Komori [30], cf. also the earlier work by Hasegawa and collaborators [31–33]. Probably, the more general limiting transitions arising from Takemura's paper [29] can again be controlled by a suitable use of the elliptic gamma function.

Appendix A. Proofs of Lemmas 3.1, 3.2, 4.1 and 4.2

The four lemmas we prove in this appendix are closely related. Indeed, the $N = 1$ lemmas in Section 3 are partially implied by their arbitrary-N generalization in Section 4, and the hyperbolic lemmas can be partly understood as the $r \downarrow 0$ limit of their elliptic counterparts. Even so, it is illuminating to proceed 'by generalization', since we can obtain a little more information in the special cases.

Proof of Lemma 3.1. Denoting the function on the rhs of (3.19) by $F(z)$, we see that $V_b(z)$ (3.17) and $F(z)$ are both even and ia_+-periodic meromorphic functions with limit 0 for $|\mathrm{Re}\, z| \to \infty$, cf. (3.18). The poles in $V(\pm z)$ at $z = 0, ia_+/2$ do not occur in $F(z)$, but their residues have opposite sign. To prove equality of $F(z)$ and $V_b(z)$, it therefore remains to verify equality of residues at the (generically) simple poles $z = ia$ and $z = ia_-/2$. Using (3.15), this is routine. □

Proof of Lemma 3.2. The definitions (3.29) imply that the functions

$$F_t(\mu; z) \equiv \frac{\mathcal{E}_t(\mu; z) - \mathcal{E}_t(\mu; z_t)}{R_+(\mu - ia_+/2)R_+(\mu - ia_- - ia_+/2)}, \quad t = 0, \ldots, 3, \tag{A.1}$$

are not only elliptic in z, but also in μ. Calculating the residues for the (generically) simple poles at $\mu = 0$ and $\mu = ia_-$, we see they vanish. By ellipticity, it follows that $F_t(\mu; z)$ has no poles for $\mu \equiv 0$ and $\mu \equiv ia_-$, so that it is constant in μ. Hence $V_b(z)$ is μ-independent as well.

It remains to prove the last assertion. Due to (3.36), the function $\Sigma(z)$ is elliptic with periods $\pi/r, ia_+$. Since $\Sigma(z)$ and $R_+(z)$ are even and π/r-periodic, the residues at the simple poles $z \equiv 0, \omega_1$ cancel. Using (2.18) and (3.36), we also verify that $\Sigma(z)$ has no poles for $z \equiv \omega_2, \omega_3$.

Next, consider the residues at the simple poles $z \equiv ia_-/2 + \omega_t$ of (3.39). Using (2.18), it is straightforward to check that they cancel. By evenness in z, no poles are present for $z \equiv -ia_-/2 - \omega_t$ either. Hence (3.39) does not depend on z. □

Proof of Lemma 4.1. The functions $\Sigma(x)$ and $\mathcal{V}(x)$ are both ia_+-periodic and even in $x_1, \ldots, x_N$, and also S_N-invariant. Comparing poles in x_1 in a period strip, let us first note that the (generically simple) poles at $x_1 = \pm x_k$, which occur in the summands of $\Sigma(x)$, are not present in $\mathcal{V}(x)$. But it is easily checked that residues at these poles cancel in pairs, so that the poles do not occur in $\Sigma(x)$ either. Thus we are left with the poles due to $V(\pm x_1)$ in $\mathcal{V}_1(x) + \mathcal{V}_1(-x)$, and due to $Q_c(x_1)$ and $Q_s(x_1)$ in $\mathcal{V}(x)$.

More specifically, by evenness it suffices to consider the poles at $x_1 = ia$ and $x_1 = ia_-/2$. Just as for $N = 1$, it now follows from (3.15) that the residues have opposite signs. Thus the function $\Sigma(x) + \mathcal{V}(x)$ has no poles in x_1. Since it is ia_+-periodic and has finite limits for $|\mathrm{Re}\, x_1| \to \infty$, it does not depend on x_1. The first assertion now follows by permutation invariance.

To verify that the constant is given by (4.26), we need only take successively $x_1 \to \infty, \ldots, x_N \to \infty$. For $\mathcal{V}(x)$ this yields limit 0, and for $\Sigma(x)$ the constant on the rhs of (4.26). □

Proof of Lemma 4.2. Due to the restriction (4.33), not only $\mathcal{V}(x)$, but also $\Sigma(x)$ is elliptic in $x_1, \ldots, x_N$. Both functions are also even in $x_1, \ldots, x_N$ and permutation invariant. As in the hyperbolic case, the poles at $x_1 \equiv \pm x_k$ in the terms of $\Sigma(x)$ do not occur in $\Sigma(x)$ due to residue cancellations. To show independence of x_1, it is therefore enough to verify residue cancellation for the x_1-poles due to $V(\pm x_1)$ in $\mathcal{V}_1(x) + \mathcal{V}_1(-x)$, and due to $\mathcal{E}_t(x_1), t = 0, \ldots, 3$, in $\mathcal{V}(x)$.

To this purpose, the previous $N = 1$ calculations in the proof of Lemma 3.2 can be used. To be specific, 'unwanted' $2\omega_2$-shifts in the R_+-arguments arising in the products over $k > 1$ in $\mathcal{V}_1(\pm x)$ can be removed via the R-AΔE (2.18). This yields extra terms (compared to $N = 1$) of the form $\exp(\pm 4ir(N-1)\mu)$, which cancel due to the restriction (4.33). □

Appendix B. Comparison with Refs. [8,9]

The symmetry under interchange of a_+ and a_-, which is built into our definitions, is not visible in the papers by van Diejen [8] and Komori/Hikami [9]. To ease our notation and for comparison purposes, we abandon our symmetric formulation in this appendix. Specifically, we change a_+ to a and R_+, s_+, to R, s, and we follow van Diejen by using the parameter

$$\gamma \equiv ia_-/2. \tag{B.1}$$

(This parameter is denoted by β in [9].)

As we shall now detail first, the analog of van Diejen's external field $w(z)$ is our field $V(-z)$, up to a certain factor. Indeed, $w(z)$ is defined by (cf. Eq. (3.20) in [8])

$$w(z) = \prod_{t=0}^{3} \frac{\sigma_t(z+\mu_t)\sigma_t(z+\gamma+\mu_t')}{\sigma_t(z)\sigma_t(z+\gamma)}. \tag{B.2}$$

Using now successively (2.28), (2.25), (2.19) and (2.18), we obtain

$$\begin{aligned} w(z) &= \exp(C_D + K_D z)[R(2z-\omega_2)R(2z+2\gamma-\omega_2)]^{-1} \\ &\quad \times R(z+\mu_0-\omega_2)R(z+\mu_2)R(z+\mu_0'+\gamma-\omega_2) \\ &\quad \times R(z+\mu_2'+\gamma)R(z+\mu_1-\omega_1-\omega_2)R(z+\mu_3-\omega_1) \\ &\quad \times R(z+\mu_1'+\gamma-\omega_1-\omega_2)R(z+\mu_3'+\gamma-\omega_1), \end{aligned} \tag{B.3}$$

$$\begin{aligned} C_D &\equiv \eta_1 r\pi^{-1} \sum_t (\mu_t^2 + \mu_t'^2 + 2[\mu_t\omega_t + \mu_t'(\omega_t+\gamma)]) \\ &\quad +(\mu_0+\mu_0')(-ir) + (\mu_1+\mu_1')(-ir-\eta_1) \\ &\quad +(\mu_2+\mu_2')(-ir-\eta_2) + (\mu_3+\mu_3')(ir-\eta_3), \end{aligned} \tag{B.4}$$

$$K_D \equiv 2\eta_1 r\pi^{-1}\sum_t(\mu_t+\mu_t'). \tag{B.5}$$

Comparing to $V(z)$ (3.26), we see that this is of the form $\exp(C_D + K_D z)V(-z)$. Since van Diejen requires $\sum(\mu_t+\mu_t') = 0$ (cf. Eq. (3.21) in [8]), his field $w(z)$ is proportional to $V(-z)$. We have written the R-factors in (B.3) in a form that makes clear how μ_t and μ_t' are related to our couplings $\mathbf{c}$, given by (3.21) and (3.43). Specifically, we get

$$\begin{aligned}&\mu_0 = ic_0, \quad \mu_2 = ic_1, \quad \mu_0' = ic_2, \quad \mu_2' = ic_3,\\ &\mu_1 = ic_4, \quad \mu_3 = ic_5, \quad \mu_1' = ic_6, \quad \mu_3' = ic_7.\end{aligned} \tag{B.6}$$

Notice that this entails (cf. also (3.45))

$$\sum_{t=0}^{3}(\mu_t+\mu_t') = 0 \Leftrightarrow \sum_{n=0}^{7} c_n = 0. \tag{B.7}$$

It is now straightforward (albeit very laborious) to verify that the Hamiltonian given by Eqs. (4.1)–(4.3) in [8] is of the form $C_1(A + C_2)$, where C_1, C_2 are constants and A is defined by (4.7). (Observe that van Diejen's parameter μ equals ours.) We point out that the need for omitting the factor $\exp(K_D z)$ in (B.3) in order to obtain an integrable $N > 1$ system *without* the constraint $\sum(\mu_t + \mu_t') = 0$ would have been hard to guess with the bare-handed approach of [8].

The integrable generalization to arbitrary couplings (obtained in [9] via elaborate Yang-Baxter machinery) leads to arbitrary-N Hamiltonians that are related by additive and multiplicative constant factors to the ones of Subsection 4.2. Again, this is not at all clear by inspection. Furthermore, as it turns out, the pertinent constants are not even S_8-invariant. The detailed comparison now follows.

With a slight adaptation of notation, the BC_N Hamiltonian arrived at in [9] reads

$$\mathcal{H}_D = \sum_{j=1}^{N}(\Psi_j(x)^{1/2}\exp(2\gamma\partial_{x_j})\Psi_j(-x)^{1/2} + (x \to -x)) + \Psi_0(x), \tag{B.8}$$

with

$$\begin{aligned}\Psi_j(x) \;=\; &\left(\prod_{t=0}^{3}\frac{\theta_{t+1}(x_j-\nu_t)}{\theta_{t+1}(x_j)}\frac{\theta_{t+1}(x_j+\gamma-\overline{\nu}_t)}{\theta_{t+1}(x_j+\gamma)}\right)\\ &\times\left(\prod_{k\neq j}\frac{\theta_1(x_j-x_k+\mu)\theta_1(x_j+x_k+\mu)}{\theta_1(x_j-x_k)\theta_1(x_j+x_k)}\right),\end{aligned} \tag{B.9}$$

$$\Psi_0(x) = \sum_{t=0}^{3} \left(\frac{\pi}{\theta_1'(0)}\right)^2 \frac{2}{\theta_1(\mu)\theta_1(\mu-2\gamma)} \prod_{u=0}^{3} \theta_{u+1}(\gamma+\nu_{\pi_t u})\theta_{u+1}(\overline{\nu}_{\pi_t u})$$
$$\cdot \prod_{j=1}^{N} \frac{\theta_{t+1}(x_j-\gamma+\mu)}{\theta_{t+1}(x_j-\gamma)} \frac{\theta_{t+1}(x_j+\gamma-\mu)}{\theta_{t+1}(x_j+\gamma)}, \qquad \text{(B.10)}$$

$$\pi_0 = \text{id}, \quad \pi_1 = (01)(23), \quad \pi_2 = (02)(13), \quad \pi_3 = (03)(12). \qquad \text{(B.11)}$$

(We have changed β, μ, z in Eqs. (1.2)–(1.3) of [9] to $\gamma, -\mu, x$, and corrected the overall sign of $\Psi_0(x)$, cf. also Komori's later paper [30].)

The connection of $\mathcal{H}_D$ to H (4.6) is as follows. To start with, our parameter r should be chosen equal to π, cf. (2.30). Next, the parameters ν_t and $\overline{\nu}_t$ are related to our couplings $\mathbf{c}$ (given by (3.21) and (3.43)) via

$$\nu_0 = -ic_0, \quad \nu_3 = -ic_1, \quad \overline{\nu}_0 = -ic_2, \quad \overline{\nu}_3 = -ic_3,$$
$$\nu_1 = -ic_4, \quad \nu_2 = -ic_5, \quad \overline{\nu}_1 = -ic_6, \quad \overline{\nu}_2 = -ic_7. \qquad \text{(B.12)}$$

With this correspondence in effect, we have

$$\Psi_j(x) = \exp(\pi C_K)\mathcal{V}_j(-x), \qquad j = 1, \ldots, N, \qquad \text{(B.13)}$$

$$\Psi_0(x) = \exp(\pi C_K)\Big[\mathcal{V}(x) + \sum_{t=0}^{3} p_t \mathcal{E}_t(\mu; z_t)^N / 2R(\mu-\omega_2)R(\mu-2\gamma-\omega_2)\Big], \qquad \text{(B.14)}$$

$$C_K \equiv -2i(N-1)\mu + c_0 + c_2 + c_4 + c_6, \qquad \text{(B.15)}$$

so that

$$\mathcal{H}_D = \exp(\pi C_K)[H + \sum_{t=0}^{3} p_t \mathcal{E}_t(\mu; z_t)^N / 2R(\mu-\omega_2)R(\mu-2\gamma-\omega_2)]. \qquad \text{(B.16)}$$

Therefore, $\mathcal{H}_D$ and H are related by multiplicative and additive constants that break most of the parameter symmetries discussed above.

We conclude this appendix with a few remarks on the verification of (B.13) and (B.14). First, (2.32) can be used to trade θ_1 for s, and then (2.36) to trade s for R. Using (2.34) and (2.35), the remaining theta-functions can also be expressed in terms of R. Then it is not hard to check (B.13). The verification of (B.14) is quite arduous, but straightforward; here, the relation (2.33) should be used as well.

Acknowledgment

We would like to thank Y. Komori for illuminating correspondence.

References

1. Ruijsenaars, S. N. M.: Factorized weight functions vs. factorized scattering, Comm. Math. Phys. **228**, 467–494 (2002)
2. Olshanetsky, M. A., Perelomov, A. M.: Quantum integrable systems related to Lie algebras, Phys. Reps. **94**, 313–404 (1983)
3. Ruijsenaars, S. N. M.: Systems of Calogero-Moser type, in *Proceedings of the 1994 Banff summer school Particles and fields*, CRM Ser. in Math. Phys., (Semenoff, G., Vinet, L., Eds.), pp. 251–352, New York: Springer (1999)
4. Inozemtsev, V. I.: Lax representation with spectral parameter on a torus for integrable particle systems, Lett. Math. Phys. **17**, 11–17 (1989)
5. Oshima, T., Sekiguchi, H.: Commuting families of differential operators invariant under the action of a Weyl group, J. Math. Sci. Univ. Tokyo **2**, 1–75 (1995)
6. Oshima, T.: Completely integrable systems with a symmetry in coordinates, Asian J. Math. **2**, 935–955 (1998)
7. Ruijsenaars, S. N. M.: Complete integrability of relativistic Calogero-Moser systems and elliptic function identities, Comm. Math. Phys. **110**, 191–213 (1987)
8. van Diejen, J. F.: Integrability of difference Calogero-Moser systems, J. Math. Phys. **35**, 2983–3004 (1994)
9. Komori, Y., Hikami, K.: Quantum integrability of the generalized elliptic Ruijsenaars models, J. Phys. A: Math. Gen. **30**, 4341–4364 (1997)
10. van Diejen, J. F.: Commuting difference operators with polynomial eigenfunctions, Comp. Math. **95**, 183–233 (1995)
11. Koornwinder, T. H.: Askey-Wilson polynomials for root systems of type BC, Contemp. Math. **138**, 189–204 (1992)
12. Askey, R., Wilson, J.: Some basic hypergeometric orthogonal polynomials that generalize Jacobi polynomials, Mem. Am. Math. Soc. **319**, (1985)
13. Gasper, G., Rahman, M.: Basic hypergeometric series, *Encyclopedia of Mathematics and its Applications*, **35**, Cambridge: Cambridge Univ. Press (1990)
14. Ruijsenaars, S. N. M.: First order analytic difference equations and integrable quantum systems, J. Math. Phys. **38**, 1069–1146 (1997)
15. Ruijsenaars, S. N. M.: A generalized hypergeometric function satisfying four analytic difference equations of Askey-Wilson type, Commun. Math. Phys. **206**, 639–690 (1999)
16. Ruijsenaars, S. N. M.: A generalized hypergeometric function II. Asymptotics and D_4 symmetry, to appear
17. Ruijsenaars, S. N. M.: A generalized hypergeometric function III. Associated Hilbert space transforms, to appear
18. Finkel, F., Gómez-Ullate, D., González-López, A., Rodríguez, M. A., Zhdanov, R.: New spin Calogero-Sutherland models related to B_N-type Dunkl operators, Nucl. Phys. **B 613**, 472–496 (2001)
19. Gómez-Ullate, D., González-López, A., Rodríguez, M. A.: Exact solutions of a new elliptic Calogero-Sutherland model, Phys. Lett. **B 511**, 112–118 (2001)
20. Whittaker, E. T., Watson, G. N.: A course of modern analysis, Cambridge: Cambridge Univ. Press (1973)
21. Ruijsenaars, S. N. M.: Special functions defined by analytic difference equations, in *Proceedings of the Tempe NATO Advanced Study Institute "Special Functions 2000"*, NATO Science Series Vol. 30, (Bustoz, J., Ismail, M., Suslov, S., Eds.), pp. 281–333, Dordrecht: Kluwer (2001)

22. Ruijsenaars, S. N. M.: Relativistic Lamé functions revisited, J. Phys. A: Math. Gen. **34**, 1–18 (2001)
23. Ruijsenaars, S. N. M.: Self-adjoint AΔOs with vanishing reflection, Theor. Math. Phys. **128**, 933–945 (2001)
24. Ruijsenaars, S. N. M.: Hilbert space theory for reflectionless relativistic potentials, Publ. RIMS Kyoto Univ. **36**, 707–753 (2000)
25. Ruijsenaars, S. N. M.: Generalized Lamé functions. I. The elliptic case, J. Math. Phys. **40**, 1595–1626 (1999)
26. Chalykh, O. A.: Macdonald polynomials and algebraic integrability, Adv. in Math. **166**, 193–259 (2002)
27. van Diejen, J. F.: Self-dual Koornwinder-Macdonald polynomials, Invent. Math. **126**, 319–339 (1996)
28. Gradshteyn, I. S., Ryzhik, I. M.: Table of integrals, series, and products, New York: Academic Press (1965)
29. Takemura, K.: Quasi-exact solvability of Inozemtsev models, J. Phys. A: Math. Gen. **35**, 8867–8881 (2002)
30. Komori, Y.: Ruijsenaars' commuting difference operators and invariant subspace spanned by theta functions, J. Math. Phys. **42**, 4503–4522 (2001)
31. Hasegawa, K.: Ruijsenaars' commuting difference operators as commuting transfer matrices, Comm. Math. Phys. **187**, 289–325 (1997)
32. Hasegawa, K., Ikeda, T, Kikuchi, T.: Commuting difference operators arising from the elliptic $C_2^{(1)}$-face model, J. Math. Phys. **40**, 4549–4568 (1999)
33. Kikuchi, T.: Diagonalization of the elliptic Macdonald-Ruijsenaars difference system of type C_2, J. Phys. A: Math. Gen. **34**, 5389–5401 (2001)

ON THE PROLONGATION OF A HIERARCHY OF HYDRODYNAMIC CHAINS

A.B. SHABAT
Landau Institute for Theoretical Physics, RAS, Moscow 117 334, Russia

L. MARTÍNEZ ALONSO
Departamento de Física Teórica II, Universidad Complutense E28040 Madrid, Spain

Abstract. The prolongation of a hierarchy of hydrodynamic chains previously studied by the authors is presented and the properties of the differential reductions of the enlarged hierarchy are derived. Several associated nonlinear integrable models are exhibited. In particular, it is found that the Camassa–Holm equation can be described as a second-order differential reduction of one of the new flows included in the hierarchy.

1. Introduction

In our previous work [1], [2] we considered a hierarchy of hydrodynamic chains arising in the theory of energy-dependent Schrödinger spectral problems [3], [8]. The methods we developed to analyze the properties and solutions of this hierarchy are rather different from those used in the standard Lax approach to the theory of integrable systems and revealed not only a rich structure but also the presence of new interesting integrable nonlinear models.

An abstract formulation of this hierarchy is provided by the system of equations

$$\Big[D_n - A_n(\lambda, \mathbf{x})D_0 \, , \, D_m - A_m(\lambda, \mathbf{x})D_0\Big] = 0, \quad n, m \geq 0, \tag{1.1}$$

where $\{D_n\}_{n\geq 0}$ denote the differentiation operators with respect to the infinite set of independent variables $\mathbf{x} := (x_0, x_1, x_2, \ldots)$ and λ is an *spectral* parameter. Equivalently, (1.1) can be written as

$$D_n A_m - D_m A_n = \langle A_n, A_m \rangle, \quad n, m \geq 1, \tag{1.2}$$

A.B. Shabat et al. (eds.), New Trends in Integrability and Partial Solvability, 263–280.

where

$$\langle U, V\rangle := U(D_0V) - (D_0U)V.$$

If we assume that the functions A_n are n-th degree monic polynomials in λ then their coefficients are interrelated through (1.2) and by using induction it can be proved that, up to unessential terms and without loss of generality, the functions A_n are determined by an infinite set of common coefficients $\{g_m(\mathbf{x})\}_{m\geq 1}$ in the form

$$A_n = \lambda^n + g_1(\mathbf{x})\lambda^{n-1} + \cdots + g_{n-1}(\mathbf{x})\lambda + g_n(\mathbf{x}). \tag{1.3}$$

Thus (1.1) defines the action of the derivations $\{D_n\}_{n\geq 1}$ on the set of dynamical variables $\{g_n(\mathbf{x})\}_{n\geq 1}$. This action is given by the system of flows

$$D_nG = \langle A_n, G\rangle, \quad n \geq 1, \tag{1.4}$$

where $G = G(\lambda, \mathbf{x})$ is the generating function

$$G = 1 + \frac{g_1(\mathbf{x})}{\lambda} + \frac{g_2(\mathbf{x})}{\lambda^2} + \cdots, \quad \lambda \to \infty. \tag{1.5}$$

In fact (1.1) is the system of compatibility conditions of (1.4). By identifying coefficients of powers of λ in (1.4) one gets the hierarchy of hydrodynamic chains

$$D_i g_n = \sum_{k=1}^{n} \langle g_{n-k}, g_{i+k}\rangle, \quad n \geq 1. \tag{1.6}$$

An alternative and useful formulation of (1.4) is obtained by introducing the generating function

$$H := \frac{1}{G} = 1 + \frac{h_1}{\lambda} + \frac{h_2}{\lambda^2} + \cdots, \quad \lambda \to \infty, \tag{1.7}$$

$$h_1 = -g_1, \quad h_2 = g_1^2 - g_2, \quad h_3 = -g_1^3 + 2g_1g_2 - g_3, \quad \ldots. \tag{1.8}$$

Thus (1.4) is equivalent to

$$D_nH = D_0\Big(A_nH\Big), \tag{1.9}$$

which means, in particular, that the coefficients $\{h_n(\mathbf{x})\}_{n\geq 1}$ supply an infinite set of conservation laws for (1.4).

A basic feature of (1.1) is its invariance under the replacement of the basic set of derivation operators $\{D_n\}_{n\geq 0}$ by

$$D_{n,1} := D_n - g_n(\mathbf{x})D_0, \quad n \geq 1. \tag{1.10}$$

Indeed, from (1.1) one deduces (see [1], [2]) that $\{D_{n,1}\}_{n\geq 1}$ forms a commutative set of operators verifying

$$\Big[D_{n,1} - A_{n-1}(\lambda, \mathbf{x})D_{1,1}\ ,\ D_{m,1} - A_{m-1}(\lambda, \mathbf{x})D_{1,1}\Big] = 0, \qquad n, m \geq 1\,. \tag{1.11}$$

Obviously, the process can be iterated to conclude that for all $p \geq 1$

$$\Big[D_{n,p} - A_{n-p}(\lambda, \mathbf{x})D_{p,p}\ ,\ D_{m,p} - A_{m-p}(\lambda, \mathbf{x})D_{p,p}\Big] = 0, \qquad n, m \geq p, \tag{1.12}$$

where

$$D_{n,p} := D_{n,p-1} - g_{n-p+1}(\mathbf{x})D_{p-1,p-1}, \quad n \geq p.$$

The goal of the present work is to enlarge the system of equations (1.1) by introducing a set of new independent variables $\{x_{-n}\}_{n\geq 1}$, its corresponding set $\{D_{-n}\}_{n\geq 1}$ of differentiation operators and a family $\{A_{-n}(\lambda, \mathbf{x})\}_{n\geq 1}$ of polynomials in λ^{-1} which depend on

$$\mathbf{x} := (\ldots, x_{-1}, x_0, x_1, \ldots).$$

It can be proved that the functions $\{A_{-n}\}_{n\geq 1}$, as well as the functions $\{A_n\}_{n\geq 1}$, are determined by an infinite set of common coefficients $\{b_n(\mathbf{x})\}_{n\geq 0}$ in the form

$$A_{-n} := \frac{b_0(\mathbf{x})}{\lambda^n} + \frac{b_1(\mathbf{x})}{\lambda^{n-1}} + \cdots + \frac{b_{n-1}(\mathbf{x})}{\lambda}. \tag{1.13}$$

Here instead of the monic condition for $\{A_n\}_{n\geq 1}$ we impose the vanishing of the constant term in the decomposition of $\{A_{-n}\}_{n\geq 1}$ in powers of λ^{-1}. The main point is that the dependence of the dynamical variables $\{g_n(\mathbf{x})\}_{n\geq 1} \bigcup \{b_n(\mathbf{x})\}_{n\geq 0}$ on $\mathbf{x}$ can be described in terms of a single generating function satisfying

$$D_n G = \langle A_n, G\rangle, \qquad n \in \mathbb{Z}, \tag{1.14}$$

as well as

$$\begin{aligned} G &= 1 + \frac{g_1(\mathbf{x})}{\lambda} + \frac{g_2(\mathbf{x})}{\lambda^2} + \cdots, \quad \lambda \to \infty, \\ G &= b_0(\mathbf{x}) + b_1(\mathbf{x})\lambda + b_2(\mathbf{x})\lambda^2 + \cdots, \quad \lambda \to 0. \end{aligned} \tag{1.15}$$

Nevertheless, we will treat the dynamical variables

$$\{g_n(\mathbf{x})\}_{n\geq 1} \bigcup \{b_n(\mathbf{x})\}_{n\geq 0}$$

as independent. Equivalently, in terms of H, the system (1.14) reads

$$D_n H = D_0\Big(A_n H\Big), \quad n \in \mathbb{Z}. \tag{1.16}$$

The paper is organized as follows: Section 2 is devoted to the introduction of several closed differential forms associated with solutions of our enlarged hierarchy. It is also analyzed how these forms are related to interesting symmetry transformations of the hierarchy. Section 3 presents several applications arising from the consideration of the new flows included in (1.14). In particular two 3-dimensional nonlinear models are exhibited. Furthermore, several 2-dimensional nonlinear models related to reductions of (1.14), including the Camassa–Holm equation, are discussed.

The following notation conventions are used in this paper. Firstly, G_∞ and G_0 stand for the expansions of G as $\lambda \to \infty$ and $\lambda \to 0$, respectively. The flows associated to the variables $\{x_n\}_{n\geq 1}$ and $\{x_{-n}\}_{n\geq 1}$ are referred to as the *positive* and *negative* flows, respectively. We notice that the negative flows of G_0 describe the Schwarzian form of the hierarchy (1.4) (see [2]), so that our enlarged hierarchy contains the two variants of (1.4) considered in [2].

Furthermore, we denote by $\mathbb{V}$ the space of formal Laurent series

$$V = \sum_{n=-\infty}^{\infty} a_n \lambda^n,$$

by $\mathbb{V}_{r,s}$ $(r \leq s)$ the subspaces of elements

$$V = \sum_{n=r}^{s} a_n \lambda^n,$$

and by $P_{r,s} : \mathbb{V} \mapsto \mathbb{V}_{r,s}$ the corresponding projectors. Given $V \in \mathbb{V}$ we will also denote

$$\Big(V\Big)_{r,s} := P_{r,s}(V).$$

In particular, notice that we can write

$$A_n = \Big(\lambda^n G_\infty\Big)_{0,+\infty}, \quad A_{-n} = \Big(\lambda^{-n} G_0\Big)_{-\infty,-1}, \quad n \geq 1.$$

2. Hodograph transformations and differential forms

The enlarged hierarchy of flows (1.14) forms a compatible system. That is to say, as a consequence of (1.14) the following consistency conditions follow

$$D_n A_m - D_m A_n = \langle A_n, A_m \rangle, \quad n, m \in \mathbb{Z}, \tag{2.1}$$

where $\{A_n\}_{n\in\mathbb{Z}}$ are given by (1.3) and (1.13). For example let us prove the compatibility between positive and negative flows. From (1.3), (1.13) and

(1.14) we deduce

$$
\begin{aligned}
D_{-m}A_n - D_nA_{-m} &= \left(\lambda^n D_{-m}G_\infty\right)_{0,+\infty} - \left(\lambda^{-m} D_n G_0\right)_{-\infty,-1} \\
&= \left(\langle A_{-m}, \lambda^n G_\infty\rangle\right)_{0,+\infty} - \left(\langle A_n, \lambda^{-m} G_0\rangle\right)_{-\infty,-1} \\
&= \left(\langle A_{-m}, A_n\rangle\right)_{0,+\infty} - \left(\langle A_n, A_{-m}\rangle\right)_{-\infty,-1} \\
&= \langle A_{-m}, A_n\rangle.
\end{aligned}
$$

The equations (2.1) represent a generalization of (1.1). Indeed they are equivalent to the system

$$
\Big[D_n - A_n(\lambda, \mathbf{x})D_0\ ,\ D_m - A_m(\lambda, \mathbf{x})D_0\Big] = 0, \quad n, m \in \mathbb{Z}. \tag{2.2}
$$

In particular this means that we can extend the abelian set of first-order differential operators (1.10) as the next theorem states.

Theorem 2.1 *The family of operators*

$$
\begin{aligned}
D_{n,1} &:= D_n - g_n(\mathbf{x})D_0, \quad n \geq 1, \\
D_{0,1} &:= b_0 D_0, \\
D_{-n,1} &:= D_{-n} + b_n(\mathbf{x})D_0, \quad n \geq 1,
\end{aligned} \tag{2.3}
$$

forms a commutative set

$$
[D_{r,1}, D_{s,1}] = 0, \quad r, s \in \mathbb{Z}. \tag{2.4}
$$

Proof. To prove (2.4) one may proceed as follows

1. For $r, s \geq 1$ it is enough to set $n = r, m = s$ and $\lambda = 0$ in (2.2)
2. For $r \geq 1$ and $s = 0$ we set to zero the coefficient of $1/\lambda$ in (2.2) with $n = r$ and $m = -1$.
3. For $r \geq 1$ and $s \leq -1$ we equate to zero the coefficient of $1/\lambda$ in (2.2) with $n = r, m = s - 1$ and find

$$
[D_r - g_r D_0, -b_{-s}D_0] + \sum_I [g_i D_0, b_j D_0] = 0,
$$

where

$$
I = \{(i, j) : i = j + r + s,\ 0 \leq i \leq r,\ 0 \leq j \leq -s - 1\}.
$$

Furthermore, by setting to zero the coefficient of λ^0 in (2.2) with $n = r$ and $m = s$ we get

$$
[D_r - g_r D_0, D_s] + \sum_I [g_i D_0, b_j D_0] = 0,
$$

so that

$$[D_r - g_r D_0, D_s + b_{-s} D_0] = 0.$$

4. Finally, for $r, s \leq -1$, equating to zero of the coefficient of $1/\lambda$ in (2.2) with $n = r, m = s - 1$ yields

$$[D_r, b_{-s} D_0] + [b_{r-1} D_0, D_{s-1}] = 0,$$

and by setting to zero the coefficient of $1/\lambda^2$ in (2.2) with

$$n = r - 1, \quad m = s - 1,$$

we deduce

$$[b_{-r} D_0, b_{-s} D_0] - [D_{r-1}, b_{-s-1} D_0] + [D_{s-1}, b_{-r-1} D_0] = 0.$$

Thus it follows at once that

$$[D_r + b_{-r} D_0, D_s + b_{-s} D_0] = 0.$$

As a consequence of this theorem we deduce

Corollary 2.2 *There is a change of variables*

$$\mathbf{y} = (\ldots, y_{-1}, y_0, y_1, \ldots) \mapsto \mathbf{x} = (\ldots, x_{-1}, x_0, x_1, \ldots),$$

determined by

$$x_0 = x_0(\mathbf{y}), \; x_n = y_n, \quad n \in \mathbb{Z} - \{0\}, \tag{2.5}$$

where

$$\begin{aligned} \omega_1 &:= \mathrm{d}x_0 = b_0 \mathrm{d}y_0 + \sum_{n\geq 1} (b_n \mathrm{d}y_{-n} - g_n \mathrm{d}y_n), \\ \hat{\omega}_1 &:= \mathrm{d}y_0 = \frac{1}{b_0}\Big(\mathrm{d}x_0 + \sum_{n\geq 1} (g_n \mathrm{d}x_n - b_n \mathrm{d}x_{-n})\Big). \end{aligned} \tag{2.6}$$

Proof. Theorem 1 implies the existence of variables $(y_n)_{n\in\mathbb{Z}}$ such that

$$\frac{\partial}{\partial y_n} = D_{n,1}, \quad n \in \mathbb{Z}. \tag{2.7}$$

Hence, by taking (2.3) into account, the statement follows at once.

We notice that (2.5) is a transformation of hodograph type as it involves dependent and independent variables. The next result shows how this transformation is related to the main symmetry property of our hierarchy.

Theorem 2.3 *Given a solution $G(\lambda, x)$ of (1.14), then it verifies*

$$D_{n,1}G = \langle A_{n-1}, G\rangle_1 \quad n \in \mathbb{Z}, \tag{2.8}$$

where

$$\langle U, V\rangle_1 := U(D_{1,1}V) - (D_{1,1}U)V.$$

Proof. For $n \geq 0$ the statement of this theorem was proved in [1], so that we concentrate on the negative flows $D_{-n,1}$, $n \geq 1$. Firstly, notice that

$$\begin{aligned} D_{-n,1}G &= D_{-n}G + b_n(D_0G) = \langle A_{-n}, G\rangle + b_n(D_0G) \\ &= \lambda A_{-n-1}(D_0G) - (D_0A_{-n})G. \end{aligned} \tag{2.9}$$

On the other hand

$$D_{1,1}G = D_1G - g_1(D_0G) = \lambda(D_0G) - (D_0g_1)G, \tag{2.10}$$

so that

$$\begin{aligned} D_0A_{-n} &= D_0\Big(\lambda^{-n}G\Big)_{-\infty,-1} = \Big(\frac{1}{\lambda^{n+1}}(D_{1,1}G + (D_0g_1)G)\Big)_{-\infty,-1} \\ &= D_{1,1}A_{-n-1} + (D_0g_1)A_{-n-1}. \end{aligned}$$

Hence, by substituting in (2.9) we deduce

$$\begin{aligned} D_{-n,1}G &= A_{-n-1}(D_{1,1}G + (D_0g_1)G) \\ &\quad -(D_{1,1}A_{-n-1} + (D_0g_1)A_{-n-1})G = \langle A_{-n-1}, G\rangle_1. \end{aligned}$$

In view of this result we may apply Theorem 1 and its corollary to the system (2.8) and conclude the existence of a further hodograph type transformation.

Corollary 2.4 *There is a change of variables*

$$\mathbf{z} = (\ldots, z_{-1}, z_0, z_1, \ldots) \mapsto \mathbf{y} = (\ldots, y_{-1}, y_0, y_1, \ldots),$$

determined by

$$y_1 = y_1(\mathbf{z}),\ y_n = z_n, \quad n \in \mathbb{Z} - \{1\}, \tag{2.11}$$

where

$$\begin{aligned} \omega_2 &:= \mathrm{d}y_1 = b_0\mathrm{d}z_1 + \sum_{n\geq 1}(b_n\mathrm{d}z_{-n+1} - g_n\mathrm{d}z_{n+1}), \\ \hat{\omega}_2 &:= \mathrm{d}z_1 = \frac{1}{b_0}(\mathrm{d}y_1 + \sum_{n\geq 1}(g_n\mathrm{d}y_{n+1} - b_n\mathrm{d}y_{-n+1}) \end{aligned}$$

Furthermore

$$D_{n,2}G = \langle A_{n-2}, G\rangle_2 \quad n \in \mathbb{Z}, \tag{2.12}$$

where

$$D_{n,2} = \frac{\partial}{\partial z_n}, \quad \langle U, V\rangle_1 := U(D_{2,2}V) - (D_{2,2}U)V.$$

Of course the process used in the above pair of theorems can be iterated to generate an infinite chain of hodograph transformations which define symmetries of (1.14).

According to the statement of Corollary 1 to each solution of (1.14) corresponds a closed differential form given by

$$\hat{\omega}_1 := \mathrm{d}y_0 = \frac{1}{b_0}\Big(\mathrm{d}x_0 + \sum_{n\geq 1}(g_n \mathrm{d}x_n - b_n \mathrm{d}x_{-n})\Big).$$

The next theorem provides another example of this type of correspondence

Theorem 2.5 *The differential form*

$$\omega := b_0 \mathrm{d}x_{-1} + \sum_{n\geq 1}(b_n \mathrm{d}x_{-n-1} - g_n \mathrm{d}x_{n-1}), \tag{2.13}$$

is closed.

Proof. We first notice that

$$b_n = \mathrm{Res}\Big(\frac{1}{\lambda^{n+1}}G_0\Big), \quad g_n = \mathrm{Res}\Big(\lambda^{n-1}G_\infty\Big).$$

In order to prove the theorem we must verify that

$$\partial_{m-1}g_n = \partial_{n-1}g_m, \quad n, m \geq 1, \tag{2.14}$$

$$\partial_{-m-1}b_n = \partial_{-n-1}b_m, \quad n, m \geq 0, \tag{2.15}$$

$$\partial_{m-1}b_n = -\partial_{-n-1}g_m, \quad n \geq 0, m \geq 1. \tag{2.16}$$

Let us start with (2.14), we have

$$\begin{aligned}
\partial_{m-1}g_n - \partial_{n-1}g_m = {}& \\
&\mathrm{Res}\Big(\langle(\lambda^{m-1}G_\infty)_{0,\infty}, \lambda^{n-1}G_\infty\rangle \\
&\quad -\langle(\lambda^{n-1}G_\infty)_{0,\infty}, \lambda^{m-1}G_\infty\rangle\Big) = \\
&\mathrm{Res}\Big(\langle\lambda^{m-1}G_\infty, (\lambda^{n-1}G_\infty)_{-\infty,-1}\rangle \\
&\quad -\langle(\lambda^{n-1}G_\infty)_{0,\infty}, \lambda^{m-1}G_\infty\rangle\Big) = \\
&\mathrm{Res}\Big(\langle\lambda^{m-1}G_\infty, \lambda^{n-1}G_\infty\rangle\Big) = 0
\end{aligned}$$

Similarly, (2.15) can be proved as follows

$$\begin{aligned}\partial_{-m-1}b_n - \partial_{-n-1}b_m \; = & \\ & \operatorname{Res}\Big(\langle(\lambda^{-m-1}G_0)_{-\infty,-1}\,,\lambda^{-n-1}G_0\rangle = \\ & \quad -\langle(\lambda^{-n-1}G_0)_{-\infty,-1}\,,\lambda^{-m-1}G_0\rangle\Big) = \\ & \operatorname{Res}\Big(\langle\lambda^{-m-1}G_0\,,(\lambda^{-n-1}G_0)_{0,\infty}\rangle \\ & \quad -\langle(\lambda^{-n-1}G_0)_{-\infty,-1}\,,\lambda^{-m-1}G_0\rangle\Big) = \\ & \operatorname{Res}\Big(\langle\lambda^{-m-1}G_0,\lambda^{-n-1}G_0\rangle\Big) = 0.\end{aligned}$$

Finally, (2.16) is proved in the form

$$\begin{aligned}\partial_{m-1}b_n + \partial_{-n-1}g_m \; = & \\ & \operatorname{Res}\Big(\langle(\lambda^{m-1}G_\infty)_{0,\infty}\,,\lambda^{-n-1}G_0\rangle \\ & \quad +\langle(\lambda^{-n-1}G_0)_{-\infty,-1}\,,\lambda^{m-1}G_\infty\rangle\Big) = \\ & \operatorname{Res}\Big(\langle(\lambda^{m-1}G_\infty)_{0,\infty}\,,(\lambda^{-n-1}G_0)_{-\infty,-1}\rangle \\ & \quad +\langle(\lambda^{-n-1}G_0)_{-\infty,-1}\,,(\lambda^{m-1}G_\infty)_{0,\infty}\rangle\Big) = 0.\end{aligned}$$

3. Applications

3.1. 3-DIMENSIONAL INTEGRABLE MODELS

The differential forms introduced in the above section reveal the presence of interesting 3-dimensional integrable models in the hierarchy (1.14) as the next examples show.

Example 1

The x_1-flow of (1.14) implies at once that

$$D_1 g_1 = D_0 g_2.$$

Moreover, from (2.5) and (2.6) we have that the potential function

$$y_0 = Q(\mathbf{x}),$$

satisfies

$$g_1 = \frac{D_1 Q}{D_0 Q}, \quad g_2 = \frac{D_2 Q}{D_0 Q},$$

and therefore, if we denote $x := x_0, y := x_2$ and $z := x_1$ we get

$$\left(\frac{Q_z}{Q_x}\right)_z = \left(\frac{Q_y}{Q_x}\right)_x. \tag{3.1}$$

Example 2

By identifying the coefficient of λ^0 in the equation for the first negative flow

$$D_{-1}G = \langle\frac{b_0}{\lambda}, G\rangle,$$

we get

$$D_{-1}b_0 = \langle b_0, b_1\rangle.$$

On the other hand , from Theorem 3 if we introduce the function

$$Q = Q(\mathbf{x})$$

which satisfies

$$\mathrm{d}Q = \omega = b_0 \mathrm{d}x_{-1} + \sum_{n\geq 1}(b_n \mathrm{d}x_{-n-1} - g_n \mathrm{d}x_{n-1}),$$

then

$$b_0 = D_{-1}Q, \quad b_1 = D_{-2}Q.$$

Hence, by denoting $x := x_0, y := x_{-2}$ and $z := x_{-1}$, we obtain

$$Q_{zz} = Q_z Q_{xy} - Q_{xz} Q_y. \tag{3.2}$$

3.2. 2-DIMENSIONAL INTEGRABLE MODELS

In [1],[2] we developed a theory of reductions of the hierarchy (1.4) based on imposing differential constraints on $G \approx G_\infty$ of the form

$$\Big(\mathcal{F}(\lambda, G, G_x, G_{xx}, \ldots)\Big)_{-\infty,-1} = 0, \quad x := x_0. \tag{3.3}$$

In particular the following three classes of reductions associated to arbitrary polynomials $a = a(\lambda)$ in λ were characterized:

Zero-order reductions

They are defined by constraints of the form

$$a(\lambda)G = U(\lambda, \mathbf{x}), \quad U := \Big(a(\lambda)G\Big)_{0,+\infty}, \tag{3.4}$$

First-order reductions

They are characterized by the differential constraints

$$G_x + a(\lambda) = U(\lambda, \mathbf{x})G, \quad U := \Big(\frac{a}{G}\Big)_{0,+\infty}, \tag{3.5}$$

or, equivalently,

$$-H_x + a(\lambda)H^2 = U(\lambda, \mathbf{x})H, \quad U := \Big(aH\Big)_{0,+\infty}. \tag{3.6}$$

Second-order reductions

They are determined by constraints of the form

$$\frac{1}{2}GG_{xx} - \frac{1}{4}G_x^2 + a(\lambda) = U(\lambda, \mathbf{x})G^2, \quad U := \Big(\frac{a}{G^2}\Big)_{0,+\infty}. \tag{3.7}$$

Equivalently, in terms of the function H, we may rewrite (3.7) as

$$\{D_x, H\} + a(\lambda)H^2 = U(\lambda, \mathbf{x}), \quad U := \Big(aH^2\Big)_{0,+\infty}. \tag{3.8}$$

Here we are denoting

$$\{D_x, H\} := \frac{3}{4}\Big(\frac{H_x}{H}\Big)^2 - \frac{1}{2}\frac{H_{xx}}{H}.$$

Under the differential constraints (3.8) the hierarchy (1.14) describes the KdV hierarchy and its generalizations associated to energy-dependent Schrödinger spectral problems. Indeed, if we define the functions $\psi(\lambda, \mathbf{x})$ by

$$\psi(\lambda, \mathbf{x}) := \exp(D_x^{-1}\phi), \quad \phi := -\frac{1}{2}\frac{H_x}{H} \pm \sqrt{a(\lambda)}\, H, \tag{3.9}$$

then from (1.16) and (3.8) it is straightforward to deduce that

$$\begin{aligned} D_n\psi &= -\frac{1}{2}(D_n \log H)\psi \pm \sqrt{a}A_nH\psi \\ &= A_n\Big(-\frac{1}{2}D_x \log H \pm \sqrt{a}H\Big)\psi - \frac{1}{2}A_{n,x}\psi \\ &= A_n\psi_x - \frac{1}{2}A_{n,x}\psi, \\ \psi_{xx} &= (\phi_x + \phi^2)\psi = (\{D_x, H\} + aH^2)\psi = U\psi. \end{aligned}$$

In other words, the functions ψ are wave functions for the integrable hierarchies associated to energy-dependent Schrödinger problems. The evolution law of the potential function U under the flows (1.14) can be determined from the equation

$$D_n U = -\frac{1}{2} A_{n,xxx} + 2U A_{n,x} + U_x A_n, \tag{3.10}$$

which arise as an straightforward consequence of (1.14) and (3.7).

Theorem 3.1 *The differential constraints (3.4) (for a given $a(\lambda)$ such that $a(0) \neq 0$), (3.5) and (3.7) are left invariant by the enlarged hierarchy (1.14).*

Proof. Let us indicate the proof by considering the second-order case

$$\mathcal{F} := \frac{1}{2}\frac{G_{xx}}{G} - \frac{1}{4}\frac{G_x^2}{G^2} + \frac{a(\lambda)}{G^2}, \quad x := x_0, \tag{3.11}$$

where $a(\lambda)$ is a N-degree polynomial in λ. After the analysis of [1], [2], it remains to prove that this constraint is invariant under the negative flows of (1.14). To this end, let us consider

$$D_t G = \langle A_{-n}, G \rangle, \quad t := x_{-n}, \; n \geq 1.$$

Then, one finds

$$\begin{aligned} D_t \mathcal{F} &= \frac{\partial \mathcal{F}}{\partial G} D_t G + \frac{\partial \mathcal{F}}{\partial G_x} D_x D_t G + \frac{\partial \mathcal{F}}{\partial G_{xx}} D_x^2 D_t G \\ &= -\frac{1}{2} D_x^3 A_{-n} + 2(D_x A_{-n})\mathcal{F} + A_{-n} D_x \mathcal{F}. \end{aligned}$$

Now, the constraint associated with (3.11) implies

$$\mathcal{F} = U = \lambda^N + \lambda^{N-1} u_{N-1}(\mathbf{x}) + \cdots u_0(\mathbf{x}), \tag{3.12}$$

where

$$U(\lambda, \mathbf{x}) := \left(\frac{a(\lambda)}{G^2} \right)_{0,+\infty}.$$

In this way we have

$$D_t \mathcal{F} = -\frac{1}{2} D_x^3 A_{-n} + 2U D_x A_{-n} + U_x A_{-n}. \tag{3.13}$$

Notice that despite of the fact that these differential constraints are formulated for $G \approx G_\infty$, because of the polynomial character of the right-hand side of (3.12) we may analytically extend (3.12) to become a constraint for both $G \approx G_\infty$ or $G \approx G_0$. Thus, (3.12) implies

$$\frac{1}{2} G_0 G_{0,xx} - \frac{1}{2} G_{0,x}^2 + a(\lambda) = U G_0^2,$$

and by differentiating with respect to x one obtains

$$-\frac{1}{2}G_{0,xxx} + 2UG_{0,x} + U_x G_0 = 0,$$

which by taking into account that

$$A_{-n} = \left(\lambda^{-n} G_0\right)_{-\infty,-1},$$

yields

$$\begin{aligned} D_t \mathcal{F} &= -\frac{1}{2} D_x^3 A_{-n} + 2UD_x A_{-n} + U_x A_{-n} \\ &= \frac{1}{2}\left(\lambda^{-n} D_x^3 G_0\right)_{0,+\infty} - 2U\left(\lambda^{-M} D_x G_0\right)_{0,+\infty} - U_x\left(\lambda^{-n} G_0\right)_{0,+\infty}. \end{aligned}$$

It proves that

$$\left(D_t \mathcal{F}\right)_{-\infty,-1} = 0.$$

It is important to analyze the action of the transformation (2.5) on the reductions of our hierarchy. To this end we observe that from (2.10) we have ($x := x_0$)

$$\begin{aligned} G_x &= \frac{1}{\lambda}(D_{1,1}G + g_{1,x}G), \\ G_{xx} &= \frac{1}{\lambda^2}\left(D_{1,1}^2 G + g_{1,x} D_{1,1} G + (\lambda g_{1,xx} + g_{2,xx} - g_1 g_{1,xx})G\right), \end{aligned}$$

and as a consequence one readily finds that the elements of the classes of first-order and second-order reductions transform under (2.5) as

$$\begin{aligned} &-H_x + a(\lambda)H^2 = UH \mapsto -D_{1,1}H + \tilde{a}(\lambda)H^2 = \tilde{U}H, \\ &\tilde{U} := \left(\tilde{a}H\right)_{0,+\infty} = \lambda U + h_{1,x}, \end{aligned} \tag{3.14}$$

where

$$\tilde{a}(\lambda) := \lambda a(\lambda),$$

and

$$\begin{aligned} &\{D_x, H\} + a(\lambda)H^2 = U \mapsto \{D_{1,1}, H\} + \tilde{a}(\lambda)H^2 = \tilde{U}, \\ &\tilde{U} := \left(\tilde{a}H^2\right)_{0,+\infty} = \lambda^2 U + \frac{1}{2}\lambda h_{1,xx} + \frac{1}{2}h_{2,xx} - \frac{3}{4}(h_{1,x})^2 - \frac{1}{2}h_1 h_{1,xx}, \end{aligned} \tag{3.15}$$

where

$$\tilde{a}(\lambda) := \lambda^2 a(\lambda),$$

respectively. This means that there are three basic cases, namely: those corresponding to linear $a(\lambda)$ for first-order reductions, and linear and quadratic $a(\lambda)$ for second-order reductions. They are associated with the Burgers, KdV (Korteweg–deVries) and NLS (Nonlinear Schrödinger) hierarchies, respectively.

The Burgers hierarchy

If we impose the differential constraint

$$-D_0 H + \lambda H^2 = UH, \quad U = \lambda + h_1, \tag{3.16}$$

then the *positive flows* of the hierarchy (2.1) reduce to the Burgers hierarchy [2]. For example, by inserting the expansion of H as $\lambda \to \infty$ in the x_1 flow

$$D_1 H = D_0\Big((\lambda - h_1)H\Big), \tag{3.17}$$

it follows

$$D_1 h_1 = D_0 h_2 - D_0(h_1^2).$$

Moreover (3.16) yields $h_2 = D_0 h_1$, so that we get Burgers equation

$$u_t = u_{xx} - (u_x)^2, \;\; t := x_1, \; x := x_0, \; u := h_1.$$

Notice that by inserting the expansion of H at $\lambda = 0$ in (3.17) and (3.16) we get

$$h_1 = \frac{a_x}{a}, \qquad a := \frac{1}{b_0},$$

and

$$a_t = a_{xx},$$

which constitutes the Hopf-Cole transformation between the Heat equation and Burgers equation.

According to (3.14) the change of variables (2.5) transforms (3.16) into

$$-D_{1,1} H + \lambda^2 H = (\lambda^2 + \lambda h_1 + h_2)H, \tag{3.18}$$

and the first positive flow of the transformed hierarchy becomes

$$D_{2,1} H = D_{1,1}\Big((\lambda - h_1)H\Big). \tag{3.19}$$

Thus, by substituting the expansions of H as $\lambda \to 0$ and $\lambda \to \infty$ in these equations we get

$$D_{1,1} a = -h_2 a, \quad D_{2,1} a = -D_{1,1}(h_1 a), \quad D_{2,1} h_1 = D_{1,1}(h_2 - h_1^2).$$

Hence, the following nonlinear model results

$$\begin{cases} u_t = -(uv)_x, \\ v_t = -(\log u)_{xx} - 2vv_x \end{cases} \tag{3.20}$$

Here

$$x := y_1,\ t := y_2,\ u := a,\ v := h_1.$$

It is interesting to rewrite the system (3.20) in terms of the potential functions associated to the differential forms (2.7) and (2.12). Indeed, from (2.7) if we take $x_0(\mathbf{y}) := -Q(\mathbf{y})$ it follows that

$$h_1 = -Q_x,\ (\log a)_x = -h_2 = Q_t - (Q_x)^2.$$

Hence the second equation of the system (3.20) becomes trivial and the first equation adopts the form

$$Q_{tt} = Q_{xxx} + 3Q_x Q_{xt} + (Q_t - 3Q_x^2)Q_{xx}. \tag{3.21}$$

This nonlinear model is the first member of a family of integrable systems with unusual dispersion law [9]. Alternatively, from (2.12) if we introduce the function $z_1 = Q(\mathbf{y})$ we have

$$a = Q_x, \quad h_1 = -\frac{Q_t}{Q_x},$$

so that the first equation of (3.20) becomes trivial, while the second equation writes

$$\left(\frac{Q_t}{Q_x}\right)_t = (\log Q_x)_{xx} + \left(\left(\frac{Q_t}{Q_x}\right)^2\right)_x. \tag{3.22}$$

The KdV hierarchy

Let us consider the differential constraint (3.8) with

$$a(\lambda) = \lambda + \epsilon.$$

It can be formulated as

$$\{D_x, H\} + (\lambda + \epsilon)H^2 = U, \quad U := \lambda + \epsilon + 2h_1. \tag{3.23}$$

By using (3.10) it follows that the first positive flow is given by the KdV equation

$$u_t = \frac{1}{4}u_{xxx} - \frac{3}{2}uu_x + \frac{1}{2}\epsilon\, u_x, \quad u := 2h_1. \tag{3.24}$$

Now, if we use the variable $y := y_0$ of (2.5), (2.6)

$$D_y = D_{01} = b_0 D_0,$$

and denote

$$\widehat{H} := b_0\, H, \tag{3.25}$$

then (3.23) implies at once that

$$\{D_y, \widehat{H}\} + (\lambda + \epsilon)\widehat{H}^2 = \lambda b_0^2 + \epsilon. \tag{3.26}$$

On the other hand, from the x_{-1}-flow

$$D_{-1}G = \langle \frac{b_0}{\lambda}, G\rangle,$$

we deduce

$$D_{-1}H = \Big(\frac{b_0}{\lambda}H\Big)_x, \quad D_{-1b} = \langle b_0, b_1\rangle.$$

Thus, it readily follows that

$$D_t\widehat{H} = D_y\left(\Big(\frac{b_0}{\lambda} + b_1\Big)H\right), \tag{3.27}$$

where

$$t := y_{-1}, \quad D_t = D_{-1,1} = D_{-1} + b_1 D_0.$$

By identifying the coefficient of λ in (3.26) as $\lambda \to 0$ we get

$$b_0^2 = -\frac{1}{2}u_{yy} + 2\epsilon u + 1, \quad u := -\frac{b_1}{b_0}. \tag{3.28}$$

Moreover, identifying the coefficient of λ^0 in (3.27) as $\lambda \to \infty$ yields

$$b_{0,t} = b_{1,y}. \tag{3.29}$$

Hence, if we differentiate (3.28) with respect to t and use (3.28), (3.29) we get

$$2b_0 b_{0,t} = 2b_0 b_{1,y} = -2b_0 b_{0,y} u - 2b_0^2 u_y,$$

and

$$2\epsilon\, u_t - \frac{1}{2}u_{yyt} + 6\epsilon\, u u_y - \frac{1}{2}u u_{yyy} - u_{yy}u_y + 2u_y = 0, \tag{3.30}$$

which is the Camassa–Holm equation [10]

The NLS hierarchy

The quadratic case of the second-order differential reduction (3.8)

$$\{D_x, H\} + (\lambda^2 + \epsilon)H^2 = U(\lambda, \mathbf{x}), \quad U := \lambda^2 + 2h_1\lambda + 2h_2 + h_1^2 + \epsilon, \tag{3.31}$$

corresponds to the NLS hierarchy [4]. This fact can be seen by noticing that the Zakharov–Shabat spectral problem

$$\mathbf{\Psi_x} = \begin{pmatrix} \lambda & u \\ v & -\lambda \end{pmatrix} \mathbf{\Psi}, \quad \mathbf{\Psi} := \begin{pmatrix} \psi_1 \\ \psi_2 \end{pmatrix}, \tag{3.32}$$

leads to a pair of Schrödinger spectral problems with quadratic dependence on λ

$$\begin{aligned} \phi_{xx} &= (\lambda^2 - \lambda D_x(\log u) + \{D_x, u\} + uv)\phi, \quad \psi_1 = \sqrt{u}\phi, \\ \psi_{xx} &= (\lambda^2 + \lambda D_x(\log v) + \{D_x, v\} + uv)\psi, \quad \psi_2 = \sqrt{v}\psi. \end{aligned} \tag{3.33}$$

Thus there are two different ways of relating (3.31) and (3.32), which are given by

$$2h_1 = -D_x \log u, \quad 2h_2 - h_{1,x} + \epsilon = uv, \tag{3.34}$$

and

$$2h_1 = D_x \log v, \quad 2h_2 + h_{1,x} + \epsilon = uv. \tag{3.35}$$

However, in both cases the first positive flow of the hierarchy (1.14) constrained by (3.31)

$$D_1 H = D_0\Big((\lambda - h_1)H\Big),$$

yields the coupled NLS system ($t := x_1$, $x := x_0$)

$$\begin{cases} u_t = \frac{1}{2}u_{xx} - u^2 v + \epsilon\, u, \\ v_t = -\frac{1}{2}v_{xx} + v^2 u - \epsilon\, v \end{cases} \tag{3.36}$$

Now let us proceed as in the KdV case by using again the variable $y := y_0$ of (2.5), (2.6)

$$D_y = D_{01} = b_0 D_0,$$

and denoting $\widehat{H} := b_0\, H$. Then (3.31) implies

$$\{D_y, \widehat{H}\} + (\lambda^2 + \epsilon)\widehat{H}^2 = \widehat{U}, \quad \widehat{U} := \lambda^2 b_0^2 + 2\lambda b^2 h_1 + \epsilon. \tag{3.37}$$

Notice that by identifying the coefficient of λ in (3.37) as $\lambda \to 0$ we get

$$2b_0^2 h_1 = -\frac{1}{2}u_{yy} + 2\epsilon u, \quad u := -\frac{b_1}{b_0}. \tag{3.38}$$

On the other hand we have

$$D_t \widehat{H} = D_y\left(\left(\frac{b_0}{\lambda} + b_1\right) H\right), \tag{3.39}$$

where

$$t := y_{-1}, \quad D_t = D_{-1,1} = D_{-1} + b_1 D_0,$$

which leads to the following equation for $\widehat{U}$

$$\widehat{U}_t = -\frac{1}{2}A_{yyy} + 2\widehat{U}A_y + \widehat{U}_y A, \quad A := \frac{1}{\lambda} - u. \tag{3.40}$$

Then, by identifying the coefficients of λ and λ^2 in (3.41) we obtain the following analogue of the Camassa–Holm equation for the quadratic case of the second-order constraints

$$\begin{cases} v_t &= -(vu)_y, \\ w_t &= -w_{yyy} + (v^2)_y - uw_y + 2wu_y, \end{cases} \tag{3.41}$$

Here we are denoting

$$v := b_0, \quad w := -\frac{1}{2}u_{yy} + 2\epsilon u.$$

Acknowledgements

A.B. Shabat was supported by the Russian Foundation for Basic Research (Grant Nos. 00-15-96007-L and 01-01-00874-A). L. Martínez Alonso was supported by the DGCYT project BFM2002-01607.

References

1. L. Martínez Alonso and A. B. Shabat, Phys.Lett. **A 300**, 58 (2002).
2. L. Martínez Alonso and A. B. Shabat, J. Non. Math. Phys. **10**, 229 (2003).
3. V. G. Mikhalev, Funct. Anal. Appl. **26** No 2, 140 (1992).
4. M. Jaulent and I. Miodek, Lett. Math. Phys. **1**, 243 (1976); Lett. Nuovo Cimento **20**, 655 (1977).
5. L. Martínez Alonso, J. Math. Phys. **21**, 2342 (1980).
6. M. Antonowicz and A. P. Fordy, Physica **28 D**, 345 (1987).
7. A. Shabat, *Universal models of solitonic hierarchies.* To appear in Theor. Math. Phys., June 2003.
8. A. N. W. Hone, Phys. Lett. **A 249**, 46 (1998).
9. R. Heredero, V. Sokolov and A. B. Shabat, *A new class of linearizable equations*, to be published.
10. R. Camassa and D. Holm, Phys. Rev. Lett. **71**, 1661 (1993).

SUPERINTEGRABLE SYSTEMS IN CLASSICAL AND QUANTUM MECHANICS

P. WINTERNITZ (wintern@crm.umontreal.ca)
Centre de recherches mathématiques et
Département de mathématiques et statistique,
Université de Montréal, C.P. 6128, succ. Centre-Ville,
Montréal, QC, H3C 3J7, Canada

Abstract. A brief review is given of the status of superintegrability, i.e., the theory of classical and quantum mechanical finite-dimensional systems with more integrals of motion than degrees of freedom. Typically, in classical mechanics such systems are characterized by periodic motion, in quantum mechanics their energy levels can be calculated algebraically.

1. Introduction

Conceptually speaking, an *integrable* system is a system for which one can make *global,* or *long term,* predictions. To express this in more technical terms, let us first consider a Hamiltonian system in classical mechanics, described by a Hamiltonian function $H(\vec{x},\vec{p})$, depending on n coordinates $\vec{x}=(x_1,\ldots,x_n)$ and n canonically conjugated momenta $\vec{p}=(p_1,\ldots,p_n)$. Such a system is called integrable, or Liouville integrable, if there exist n well defined functions on phase space $X_i(\vec{x},\vec{p})$ (including the Hamiltonian H) satisfying the following properties:

$$\frac{\mathrm{d}X_i}{\mathrm{d}t}=\{X_i,H\}=0 \tag{1.1}$$

$$\{X_i,X_j\}=0 \tag{1.2}$$

$$\operatorname{rank}\frac{\partial(X_1,\ldots,X_n)}{\partial(x_1,\ldots,x_n,p_1,\ldots,p_n)}=n, \tag{1.3}$$

where $\{\,,\,\}$ denotes the Poisson bracket. In other words, a Hamiltonian system with n degrees of freedom is integrable if there exist n functionally independent well defined integrals of motion in involution [1, 2].

A.B. Shabat et al. (eds.), New Trends in Integrability and Partial Solvability, 281–297.

A Hamiltonian system is *superintegrable* if it is integrable and there exist k further integrals $Y_a(\vec{x}, \vec{p})$ (with $1 \le k \le n-1$) satisfying

$$\frac{\mathrm{d}Y_a}{\mathrm{d}t} = \{Y_a, H\} = 0, \quad a = 1, \dots, k; \tag{1.4}$$

$$\operatorname{rank} \frac{\partial(X_1, \dots, X_n, Y_1, \dots, Y_k)}{\partial(x_1, \dots, x_n, p_1, \dots, p_n)} = n + k. \tag{1.5}$$

In general we have $\{X_i, Y_a\} \neq 0$, i.e., only n of the $n + k$ integrals are in involution. If the total number of independent integrals of motion is $N = 2n - 1$, the system is called *maximally superintegrable.* If we have $N = n + 1$, it is *minimally superintegrable.*

Integrability and superintegrability have physical consequences. Indeed, if a system is integrable, the equations

$$X_1 = H = E, \quad X_i = c_i, \quad i = 2, \dots, n \tag{1.6}$$

confine the motion to an n-dimensional torus in phase space (possibly a noncompact version of a torus). If a system is maximally superintegrable the additional equations $Y_a = k_a$ ($a = 1, \dots, n-1$) further constrain the motion to a line in phase space (E, k_a, c_i are constants), in particular the trajectories are completely determined. It follows that for a classical maximally superintegrable system all finite trajectories are periodic.

The best known maximally superintegrable Hamiltonian systems are the Kepler–Coulomb system and the isotropic harmonic oscillator, described by the potentials

$$V_K(r) = \frac{\alpha}{r}, \quad V_H(r) = \omega^2 r^2, \tag{1.7}$$

respectively. Indeed, in addition to the angular momentum $\vec{L}$, the Kepler–Coulomb problem admits the Laplace–Runge–Lenz vector

$$\vec{A} = \vec{L} \times \vec{p} + \frac{\alpha}{r^3}\vec{r} \tag{1.8}$$

and the harmonic oscillator admits the tensor

$$T_{ik} = \frac{1}{2} p_i p_k + \omega^2 x_i x_k, \tag{1.9}$$

as integrals of motion.

A classical theorem due to Bertrand [3] states that the potentials (1.7) are the only two rotationally invariant ones for which all finite trajectories are closed.

In quantum mechanics the situation is quite analogous. A system described by the Schrödinger equation

$$H\psi = E\psi, \quad H = -\frac{1}{2}\hbar^2 \Delta + V(\vec{x}) \tag{1.10}$$

is superintegrable if there exists a set of Hermitian operators $\{X_1, \dots, X_n, Y_1, \dots, Y_k\}$ (with $H = X_1$ and $1 \leq k \leq n-1$) satisfying

$$[H, X_i] = 0, \quad [H, Y_a] = 0\,, \tag{1.11}$$
$$[X_i, X_j] = 0\,; \tag{1.12}$$
$$i = 1, \dots, n, \quad a = 1, \dots k\,.$$

The operators $\{X_i, Y_a\}$ must satisfy an independence condition (that needs a rigorous definition with the correct classical limit). The physical consequences of superintegrability in quantum mechanics are interesting. In particular, since in general we have $[X_i, Y_a] \neq 0$, the integrals of motion generate a non-Abelian Lie algebra (finite or infinite-dimensional). This in turn implies that the energy levels of such systems will be degenerate. Indeed, the *accidental* degeneracy of the hydrogen atom and the harmonic oscillator, described respectively by their O(4) and SU(3) symmetries, are examples of this phenomenon [4–6].

Another consequence of superintegrability is at this stage conjectured, rather than proven. The conjecture is that all maximally superintegrable quantum systems are exactly solvable, i.e., their energy levels can be calculated algebraically and their bound state wave functions, in an appropriate gauge and in appropriate coordinates, are polynomials. This conjecture is borne out by many examples [7–9].

We see that superintegrable systems in both classical and quantum mechanics have interesting mathematical and physical properties. Such systems are rare and it is worthwhile to search for them systematically.

To formulate the problem of finding superintegrable systems mathematically, some restrictions must be imposed and some choices made.

First of all one must choose the space one is working in: a Euclidean space E_n, a space of constant nonzero curvature, or a more general Riemannian, pseudo Riemannian, or complex Riemannian space.

Secondly, the form of the Hamiltonian must be specified, involving a scalar potential, or a scalar and vector one. Alternatively, one could consider a many-particle Hamiltonian, conceivably involving particles with spin.

Finally restrictions must be imposed on the form of the integrals of motion. Usually they are postulated to be polynomials of a definite order in the momenta, often just second-order polynomials.

The purpose of this report is to provide a brief review of the status of superintegrability in classical and quantum mechanics. In Section 2 we concentrate on *quadratic integrability,* i.e., the case of a standard Hamiltonian and integrals of motion, all of second order in the momenta. The case of third-order integrals is discussed in Section 3. We limit ourselves to two dimensional spaces though the extension to n dimensions is immediate, at least conceptually.

2. Quadratic superintegrability

2.1. EUCLIDEAN SPACES

The systematic search for superintegrable systems started quite some time ago [10, 11]. The approach was via commuting operators in quantum mechanics in two-dimensional Euclidean space with the Hamiltonian

$$H = -\frac{1}{2}\Delta + V(x, y) \tag{2.1}$$

and the operators commuting with H were assumed to be of second order

$$X = f^{ik}(\vec{x})p_i p_k + g^i(\vec{x})p_i + \phi(\vec{x}). \tag{2.2}$$

The form of the Hamiltonian (2.1) is left invariant by the group E(2) of Euclidean transformations, which do however change the potential $V(x, y)$.

The corresponding Lie algebra e(2) has a basis given by

$$L_3 = y\partial_x - x\partial_y, \quad P_1 = \partial_x, \quad P_2 = \partial_y. \tag{2.3}$$

The commutation relation $[H, X_a] = 0$ implies very strong restrictions on the functions f_a^{ik} and g_a^i, namely we find:

$$\begin{aligned} X = {} & aL_3^2 + b(L_3P_1 + P_1L_3) + c(L_3P_2 + P_2L_3) + \\ & d(P_1^2 - P_2^2) + 2eP_1P_2 + \alpha L_3 + \beta P_1 + \gamma P_2 + \phi(x, y), \end{aligned} \tag{2.4}$$

where $a, \ldots, e$ and α, β, γ are constants. Furthermore, the potential $V(x, y)$, the function $\phi(x, y)$ and the constants in (2.4) must satisfy:

$$[\alpha(y\partial_x - x\partial_y) + \beta\partial_x + \gamma\partial_y]V = 0, \tag{2.5}$$

$$\begin{aligned} \phi_x &= -2(ay^2 + 2by + d)V_x + 2(axy + bx - cy - e)V_y, \\ \phi_y &= 2(axy + bx - cy - e)V_x + 2(-ax^2 + 2cx + d)V_y. \end{aligned} \tag{2.6}$$

The compatibility condition $\phi_{xy} = \phi_{yx}$ provides a further equation for the potential, namely

$$\begin{aligned} & (-axy - bx + cy + e)(V_{xx} - V_{yy}) + [a(x^2 - y^2) \\ & - 2by - 2cx - 2d]\, V_{xy} - 3(ay + b)V_x + 3(ax - c)V_y = 0. \end{aligned} \tag{2.7}$$

Eq. (2.5) is the condition for a first-order operator commuting with the Hamiltonian to exist. This is the same as requiring that $V(x, y)$ be invariant under rotations, or translations. For $\alpha \neq 0$ this is equivalent to putting $V = V(r)$. For $\alpha = 0$, $\beta^2 + \gamma^2 \neq 0$ this is equivalent to putting $V = V(x)$. To find potentials $V(x, y)$ allowing only a second-order operator X we put

$\alpha = \beta = \gamma = 0$. We can simplify the operator X (and Eq. (2.7)) by Euclidean transformations and linear combinations with the Hamiltonian H:

$$X \to \tilde{X} = \lambda g X g^{-1} + \mu H, \quad \lambda \neq 0, \quad g \in \mathrm{E}(2). \tag{2.8}$$

The transformation (2.8) admits two invariants

$$I_1 = a, \quad I_2 = (2ad - b^2 + c^2)^2 + 4(ae - bc)^2. \tag{2.9}$$

If we have $I_1 = I_2 = 0$, there is a third invariant, namely $I_3 = d^2 + e^2$. Depending on the values of these invariants we can transform X and the potential $V(x, y)$ into precisely one of the following forms:

1. $I_1 = I_2 = 0,\ I_3 \neq 0$

$$X_C = -\frac{1}{2}(P_1^2 - P_2^2) + f(x) - g(y), \quad V = f(x) + g(y). \tag{2.10}$$

2. $I_1 \neq 0,\ I_2 = 0$

$$X_R = L_3^2 - 2g(\theta), \quad V = f(r) + \frac{1}{r^2}g(\theta), \tag{2.11}$$
$$x = r\cos\theta, \quad y = r\sin\theta.$$

3. $I_1 = 0,\ I_2 \neq 0$

$$X_P = L_3 P_2 + P_2 L_3 + \frac{g(\eta)\xi^2 - f(\xi)\eta^2}{\xi^2 + \eta^2}, \quad V = \frac{f(\xi) + g(\eta)}{\xi^2 + \eta^2}, \tag{2.12}$$
$$x = \frac{1}{2}(\xi^2 - \eta^2), \quad y = \xi\eta.$$

4. $I_1 \neq 0,\ I_2 \neq 0$

$$X_E = L_3^2 + \frac{l^2}{2}(P_1^2 - P_2^2) - l^2 \frac{\cosh 2\rho f(\sigma) + \cos 2\sigma g(\rho)}{\cos^2\sigma - \cosh^2\rho}, \quad l > 0$$
$$V = \frac{f(\sigma) + g(\rho)}{\cos^2\sigma - \cosh^2\rho}, \tag{2.13}$$
$$x = l\cosh\rho\cos\sigma, \quad y = l\sinh\rho\sin\sigma.$$

We see that a second-order operator X commuting with the Hamiltonian H in E_2 exists if and only if the Schrödinger equation allows the separation of variables in cartesian, polar, parabolic, or elliptic coordinates. These correspond to the cases (2.10), (2.11), (2.12) and (2.13), respectively (l in (2.13) is the focal distance and we have $l^2 = I_2/I_1^2$. Thus all *quadratically integrable* systems (2.1) are also separable. They allow (multiplicative) separation of variables in the Schrödinger equation and also (additive) separation of variables in the Hamilton–Jacobi equation.

The close relation between quadratic integrability and separation of variables persists in other spaces (E_n for arbitrary n) and also in spaces of constant curvature, as well as much more general Riemannian and pseudo-Riemannian spaces.

The system will be *quadratically superintegrable* if it allows two integrals of the form (2.2). It will then also be *multiseparable,* i.e., allow the separation of variables in at least two coordinate systems. Four such systems were shown to exist in E_2. The corresponding potentials are:

$$V_{\mathrm{I}} = \omega^2(x^2+y^2)+\frac{\alpha}{x^2}+\frac{\beta}{y^2}, \tag{2.14}$$

$$V_{\mathrm{II}} = \omega^2(4x^2+y^2)+\frac{\alpha}{y^2}+\beta x, \tag{2.15}$$

$$\begin{aligned} V_{\mathrm{III}} &= \frac{\alpha}{r}+\frac{1}{2r^2}\left(\frac{\beta_1}{1+\cos\theta}+\frac{\beta_2}{1-\cos\theta}\right) \\ &= \frac{1}{\xi^2+\eta^2}\left(2\alpha+\frac{\beta_1}{\xi^2}+\frac{\beta_2}{\eta^2}\right), \end{aligned} \tag{2.16}$$

$$\begin{aligned} V_{\mathrm{IV}} &= \frac{\alpha}{r}+\frac{1}{\sqrt{2r}}\left(\beta\cos\frac{\theta}{2}+\gamma\sin\frac{\theta}{2}\right) \\ &= \frac{1}{\xi^2+\eta^2}(2\alpha+\beta\xi+\gamma\eta). \end{aligned} \tag{2.17}$$

Classical trajectories, wave functions and energy spectra for these four potentials were determined long ago [10, 11]. More recently, it was shown that these superintegrable systems are invariant under generalized Lie symmetries, and that they allow recursion operators [12].

Let us now demonstrate the exact solvability of these four superintegrable systems [7]. First of all we note that the *generalized hydrogen atom potentials* V_{III} and V_{IV} can be reduced to the case of the generalized harmonic oscillator V_{I}. Indeed, the Schrödinger equation for V_{III} in parabolic coordinates is

$$\left\{-\frac{1}{2}\frac{1}{\xi^2+\eta^2}(\partial_\xi^2+\partial_\eta^2)+\frac{1}{\xi^2+\eta^2}\left(2\alpha+\frac{\beta_1}{\xi^2}+\frac{\beta_2}{\eta^2}\right)\right\}\psi = E\psi. \tag{2.18}$$

This can be rewritten as

$$\left\{-\frac{1}{2}(\partial_\xi^2+\partial_\eta^2)-E(\xi^2+\eta^2)+\frac{\beta_1}{\xi^2}+\frac{\beta_2}{\eta^2}\right\}\psi = -2\alpha\psi. \tag{2.19}$$

Similarly, the Schrödinger equation for V_{IV} can be rewritten as

$$\begin{aligned} &\left\{-\frac{1}{2}(\partial_\xi^2+\partial_\eta^2)-E\left[\left(\xi-\frac{\beta}{2E}\right)^2+\left(\eta-\frac{\gamma}{2E}\right)^2\right]\right\}\psi = \\ &\left(-2\alpha-\frac{\beta^2+\gamma^2}{4E}\right)\psi. \end{aligned} \tag{2.20}$$

Changing notation, we write the Schrödinger equation for the potential $V_{\rm I}$ as

$$H_{\rm I}\left(x,y;\frac{\omega^2}{2},\frac{A}{2},\frac{B}{2}\right)\psi = \left\{-\frac{1}{2}(\partial_x^2+\partial_y^2)+\frac{\omega^2}{2}(x^2+y^2)+\frac{1}{2}\left(\frac{A}{x^2}+\frac{B}{y^2}\right)\right\}\psi = E\psi. \quad (2.21)$$

Equations (2.19) and (2.20) can be rewritten as

$$Q_{\rm III}\psi = -2\alpha\psi, \quad Q_{\rm III} = H_{\rm I}(\xi,\eta,-E,\beta_1,\beta_2) \quad (2.22)$$

and

$$Q_{\rm IV}\psi = \left(-2\alpha - \frac{\beta^2+\gamma^2}{4E}\right)\psi, \quad Q_{\rm IV} = H_{\rm I}\left(\xi-\frac{\beta}{2E},\eta-\frac{\gamma}{2E},-E,0,0\right). \quad (2.23)$$

This means that the Schrödinger equation for potentials $V_{\rm III}$ and $V_{\rm IV}$ has been reduced to that for potential $V_{\rm I}$, with a *metamorphosis of the coupling constant* [13]: the energy and coupling constant ω are interchanged. Thus it suffices to show that the generalized harmonic oscillator (2.21) is exactly solvable. To see this, we solve eq. (2.21) and obtain:

$$\psi_{nm}(x,y) = x^{p_1}y^{p_2}{\rm e}^{-\frac{\omega}{2}(x^2+y^2)}L_n^{(-\frac{1}{2}+p_1)}(\omega x^2)L_m^{(-\frac{1}{2}+p_2)}(\omega y^2), \quad (2.24)$$
$$A = p_1(p_1-1), \quad B = p_2(p_2-1),$$

where $L_k^{(\alpha)}(z)$ is a Laguerre polynomial. The ground state wave function

$$\psi_{00}(x,y) = x^{p_1}y^{p_2}{\rm e}^{-\frac{\omega}{2}(x^2+y^2)} \quad (2.25)$$

figures in eq. (2.24) as a universal factor for all bound states. We use it to perform a gauge rotation of the Hamiltonian $H_{\rm I}$ and simultaneously transform to new variables $t=\omega x^2$, $u=\omega y^2$. We obtain

$$h_{\rm I} = \frac{1}{\omega}\psi_{00}^{-1}H_{\rm I}\psi_{00}\Big|_{t=\omega x^2,u=\omega y^2} = -2t\partial_t^2 - 2u\partial_u^2 + 2t\partial_t + 2u\partial_u - (2p_1+1)\partial_t - (2p_2+1)\partial_u + 1 + p_1 + p_2. \quad (2.26)$$

Thus the transformed Hamiltonian $h_{\rm I}$ is an element of the enveloping algebra of the affine Lie algebra aff$(2,\mathbb{R})$, realized by the vector fields

$$J_1=\partial_t, \quad J_2=\partial_u, \quad J_3=t\partial_t, \quad J_4=u\partial_u, \quad J_5=u\partial_t, \quad J_6=t\partial_u. \quad (2.27)$$

For $h_{\rm I}$ the Schrödinger equation reduces to

$$h_{\rm I} P_N(t,u) = (n+m) P_N(t,u), \quad N = n+m, \tag{2.28}$$

where $P_N(t,u)$ are polynomials of order $n+m$. The operators commuting with $H_{\rm I}$ are

$$\begin{aligned} X_C &= -\frac{1}{2}(P_1^2 - P_2^2) + \frac{\omega^2}{2}(x^2 - y^2) + \frac{1}{2}\left(\frac{A}{x^2} - \frac{B}{y^2}\right), \\ X_R &= L_3^2 - \frac{A}{\cos^2\theta} - \frac{B}{\sin^2\theta}. \end{aligned} \tag{2.29}$$

After a gauge rotation with the gauge factor $\psi_{00}(x,y)$ and a change of variables to t and u we can write $h_{\rm I}$, $x_c = \psi_{00}^{-1} X_C \psi_{00}$ and $x_R = \psi_{00}^{-1} X_R \psi_{00}$ as:

$$\begin{aligned} h_{\rm I} &= -2J_3J_1 - 2J_4J_2 + 2J_3 + 2J_4 - (2p_1+1)J_1 - (2p_2+1)J_2, \\ x_C &= 2J_3J_1 - 2J_4J_2 - 2J_3 + 2J_4 + (2p_1+1)J_1 - (2p_2+1)J_2, \\ x_R &= 4J_3J_5 + 2J_4J_6 - 8J_3J_4 + 2(2p_1+1)J_5 - 2(2p_2+1)J_3 - \\ &\quad 2(2p_1+1)J_4 + 2(2p_2+1)J_6. \end{aligned} \tag{2.30}$$

All three integrals of motion $h_{\rm I}$, x_C and x_R are in the enveloping algebra of aff$(2, \mathbb{R})$. This algebra contains no raising operators: only lowering and reproducing ones. Hence all three operator (2.30) preserve the flag of polynomials

$$\mathcal{P}_N(t,u) = \{t^m u^n : 0 \le m+n \le N\}. \tag{2.31}$$

This is the origin of the exact solvability of the superintegrable Hamiltonian $H_{\rm I}$. The arguments for the generalized anisotropic harmonic oscillator $H_{\rm II}$ are very similar [7]. The generalized hydrogen atoms $H_{\rm III}$ and $H_{\rm IV}$ are reduced to $H_{\rm I}$ by the coupling constant metamorphosis (2.22), or (2.23), respectively.

A systematic study of quadratically superintegrable systems in three-dimensional Euclidean space E_3 was also started in the 1960s [14] and continued in the 1990s [15, 16]. Superintegrable and exactly solvable systems also exist in E_n for any n [9, 17].

2.2. SPACES OF NONZERO CONSTANT CURVATURE

Quadratic superintegrable systems have been classified and analyzed on real spheres S_n and real Lorentzian hyperboloids H_N, as well as on complex Euclidean spaces and spheres. For lack of space we shall not reproduce the results here, but merely refer to the original articles.

The two dimensional cases S_2, H_2, $E_2(\mathbb{C})$ and $S_2(\mathbb{C})$ have been treated in detail [18–20]. Higher-dimensional spaces, more sporadically [21–24].

2.3. TWO DIMENSIONAL SPACES OF NONCONSTANT CURVATURE: DARBOUX SPACES

Let us consider a two-dimensional space with infinitesimal distance

$$ds^2 = \sum_{i,j=1}^{2} g_{ij}(u)\mathrm{d}u^i\mathrm{d}u^j. \tag{2.32}$$

The classical Hamiltonian in this space is

$$H = \frac{1}{2}g^{ij}(u)p_ip_j + V(u), \tag{2.33}$$

and the Schrödinger equation is written as:

$$\hat{H}\psi = \left[-\frac{1}{2}\frac{1}{\sqrt{g}}\partial_{u^i}\sqrt{g}g^{ik}\partial_{u^k} + V(u)\right]\psi = E\psi, \quad g = \det g_{ik}. \tag{2.34}$$

In keeping with our general approach, we are looking for a metric $g_{ij}(u)$ and potentials $V(u)$ such that the Hamiltonian H admits first, or second-order integrals of motion

$$X^{(1)} = a^i(u)p_i, \quad X^{(2)} = a^{ij}(u)p_ip_j + \phi(u), \tag{2.35}$$

$$\frac{\mathrm{d}X^{(i)}}{\mathrm{d}t} = \{X^{(i)}, H\} = 0. \tag{2.36}$$

First of all we consider the case $V(u) = 0$, $\phi(u) = 0$. Eq. (2.36) then determines Killing vectors $a^i(u)$, or Killing tensors $a^{ij}(u)$. The problem of finding all complex two-dimensional Riemannian spaces with more than one quadratic integral of motion was completely solved by Darboux [25] and Koenigs [26]. They established the following results

1. A two-dimensional Riemannian space can admit 0, 1, or 3 Killing vectors. If it admits one it is a space of revolution. If it admits three, it is a space of constant curvature.
2. A two-dimensional Riemannian space can admit 0, 1, 2, 3, or 5 Killing tensors. If it admits 5, then it is a space of constant curvature and all the Killing tensors are reducible: they are bilinear expressions in the 3 Killing vectors. If the space admits 3 Killing tensors, then one of them is reducible, i.e., it is the square of a Killing vector, and the space is a space of revolution.
3. They gave a complete classification of spaces with 2 and 3 Killing tensors.

4. Four types of Riemannian spaces with 3 Killing tensors exist. The corresponding infinitesimal distances are given by:

$$\begin{aligned} \text{I.}\quad \mathrm{d}s^2 &= (x+y)\mathrm{d}x\,\mathrm{d}y, \\ \text{II.}\quad \mathrm{d}s^2 &= \left(\frac{a}{(x-y)^2}+b\right)\mathrm{d}x\,\mathrm{d}y, \\ \text{III.}\quad \mathrm{d}s^2 &= \left(a\mathrm{e}^{-(x+y)/2}+b\mathrm{e}^{-(x+y)}\right)\mathrm{d}x\,\mathrm{d}y, \\ \text{IV.}\quad \mathrm{d}s^2 &= \frac{a(\mathrm{e}^{(x-y)/2}+\mathrm{e}^{-(x-y)/2})+b}{[\mathrm{e}^{(x-y)/2}-\mathrm{e}^{-(x-y)/2}]^2}\mathrm{d}x\,\mathrm{d}y. \end{aligned} \tag{2.37}$$

We shall call these spaces *Darboux spaces* of type I,. . . ,IV. Superintegrable systems exist in all four of them [8, 27]. Here we shall discuss the first space only and consider a real form of it. We put $x = u+iv, y = u-iv$. The infinitesimal distance and Hamiltonian reduces to

$$\mathrm{d}s^2 = 2u(\mathrm{d}u^2+\mathrm{d}v^2), \quad H = \frac{1}{4u}(p_u^2+p_v^2), \tag{2.38}$$

respectively. The three classical integrals of motion (for the case of zero potential) are

$$\begin{aligned} K = p_v, \quad X_1 &= p_u p_v - \frac{v}{2u}(p_u^2+p_v^2), \\ X_2 &= p_v(vp_u - up_v) - \frac{v^2}{4u}(p_u^2+p_v^2). \end{aligned} \tag{2.39}$$

The Killing vector K generates a one-dimensional group of isometries of the space. The most general second-order integral of motion will have the form

$$X = \alpha X_1 + \beta X_2 + \gamma K^2. \tag{2.40}$$

Acting on X with the transformation $g = \exp(\alpha K)$ we find that X can be transformed into precisely one of the following three representatives of orbits:

$$L_1 = X_1 + aK^2, \quad L_2 = X_2 + aK^2, \quad L_3 = K^2. \tag{2.41}$$

Let us now find coordinates that will allow the separation of variables in the Hamilton—Jacobi equation

$$H \equiv \frac{1}{2}g^{ik}\frac{\partial S}{\partial u^i}\frac{\partial S}{\partial u^k} = E. \tag{2.42}$$

To do this we add a further equation, given by the integral X:

$$X \equiv a^{ik}\frac{\partial S}{\partial u^i}\frac{\partial S}{\partial u^k} = \lambda, \tag{2.43}$$

where X is one of the expressions (2.41). The separable coordinates are the eigenvalues ρ_1, ρ_2 of the characteristic equation

$$|a^{ik} - \rho g^{ik}| = 0, \tag{2.44}$$

under the condition that they are distinct, i.e., $\rho_1 \neq \rho_2$. Once separable coordinates are found, we can add a potential $V(\rho_1, \rho_2)$ that does not spoil the separation. In general, the classical quadratically integrable and separable system will have the form

$$H = \frac{1}{\sigma(\alpha)+\tau(\beta)}[p_\alpha^2 + p_\beta^2 + f(\alpha) + g(\beta)], \tag{2.45}$$

$$L = \frac{1}{\sigma(\alpha)+\tau(\beta)}[\sigma(\alpha)(p_\beta^2 + g(\beta)) - \tau(\beta)(p_\alpha^2 + f(\alpha))]. \tag{2.46}$$

The system will be superintegrable if it allows a further integral, independent of the pair $\{H, L\}$. By necessity it will have the form (2.40).

For all details we refer to the original article [8]. Here we just state the result, namely: there exist precisely three quadratically superintegrable systems in the Darboux space of type I. They are all separable in at least two coordinates systems. Their Hamiltonians and integrals of motion are

$$\begin{aligned} 1)\quad H_1 &= \frac{1}{4u}[p_u^2 + p_v^2 + a_1 + a_2 v + a_3(u^2+v^2)], \\ R_1 &= X_1 + \frac{1}{2u}[-a_1 v + a_2(u^2 - v^2) + a_2 v(u^2 - v^2)], \\ R_2 &= K^2 + a_2 v + a_3 v^2. \end{aligned} \tag{2.47}$$

$$\begin{aligned} 2)\quad H_2 &= \frac{1}{4u}\left[p_u^2 + p_v^2 + b_1(4u^2+v^2) + b_2 + \frac{b_3}{v^2}\right], \\ R_1 &= X_2 - \frac{1}{4u}\left[b_1 v^4 + b_2 + b_3\frac{4u^2+v^2}{v^2}\right], \\ R_2 &= K^2 + b_1 v^2 + \frac{b_3}{v^2}. \end{aligned} \tag{2.48}$$

$$\begin{aligned} 3)\quad H_3 &= \frac{1}{u}(p_u^2 + p_v^2 + a), \\ R_1 &= X_1 - \frac{av}{2u}, \\ R_2 &= X_2 - \frac{av^2}{4u}, \\ R_3 &= K. \end{aligned} \tag{2.49}$$

The quantum integrable systems are obtained by substituting

$$p_u \to -i\hbar\frac{\partial}{\partial u}, \quad p_v \to -i\hbar\frac{\partial}{\partial v}$$

in all of the above expressions.

All of the corresponding quantum systems are exactly solvable. Indeed, if we multiply the Schrödinger equation by the denominator u and interchange the energy E with a_1 for H_1, with b_2 for H_2 and with a for H_3, we obtain a shifted isotropic harmonic oscillator, a generalized anisotropic harmonic oscillator and a linear potential, respectively. Thus the above simple metamorphosis of the coupling constants takes each of the above systems into a superintegrable system in E_2 that has been shown to be exactly solvable [7].

3. Superintegrability with a third-order integral of motion

Quadratic superintegrability is by now rather well understood, at least for scalar potentials. The main features that have emerged are the following

1. Integrable and superintegrable systems with integrals of motion that are linear, or quadratic in the momenta coincide in classical and quantum mechanics.
2. Integrable systems in n dimensions with n quadratic (or linear) commuting integrals of motion are separable: the Hamilton-Jacobi and Schrödinger equations allow the separation of variables. Superintegrable systems are multiseparable.
3. All known maximally superintegrable systems are exactly solvable

Much less is known about integrable systems with higher-order integrals of motion. One systematic study is due to Drach, [28] who considered classical integrable systems in two-dimensional complex Euclidean space $E(2, C)$. He found 10 such systems with third-order integrals (in classical mechanics). It was shown by Rañada [29] (see also Tsiganov [30]) that 7 of these systems are actually *reducible*. They are quadratically superintegrable and the third-order integral is the commutator of two second-order ones.

A classification of all integrable classical and quantum systems with third-order integrals is a difficult task, even in two-dimensional Euclidean space. Indeed, let us start from the quantum case, setting

$$\begin{aligned} H &= -\frac{1}{2}\hbar^2\Delta + V(x,y), \\ X &= \sum_{0\leq j+k\leq 3} [f_{jk}(x,y)p_1^j p_2^k + p_1^j p_2^k f_{jk}(x,y)], \quad p_k = -\mathrm{i}\hbar\frac{\partial}{\partial x_k}. \end{aligned} \tag{3.1}$$

The operator X contains terms of order 3, 2, 1 and 0 in the momenta, and has been symmetrized to ensure that it is Hermitian. The commutativity condition $[H, X] = 0$ implies that even and odd terms in the momenta must commute separately with H. Thus, if we require integrability with a third-order integral, rather than superintegrability, we must put $f_{jk} = 0$ for $j + k = 2$ and also $f_{00} = 0$. Furthermore, the third-order terms must lie in the enveloping algebra of e(2). This is true for any potential $V(x, y)$. The potential will figure in four remaining determining equations. Explicitly, we obtain [31]

$$\begin{aligned} X = & \sum_{i+j+k=3} A_{ijk}(L_3^i p_1^j p_2^k + p_1^j p_2^k L_3^i) + g_1(x,y)p_1 + p_1 g_1(x,y) \\ & + g_2(x,y)p_2 + p_2 g_2(x,y), \end{aligned} \tag{3.2}$$

where A_{ijk} are constants. The remaining determining equations are

$$g_{1,x} = 3f_1 V_x + f_2 V_y, \tag{3.3}$$

$$g_{2,y} = f_3 V_x + 3f_4 V_y, \tag{3.4}$$

$$g_{1,y} + g_{2,x} = 2(f_2 V_x + f_3 V_y), \tag{3.5}$$

$$\begin{aligned} & g_1 V_x + g_2 V_y - \frac{\hbar^2}{4}[f_1 V_{xxx} + f_2 V_{xxy} + f_3 V_{xyy} + f_4 V_{yyy} \\ & + 8A_{300}(xV_y - yV_x) + 2(A_{210}V_x + A_{201}V_y)] = 0. \end{aligned} \tag{3.6}$$

The functions f_i are given in terms of the constants A_{ijk} as

$$\begin{aligned} f_1 &= -A_{300}y^3 + A_{210}y^2 - A_{120}y + A_{030}, \\ f_2 &= 3A_{300}xy^2 - 2A_{210}xy + A_{201}y^2 + A_{120}x - A_{111}y + A_{021}, \\ f_3 &= -3A_{300}x^2y + A_{210}x^2 - A_{201}xy + A_{111}x - A_{102}y + A_{012}, \\ f_4 &= A_{300}x^3 + A_{201}x^2 + A_{102}x + A_{003}. \end{aligned} \tag{3.7}$$

Thus we have an overdetermined system of 4 partial differential equations for 3 functions $V(x, y)$, $g_1(x, y)$ and $g_2(x, y)$.

The first thing to notice about the system (3.3–(3.6) is that the Planck constant $\hbar$ figures in eq. (3.6). Hence the classical and quantum case differ, contrary to the case of quadratic integrability. Indeed, in the classical case eqs. (3.3), (3.4) and (3.5) are the same, but eq. (3.6) reduces to

$$g_1 V_x + g_2 V_y = 0. \tag{3.8}$$

The difference between classical and quantum integrability was already noticed by Hietarinta [32, 33].

Another difference between the cases of second- and third-order integrals of motion is that the system (3.3)–(3.6) (or (3.7)) is overdetermined, and hence gives rise to compatibility conditions. Eqs. (3.3), (3.4) and (3.5) imply a linear equation for the potential, namely

$$f_3 V_{xxx} + (3f_4 - 2f_2)V_{xxy} + (3f_1 - 2f_2)V_{xyy} + f_2 V_{yyy}$$

$$+ 2(f_{3,x} - f_{2,y})V_{xx} + 2(3f_{1,y} + 3f_{4,x} - f_{2,x} - f_{3,y})V_{xy}$$
$$+ 2(f_{2,y} - f_{3,x})V_{yy} + (3f_{1,yy} + 3f_{3,xx} - 2f_{2,xy})V_x$$
$$+ (f_{2,yy} + 3f_{4,xx} - 2f_{3,xy})V_y = 0. \quad (3.9)$$

Further compatibility conditions for the potential are nonlinear. This is what makes the problem difficult to solve, even in the classical case. The nonlinear compatibility conditions are third-order equations in the classical case, fifth-order ones in quantum mechanics.

Let us now look at a simpler problem, namely that of superintegrability with one first-order integral of motion and one third-order one. We consider the classical and quantum cases separately. In both cases, the first-order integral can be either $X = P_2$ or $X = L_3$. The corresponding potential satisfies $V = V(x)$, or $V = V(r)$ respectively.

In the classical case, no new superintegrable systems emerge. For $X = P_2$, we obtain two potentials

$$V_1 = ax, \quad V_2 = \frac{a}{x^2}. \quad (3.10)$$

Both allow second-order integrals, in addition to P_2. The third-order integrals are products of P_2 and the second-order ones.

For $X = L_3$ we once again get the Kepler–Coulomb potential and the harmonic oscillator, as in eq. (1.7). The third-order integrals are again reducible.

In the quantum case, the terms in eq. (3.6) proportional to $\hbar^2$ make all the difference. Indeed, taking $X_1 = P_2$ and a third-order integral of the form

$$X_2 = L_3P_1^2 + P_1^2L_3 + g_1P_1 + P_1g_1 + g_2P_2 + P_2\,g_2, \quad (3.11)$$

we find that these two integrals are allowed by any potential $V(x)$ satisfying

$$\hbar^2V_x^2 = 4(V - V_1)(V - V_2)(V - V_3), \quad (3.12)$$

where V_i are constants. If they are all different, we obtain a finite or a singular solution in terms of Jacobi elliptic functions:

$$V(x) = \hbar^2\omega^2k^2\,\mathrm{sn}^2(\omega x, k), \quad V(x) = \frac{\hbar^2\omega^2}{2(1 + \mathrm{cn}(\omega x, k))},$$
$$\text{or} \quad V(x) = \frac{\hbar^2\omega^2}{\mathrm{sn}^2(\omega x, k)}. \quad (3.13)$$

In the case of multiple roots, $V_1 < V_2 = V_3$, or $V_1 = V_2 < V_3$ we obtain solutions in terms of elementary functions:

$$V(x) = \frac{\hbar^2\omega^2k^2}{\cosh^2 \omega x}, \quad V(x) = \frac{\hbar^2\omega^2}{\sin^2 \omega x}. \quad (3.14)$$

For $V_1 = V_2 = V_3$ we reobtain a known potential, $V(x) = \hbar^2/x^2$, however, with additional integrals of motion. In general, the third-order integral (3.11) is equal to

$$\begin{aligned} X_1 &= L_3 P_1^2 + P_1^2 L_3 + \{-y(3V-\alpha), P_1\} + \\ &\quad \{2xV - \alpha x + \int V \mathrm{d}x, P_2\}, \end{aligned} \tag{3.15}$$

where $\{\cdot\,,\,\cdot\}$ denotes the anticommutator. Commuting X_1 with X_2 we obtain a further integral

$$X_3 = P_1^3 + \{3V - \alpha, P_1\}. \tag{3.16}$$

For $X_1 = L_3$, i.e., rotationally invariant potentials, the classical and quantum cases coincide. In fact, no new superintegrable potentials emerge. The only ones are $V_K = \alpha/r^2$ and $V = \omega^2 r^2$, and the third-order quantum integrals are commutators of lower-order ones.

4. Conclusions

Finite-dimensional superintegrable systems are of considerable interest in their own right, because of their mathematical and physical properties. Moreover, an intriguing relationship with soliton theory, i.e., the theory of infinite-dimensional integrable systems, is emerging. First of all, a superintegrable system can be characterized by the fact that it allows a non-Abelian algebra of integrals of motion with at least one n-dimensional Abelian subalgebra. The situation is similar for soliton equations, e.g., the Korteweg-de Vries and the Kadomtsev-Petviashvili equations. Their integrals of motion form infinite-dimensional non-Abelian algebras. These have Abelian subalgebras that are also infinite-dimensional. In this sense, soliton equations are superintegrable, rather than just integrable.

A further relation between soliton theory and superintegrable systems appears in the study of quantum superintegrability in E_2 space with third-order integrals of motion. In Section 3 we saw that elliptic function (3.13) and soliton (3.14) solutions of the KdV equation figure as superintegrable potentials. The relation between solutions of the KdV and superintegrability in E_2 goes further. When superintegrable systems allowing one third-order and one second-order integral are considered, Painlevé transcendents emerge as the corresponding potentials [34].

Acknowledgements

This text was written during the author's visit to the Departamento de Física Teórica II of the Universidad Complutense of Madrid. The author thanks the Department for its hospitality, and specially Miguel A. Rodríguez for helpful discussions. The author's research is partially supported by research grants from NSERC of Canada, FQRNT du Québec and the NATO collaborative grant PST.CLG 978431.

References

1. H. Goldstein, Classical Mechanics (Addison-Wesley, Reading, MA, 1990).
2. V.I. Arnold, Mathematical Methods of Classical Mechanics, Graduate texts in Mathematics, 60 (Springer–Verlag, New York, 1978).
3. J. Bertrand, *Théorème relatif au mouvement d'un point attiré vers un centre fixe,* C. R. Ac. Sci. **77**, 849–853 (1873).
4. V.A. Fock, *Zur Theorie des Wasserstoffatoms,* Z. Phys. **98**, 145–154 (1935).
5. V. Bargmann, *Zur Theorie des Wasserstoffatoms,* Z. Phys. **99**, 576–582 (1936).
6. J. Jauch and E. Hill, *On the problem of degeneracy in quantum mechanics,* Phys. Rev. **51**, 641–645 (1940).
7. P. Tempesta, A.V. Turbiner, and P. Winternitz, *Exact solvability of superintegrable systems,* J. Math. Phys. **42**, 4248–4252 (2001).
8. E.G. Kalnins, J.M. Kress, and P. Winternitz, *Superintegrability in two-dimensional spaces of nonzero curvature,* J. Math. Phys. **43**, 970–983 (2002).
9. M.A. Rodríguez and P. Winternitz, *Quantum superintegrability and exact solvability in n dimensions,* J. Math. Phys. **43**, 1309–1322 (2002).
10. I. Friš, V. Mandrosov, J. Smorodinsky, M. Uhlíř, and P. Winternitz, *On higher symmetries in quantum mechanics,* Phys. Lett. **16**, 354–356 (1965).
11. P. Winternitz and Ya. A. Smorodinsky, M. Uhlíř, and I. Friš, *Symmetry groups in classical and quantum mechanics,* Sov. J. Nucl. Phys. **4,** 444–450 (1967).
12. M.B. Sheftel, P. Tempesta, and P. Winternitz, *Superintegrable systems in quantum mechanics and classical Lie theory,* J. Math. Phys. **42**, 659–673 (2001).
13. J. Hietarinta, B. Grammaticos, B. Dorizzi, and A. Ramani, *Coupling constant metamorphosis and duality between integrable Hamiltonian systems,* Phys. Rev. Lett. **53**, 1707–1710 (1984).
14. A. Makarov, Ya. Smorodinsky, Kh. Valiev, and P. Winternitz, *A systematic search for non-relativistic systems with dynamical symmetries,* Nuovo Cim. A **52**, 1061–1084 (1967).
15. N.W. Evans, *Superintegrability in classical mechanics,* Phys. Rev. A **41**, 5666–5676 (1990).
16. N.W. Evans, *Group Theory of the Smorodinsky-Winternitz system,* J. Math. Phys. **32**, 3369–3375 (1991).
17. N.W. Evans, *Superintegrability of the Winternitz system,* Phys. Lett. A **147**, 483–486 (1990).
18. E.G. Kalnins, W. Miller Jr, and G.S. Pogosyan, *Superintegrability and associated polynomial solutions: Euclidean space and the sphere in two dimensions,* J. Math. Phys. **37**, 6439–6467 (1996).
19. E.G. Kalnins, W. Miller Jr, and G.S. Pogosyan, *Superintegrability on the two-dimensional hyperboloid, I and II,* J. Math. Phys. **38**, 5416–5433 (1997); **40**, 2291–2306 (1999).

20. E.G. Kalnins, J.M. Kress, G.S. Pogosyan, and W. Miller Jr, *Completeness of superintegrability in the two-dimensional constant-curvature spaces,* J. Phys. A **34**, 4705–4720 (2001).
21. C.P. Boyer, E.G. Kalnins, and P. Winternitz, *Completely integrable relativistic Hamiltonian systems and separation of variables in Hermitian hyperbolic spaces,* J. Math. Phys. **24**, 2022–2034 (1983).
22. C.P. Boyer, E.G. Kalnins, and P. Winternitz, *Separation of variables for the Hamilton-Jacobi equation on complex projective spaces,* SIAM J. Math. Anal. **16**, 93–109 (1985).
23. M.A. del Olmo, M.A. Rodríguez, and P. Winternitz, *Integrable systems based on SU(p,q) homogenous manifolds,* J. Math. Phys. **34**, 5118–5139 (1993).
24. M.A. del Olmo, M.A. Rodríguez, and P. Winternitz, *The conformal group and integrable systems on a Lorentzian hyperboloid,* Fortschr. der Physik **44**, 90–125 (1996).
25. G. Darboux, *Leçons sur la théorie générale des surfaces,* Vols.1-4 (Chelsea, New York, 1972).
26. G. Koenigs, *Sur les géodésiques à intégrales quadratiques* in Leçons sur la théorie générale des surfaces, by G. Darboux (Chelsea, New York, 1972), Vol. 4, pp 368–404.
27. E.G. Kalnins, J.M. Kress, W. Miller Jr, and P. Winternitz, *Superintegrable systems in Darboux spaces,* to be published.
28. J. Drach, *Sur l'intégration logique des équations de la dynamique à deux variables: forces conservatrices. Intégrales cubiques. Mouvements dans le plan,* C. R. Ac. Sci. III, **200**, 22–26 (1935).
29. M.F. Rañada, *Superintegrable $n = 2$ systems, quadratic constants of motion and the potentials of Drach,* J. Math. Phys. **38**, 4165–4178 (1997).
30. A.V. Tsiganov, *The Drach superintegrable systems,* J. Phys. A **33** 7407–7422 (2000).
31. S. Gravel and P. Winternitz, *Superintegrability with third order integrals in quantum and classical mechanics,* J. Math. Phys. **43**, 5902–5912 (2002).
32. J. Hietarinta, *Classical vs. quantum integrability,* J. Math. Phys. **25**, 1833–1840 (1984).
33. J. Hietarinta and B. Grammaticos, *On the $\hbar^2$ correction in quantum mechanics,* J. Phys. A **22**, 1315–1322 (1989).
34. S. Gravel, *Hamiltonians separable in cartesian coordinates and third order integrals of motion,* preprint math-ph/030228.